£8.95

Battle Tactics of the Western Front

Battle Tactics of the Western Front

The British Army's Art of Attack, 1916–18

Paddy Griffith

Yale University Press
New Haven & London · 1994

For Françoise Beucher

Copyright © 1994 by Paddy Griffith

All rights reserved. This book may not be reproduced in whole or in part, in any form (beyond that copying permitted by Sections 107 and 108 of the U.S. Copyright Law and except by reviewers for the public press), without written permission from the publishers.

Set in Monotype Bembo
Printed and bound in Great Britain by St Edmundsbury Press

Library of Congress Cataloging-in-Publication Data

Griffith, Paddy.
    Battle tactics of the western front: the British Army's art of attack, 1916–18 / Paddy Griffith.
       p.   cm.
    Includes bibliographical references (p.258) and index.
    ISBN 0–300–05910–8
    1. World War, 1914–18—Campaigns—Westen. 2. Tactics.
3. Strategy. I. Title.
D756.G75   1994
940.4'144—dc20   93–42656
                                                                                         CIP

A catalogue record for this book is available from the British Library.

Contents

List of Figures		viii
List of Tables		viii
Preface		ix
Abbreviations		xiii
Part One:	*Setting the Scene*	1
1	*Introduction*	3
	Competence and incompetence	3
	The larger second half of the war	10
2	*The Tactical Dilemma*	20
	The nature of tactics	20
	Lines, densities and timescales	29
	The storm of steel	38
Part Two:	*Infantry*	45
3	*Infantry during the First Two Years of the War*	47
	Tactics of the Old Contemptibles	48
	The New Armies arrive	52
4	*The Lessons of the Somme*	65
	Rifles, bayonets and the cult of the bomb	65
	The importance of careful preparation	74
	The assault spearhead of the BEF	79

5	The Final Eighteen Months	84
	The formal Battles of 1917 and the chaotic battles of 1918	84
	Flexible formations for mobile war	93

Part Three: Heavier Weapons — 101

6	The Search for New Weapons	103
	The mobilisation of invention	104
	Bombs, smoke and gas	112
7	Automatic Weapons	120
	The struggle to control automatic fire	120
	The rise of the light machine gun	129
8	Artillery	135
	The evolution of precision munitions	135
	The shift from destructive to neutralising fire	142
	The emergence of the deep battle	153
9	Controlling the Mobile Battle	159
	Cavalry and armour	159
	Signals and command	169

Part Four: The BEF's Tactical Achievement — 177

10	Doctrine and Training	179
	Captain Partridge and the dissemination of doctrine	179
	Training schools and other exhortations	186
11	Conclusion	192

Appendix 1	Some limitations in the university approach to military history	201
Appendix 2	A Great War perspective on the American Civil War	204
Appendix 3	Armies, corps and divisions of the BEF	208
Notes		220
Bibliography		258
Index		277

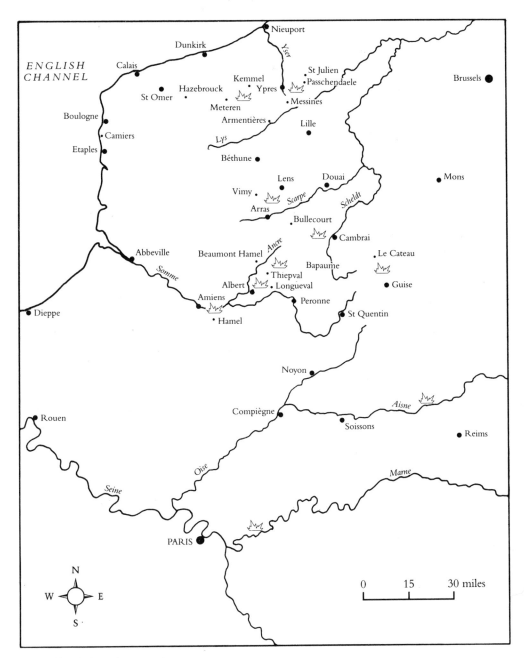

Some of the chief British sites on the Western Front

List of Figures and Tables

Figures

1	(Notional) Comparison between the loss of trained tactical leaders and the need for them	23
2	Variants of the wave attack by the 9th Division, 1915-16	55
3	Platoon tactics, February 1917	78
4	Successive fragmentation of XVIII Corps in March 1918	91
5	Advances of the BEF in the Hundred Days, 1918	94
6	The flexibility of diamonds: the BEF's 1918 concept of a fluid 'infiltration' attack	97
7	Schematic layout of BEF cable grid, as from late 1916	171
8	Armies in autumn 1916	210
9	Armies in autumn 1917	211
10	Armies in spring 1918	212
11	Armies in autumn 1918	214

Tables

1	Approximate number of 'divisional battles'	18
2	Length of the British line in France	37
3	Effective combat performance of British infantry weapons	115
4	Organisation of motor machine gun brigades	129
5	British artillery weapons	136
6	Total British shell production per month	139
7	Proportion of shell types in selected creeping barrages	141
8	Speed of creeping barrages in some 9th Division attacks	144
9	Speed of some creeping barrages in the Hundred Days	146
10	Expenditure of artillery ammunition, by weeks	148
11	Yards of front per gun on 'first days' of battle	150
12	Tanks committed to action during the Hundred Days	167

Preface

The present book began life during a stimulating conversation with my friend and former colleague G. D. Sheffield, a 'War and Society' historian who is best known to the wider world for his scholarship in Great War studies. This in turn recalled a series of pertinent thoughts on tactics that had been provided some years earlier by another scholarly observer of twentieth-century warfare, Andrew Grainger. In the autumn of 1990, therefore, I discovered that between them these two had somehow contrived to awaken an interest within me that I had for many years sworn seriously to abjure.

Throughout my life as a Briton and a military historian, the Western Front had always seemed to represent something of a blighted absolute. It was the British Army's biggest campaign ever, but apparently also its biggest disaster. Technically a victory, the war continued so indecisively and for so long that its commanders should surely shoulder a whole mountain of blame for their failure to modernise tactics — all the alluring protestations of apologists notwithstanding. Merely to visit the Ypres Salient today, and to trace the fields no bigger than football pitches where entire brigade attacks were halted almost before they had begun, is to suffer a cold depression of the soul.

In historiographical terms the Great War also seemed to be something of a disaster area, since it has become the favoured playground of popular authors who love to dwell on the horrors and the incompetence without stopping to make a proper military analysis. This is annoying in itself; but for me it has had the additional effect of helping turn the academic world away from true military history (see Appendix 1). During my working life I have encountered considerable resistance to serious military study among the 'politically correct' members of the university, and it has often been the Great War which they have used as the reference point for their prejudices.

All these unpromising features of Great War historiography conspired together for many years to warn me away from the subject, and I tried to confine my studies to the conflicts that came either before 1914 or after 1918. However, I have now at last come to swallow my reservations and have attempted to give the Great War the long hard gaze that it had previously tried so hard to avert...and I have been fascinated to discover just how richly it has rewarded me for my effort.

As a first step I saw some illuminating parallels and contrasts with the American Civil War, which was a conflict that I had studied in the past (see Appendix 2). Beyond those, it came as something of a surprise to find that much Great War historiography tended to be dismissive of the many technical achievements of the infantry, as opposed to its awesome and oft-repeated ability to suffer and endure. It is therefore around this subject that I have tried to construct the present volume. While I must certainly apologise that it does not look in very great detail into the actions of the artillery, the tanks or the air forces on the Western Front, I would respectfully point out that each of these branches has received many excellent modern treatments, and each even still maintains its own active network of propagandists. I hasten to add that I have not yet met or read a single *unfair* modern propagandist for any of them; but I suspect that there does nevertheless remain a deep structural unfairness in the literature, insofar as many of the other branches that were essential to the Great War fighting have now apparently fallen completely silent. The Machine Gun Corps is the most obvious example, since it was disbanded immediately after the armistice and has left no successor apart from a fading Old Comrades association. By the same token there is no longer an identifiable trench-mortar lobby or a hand-grenade lobby — and both the gas debate and the signals debate have long ago moved forward to very much higher manifestations of their respective black arts. But above all it seems to be the infantry itself, which so horrifically bore so much more than the lion's share of the war, that has most seriously been denied a proper account of its tactical methods in modern times. In the present book I therefore hope to offer something of a corrective to this trend.

I am nevertheless conscious that I am a newcomer to Great War studies, and that time has not been granted for me to read more than a tiny percentage of the books and manuscripts available. Such an undertaking would demand at least a lifetime of specialisation, and even then it remains true that 'the more you know, the more you know you don't know'. My sources are therefore necessarily limited, and attentive readers will observe a certain recurrent gravitational spin towards the 2nd Royal Welch Fusiliers at the battalion level, the Ninth Scottish Division at the divisional level, and the official histories in much of the rest.

Preface xi

However, I am encouraged to find that a number of the absolutely most obvious sources for tactics have apparently never been properly explained to the public — for example, the papers of Maxse, Lindsay, Laffargue and even Partridge, not to mention Liddell Hart's actual scribblings during the period 1916–18. It seems incredible that such a major war should have been so badly served by previous historians, and that as a result so few of its most simple tactical assumptions have ever been systematically probed. Admittedly in 1985 I found that precisely the same phenomenon was observable in relation to the American Civil War, so I ought to have been forewarned: but really I am today astonished to find just how neglected the whole field of tactical history still appears to be for the Great War.

Quite apart from such technical criticisms, I am nevertheless very fortunate that I have been able to patch over some of my original ignorance with profuse help given freely by a large number of genuine experts in the field. They are of course in no sense responsible for the doubtless numerous mistakes and twisted opinions which I present in my text; but I am still extremely grateful to them all for their efforts on my behalf. I would therefore like to take this opportunity to register my deep appreciation to those of them who are named below, and my apologies to any I have missed.

Apart from their inspirational rôle in starting off this study, G. D. Sheffield and Andy Grainger both offered much valuable advice and information throughout its course, not all of which I have followed. For example, the former shook his head vigorously and sadly over my protest against the 'War and Society' school, while the latter took his life in his hands for an ice-bound 'trench raid' in February 1991, in which my car succumbed to the rigours of the Passchendaele Road as so many others had done before it. Andy also directed me decisively towards the epic qualities of the fighting 9th (Scottish) New Army Division.

Greg McCauley and Bill Rogan took me on other revealing battlefield visits — to Ypres in 1989 and the Somme in 1992 respectively — while Dr Paul Harris, who must be classed as the leading contemporary expert on British armoured warfare doctrines before 1945, was especially helpful with ideas, comments and photocopies, not to mention sanctuary and hospitality, over a very long period.

Apart from my family, who have shown amazing forbearance during two years living with a ratty and distracted author, I would also like to thank Graham Evans for his grandfather's memoirs, as well as my other friends who participated in the '4th Warwickshire' series of rôle plays — Joe Lawley, Richard Shields, Chris Kemp, Mark Hone and Peter McManus, not to mention Mrs Shirley Barkes and the 1991 fourth year of Weddington Middle School for their original encouragement in that direction. I thank all past and present members of the Mercia Military

Society; Dennis Gillard — who died tragically just as this book was being completed — Dr Geoffrey Noon, and other members of the Western Front Association; Peter Scott and his, now alas defunct, *The Great War* journal; the Ordnance Society; all the contributors to the forthcoming Frank Cass book of essays, *British Fighting Methods in the Great War*; Peter Simkins, Simon Robbins and other members of staff at the Imperial War Museum, and Peter and Penny Thompson for their nearby hospitality; David Fletcher at the Tank Museum, and Diana and Roger Ashdown for the loan of their caravan at Bovington; Mr Woodends and Mr Ellis at the MOD Pattern Room in Nottingham; the staff of the Liddell Hart Archives at King's College, London, of the Public Record Office at Kew, and of the Nuneaton public library. Above all I am indebted to Andrew Orgill and the staff of the central library of the Royal Military Academy, Sandhurst, without whose professionalism and tolerance over a very long period this book — and indeed most of my books — would have been entirely impossible to write.

Finally, I am particularly indebted to Françoise Beucher — a modern-day neighbour of what used to be 'Infantry Base Dépôt No.5' in Rouen, and an occasional diner in what is still the Hôtel de la Poste (albeit now thankfully stripped of its notorious British 'Officers' Club') — to whom this book is affectionately dedicated.

Not only have she and her husband Guy given my family quite irreplaceable help, support and generosity over many years, but she once very tellingly put the whole BEF experience into what I suspect may be rather more of its true context than many Anglo-Saxons would easily accept. When I told her I was about to visit the Great War battlefields, she shook her head at the great distances I would have to travel to reach the Chemin des Dames and Verdun. She could not understand how I could possibly hope to find any such battlefields as close as Picardy or Flanders since the war, as everyone knew, had been fought exclusively in Eastern France. She would therefore doubtless have wished me to turn this book into an English-language rehabilitation of the French contribution to victory, and such a work is in fact long overdue. Alas, the present volume is not that book, but merely yet another Anglo-centric collection of trench stories...

Abbreviations

ADC   Aide De Camp (a relatively junior officer serving as personal assistant to a general, and basking in many privileges — and responsibilities — as a result).
Adjutant   In effect the chief of staff in a battalion; normally a captain.
ADVS   Assistant Director of Veterinary Services
ANZAC   Australian and New Zealand Army Corps, the memorable code name devised by an English clerk in Cairo to denote the Australasian forces deployed to Gallipoli in 1915. It was retained in the titles of their two army corps until these were amalgamated into one corps — without the New Zealanders — at the start of 1918, and redesignated as 'The Australian Corps'.
AR   Automatic Rifle (i.e. a non-tripod-mounted machine gun, e.g. the Lewis).
Army   A formation made up of a number of army corps, plus additional 'army troops'.
Army Corps A formation made up of a number of divisions, plus additional 'corps troops'.
ASC   Army Service Corps. A body of men who were responsible for transport, envied for their high levels of pay and low levels of exposure to danger, and mocked in the wartime song 'Fred Karno's Army', which suggested they were followers of the music hall impresario of that name.
Bde   Brigade (i.e. a unit usually containing four battalions, reduced to three at the start of 1918 except in colonial and 'newly returned from Italy' divisions).
BEF   British Expeditionary Force, a name that was later officially changed to 'BAF' or 'British Armies in France', although I have not followed that practice.
BGC   Brigadier General Commanding (usually a brigade commander).

BGGS	Brigadier General General Staff (i.e. the chief of staff in a corps).
BGRA	Brigadier General Royal Artillery (i.e. the artillery commander — or until the Somme battle merely an adviser — in a division or corps).
BM	Brigade Major (i.e. the chief of staff in a brigade).
Bn	Battalion (i.e. a unit made up of a number of companies, squadrons or batteries). In the British Army this was often referred to as a 'Regiment', since although most regiments comprised more than one battalion — some numbered more than 20 by 1918, or almost the size of a pukka army corps — the battalions would usually be brigaded alongside a varied selection of battalions from different regiments, and hence would retain their own distinct regimental identities within their particular brigade or division.
CDS	Central Distribution Section, for publications, at the War Office.
CO	Commanding Officer (usually the lieutenant colonel commanding a battalion).
CRE	Commander Royal Engineers.
DA	Deputy Adjutant.
Div	Division (i.e. a formation of several brigades plus additional 'divisional troops').
FSR	Field Service Regulations.
GHQ	General Headquarters (i.e. the field HQ of the BEF).
GOC	General Officer Commanding (i.e. 'commander' of a Division, Corps or Army).
GOCRA	General Officer Commanding Royal Artillery; the formal title of a BGRA during the second half of the war, when he had more authority.
GPMG	General Purpose Machine Gun (a 1960s term for successors to the 'light' but belt-fed German MG 08/15 Maxim).
GSO1	General Staff Officer, Grade 1 (i.e. the chief of staff in a division).
HE	High Explosive.
HQ	Headquarters.
IGC	Inspector General of Communications (the GHQ department responsible for the line of communication).
IT	Inspectorate of Training (set up under General Ivor Maxse from July 1918).
IWM	Imperial War Museum.
K	'Kitchener Army', or 'New Army'.
LMG	Light Machine Gun (originally the Vickers; later the Lewis and then the Bren!).
Lt Col	Lieutenant Colonel (usually a battalion commander).
MG	Machine Gun.
MGC	Machine Gun Corps.
MMG	Medium Machine Gun (normally the Vickers).

OC	Officer Commanding (usually applies to a company or platoon commander).
OH	Official History (e.g. Edmonds, *Military Operations, France and Belgium*; Becke, *Order of Battle*, etc.: see bibliography for full references).
OP	Observation Post.
PBI	'Poor Bloody Infantry' (i.e. what Wellington had called 'That Article', or 'The Scum of the Earth', but which rose a little in the public estimation with the advent of citizen armies).
PRO	Public Record Office, Kew.
QMG	Quarter Master General.
RA	Royal Artillery (or occasionally, within the camouflage service, 'Royal Academicians'!).
RAMC	Royal Army Medical Corps (laughingly known in some quarters as 'Rob All My Comrades').
RE	Royal Engineers.
RFA	Royal Field Artillery (responsible for guns within infantry divisions).
RFC	Royal Flying Corps (converted into the RAF on 29 November 1917).
RGA	Royal Garrison Artillery (responsible for medium and some heavy guns).
RHA	Royal Horse Artillery (responsible for guns in cavalry divisions).
RN	Royal Navy.
RS	Royal Scots regiment.
RSM	Regimental Sergeant Major (i.e. the most powerful 'executive' at a battalion commander's disposal).
RWF	Royal Welch Fusiliers, the Liverpool-based regiment which (despite strong Australian competition) has managed to dominate the literature of this war, although in the event it did not quite manage to win it single-handedly. The many RWF officers who passed through the Adelphi Hotel (Sassoon's 'Olympic') during the Great War may thus be legitimately regarded as the direct but opposite precursors to the many naval officers who passed through the same hotel in the Second World War on their way to making an undoubtedly lesser impact on the literature — but nevertheless winning the genuinely decisive battle of the Western Approaches. (My own interest in this hotel is simply that I attended some children's parties there, about forty years before I began to suspect its deep military connections.)
RWK	Royal West Kent regiment.
SOS	Call or coloured flare signalling for immediate help, especially preplanned artillery defensive fire (originally from the nautical signal 'Save Our Souls', which now seems to have turned into 'Mayday' after the French 'M'aidez!').

SS	Stationery Services (a prefix often attached to official manuals and other publications).
TF	Territorial Forces.
TM	Trench Mortar (various sizes).
VC	Victoria Cross: gallantry decoration awarded most often to colonials, British regulars and other formations felt to be in need of encouragement.
WO	War Office (including its series classification code in the PRO).

Names of Battles

I have not generally used the elaborately pedantic official system of nomenclature (as found, *inter alia,* at the start of every volume of *Military Operations, France and Belgium),* but have followed the far more straightforward popular usage by which, for example, 'The Somme' is the whole battle fought between June and November 1916, roughly between Chaulnes in the south and Gommecourt in the north; 'Arras' is the whole battle of April and May 1917, including the capture of Vimy Ridge and the Fifth Army operations around Bullecourt etc.; 'Passchendaele' or 'Third Ypres' is the whole of the autumnal mud bath of the same year, and 'The Hundred Days' is the 'semi-open warfare' advance from 8 August 1918 to the end of the war.

Note that according to this informal system of nomenclature the BEF fought only about half a dozen battles in all, between 1 July 1916 and the end of the war, rather than the several dozen that the official nomenclature would suggest. Contrary to Edmonds's face-lifting instinct in his second 1917 volume, however, I have kept the successful attack on Messines distinct from the highly contentious one on Passchendaele, and have counted it as a quite separate battle. I was admittedly deeply tempted to lump 'Cambrai' into the face-saving butt end of 'Passchendaele', but ultimately I have closed ranks with Edmonds by agreeing that it was indeed a separate battle...or perhaps just a separate 'raid'?

Part One
Setting the Scene

1

Introduction

> ...it would seem that events in the earlier part of the war made greater impression than did later adventures.
>
> *Official History 1918,* vol. 5, p. viii

In the two chapters of this initial 'Setting the Scene' section, we shall review the general thrust of the modern historiography of Great War tactics, and locate a certain failure to appreciate the full extent of the British achievement, especially in the very important second half of the war. We shall then attempt to define just what the tactical problem actually was, before we move on in subsequent sections to examine the ways in which the BEF — usually successfully — tried to solve it.

Competence and incompetence

The Great War on the Western Front has long been a synonym for futile industrialised slaughter. It was a war of barbed wire, poison gas, impersonal massed bombardments and all-embracing mud, trench foot, stench, rats and lice. The casualty list was longer than for any previous British war, made doubly horrific by being concentrated into such a short period of time. Nor did the bloodletting seem to lead to any great military result. There was no decisive breakthrough nor — the quite rapid advances of late 1918 notwithstanding — any true restoration of 'open warfare'. Admittedly the Kaiser's army was completely beaten and his people starved into submission, but there nevertheless remained an ineradicable 'feeling of failure' about the allied victory.[1] The British public was numbed by victory's unthinkable cost while the army was both physically and mentally exhausted — and was finally even denied the victor's satisfaction of blowing up the towns in Germany.[2]

The Somme attack on 1 July 1916 has especially been seen as the absolute all-time nadir of British military enterprise: not only by the middle-class television-watching public of modern times, but even by the BEF within a very few days of the fatal moment itself.[3] Before the event, this attack had been widely heralded as a war-winning offensive whose success would be absolutely guaranteed by an unprecedented mass of men and munitions concentrated on an awesomely short frontage.[4] Yet some 57,000 casualties were sustained in the first morning, and at many parts of the line — at Serre, Beaumont Hamel, Thiepval and La Boisselle — the progress made can be described as at best 'negligible'. On the right flank of the British line (and in the far from negligible French sector), progress was admittedly much less disgraceful; nevertheless, the successes have been obliterated from the public memory by a fatally disillusioning combination of breezy official optimism before the event, mixed with unacceptably localised Accrington, Ulster (and many other) death tolls on one fatal day.

Then again there is the peculiar case of the Ypres Salient — an evilly waterlogged and overlooked spot that had been made notorious throughout the BEF as early as October–November 1914 by Haig's first — desperate but, for him personally, formative — battle. 'Wipers' would become still more notorious in every succeeding year. In 1915 it was the scene of the first major military use of poison gas, leading to one of the most precipitate allied retreats of the war. This was followed by a winter of what even an élite formation remembered as 'almost unmitigated gloom and discomfort'.[5] In 1917 the Salient was nevertheless carefully selected to be the site of what was surely the most futile and foredoomed of all 'the mature Haig offensives'. The very name of Passchendaele quickly entered the language as a synonym for both 'Slaughter' in the commercial wholesale butcher's sense, and for 'Passion' in the Eastertide theological usage of that term. It also has a peculiarly liquid resonance about it, reminding us that many of the wounded were destined to drown where they lay, and many of the tactical movements sank literally waist deep in the mud. Overall there were no fewer than five battles around Ypres, at least three of which were truly awful; and the place has been blighted with a particularly bad memory forever afterwards.

That the war involved enormous frustration and sacrifice is a self-evident fact; but more controversial is the underlying question of whether or not the disasters could have been avoided — whether or not 'someone had blundered' in the most massive and odious possible way. The majority opinion since at least the late 1920s has been that it was largely the stupidity or inflexibility of the high command that prevented a timely adaptation to new conditions, thereby transforming an 'ordinary' modern

Introduction

war into 'the war to end all wars'.[6] According to this interpretation the war was fought by 'Lions led by Donkeys',[7] or in other words by courageous soldiers who deserved far better leadership than they actually received. Thus Haig is supposed to have filled his high command with cavalrymen who wanted a glorious mounted breakout, and could not see that trench warfare was really, in General Monash's immortal phrase, 'simply a problem of engineering'.[8] Haig's allegedly hidebound nineteenth-century mind could not cope with a twentieth-century war, and the arrogant ruling caste from which he came could rise no higher than dismissing someone like Monash as just 'a typical old Jew' who had little to offer.[9]

There have doubtless been many different motives behind these allegations, ranging from populist mockery of the ruling class to a widespread belief that if something ever goes badly wrong in any great enterprise it must automatically be the fault of the man at the top. Haig himself held to the doctrine that 'a commander must be allowed to command', and therefore that such an officer must be ruthlessly stellenbosched — sent home — if his efforts finally end in failure.[10] In the specific case of Haig himself, however, it has always been notoriously difficult to identify just who should have been responsible for stellenbosching him and, if so, precisely when. Even if it may be easy to see with hindsight that Kitchener should have substituted Smith-Dorrien for French immediately after Le Cateau in August 1914, with Haig there was never any equivalent moment when he actually panicked — and some would say no moment when he actually betrayed any emotion at all. In any event, Kitchener drowned on 5 June 1916, only three weeks before Haig's generalship was due to undergo its first and most decisive test in the Somme offensive of 1 July. This left inadequate time for Kitchener's successors to form themselves into a jury that was properly competent to judge the battlefield performance of its legally appointed field commander.

Following the loss of Kitchener, the War Office passed into the hands of career politicians who lacked sufficient leverage to get Haig sacked for the 1 July débâcle, and later even for Passchendaele itself. The clear chain of military command had been broken, and Lloyd George found himself faced by an almost united 'Army' front against which he could proceed only very cautiously. Instead of a clear-cut distribution of power, therefore, the whole question of Haig's appointment rapidly degenerated into a stinking miasma of speculation, gossip, intrigue and faction-fighting, which refuses to go away even today.[11] The problem seems to be that although there was a clear chain of responsibility within the BEF itself, by 1916 its commander had become the servant of an over-diffuse and probably ultimately incoherent group of 'statesmen' working in London, who tended to be exposed to amazingly few clues about the underlying military realities. Much of the received wisdom about Haig therefore comes from tainted sources. His two leading critics, Lloyd George and Winston Churchill, eventually even managed to argue themselves into a most peculiar position, according to which most of the decisive fighting should apparently have been entrusted to a combination of the Tank Corps plus the Rumanian and Italian armies.[12] Such a policy would certainly have left the BEF's infantry with little more to do than hold quiet sectors of the line, but it would also surely have led to total defeat in the West.[13]

Although Lloyd George never did manage to get rid of Haig, by the end of the war he had undoubtedly emerged as by far the most prominent British politician, and a version of his general outlook has very naturally found its way into many of our present history books.[14] However, it is important to remember that through much of the war he was consistently seeking to further his position by searching high and low for absolutely any voices of dissent he could possibly find, for use as ammunition against the army high command. Once he had collected a number of such criticisms, his technique was to cobble them all together into an artificially unified alternative polemic of how the war as a whole ought to have been fought. Churchill too, albeit perhaps in a slightly different and more 'Westerner' vein, seems to have adopted much the same approach.

From today's perspective none of this at first sight seems to be illegitimate, and we can agree that the combined efforts of Lloyd George and Churchill did add a certain dynamism to several aspects of the war's prosecution. Both the tank and the Stokes mortar were given significant boosts in 1915, as perhaps was the adoption of the convoy system in 1917. Nevertheless this gadfly outsider's approach did suffer from very grave weaknesses, insofar as it never fully came to grips with the real issues of command at the operational or tactical levels. For example, Lloyd George himself had accepted Haig's misguided idea of an offensive at Ypres in 1917, just as he was directly responsible for restricting the flow of reinforcements which might have made Fifth Army's task considerably easier in March 1918. He was personally very deeply implicated in the outcome of all these events. In technological terms, equally, it may have been perfectly wonderful to hold up the Stokes mortar, the machine gun or the tank as 'neglected war-winners' and symbols of a golden future age, but that scarcely advanced the far more important practical question of how the BEF should adapt itself to an 'all-arms' doctrine of tactics which could successfully accommodate such inventions. That was really the deeper point at issue, and in the event Haig eventually managed to push it rather further forward than either Lloyd George or Churchill ever could.

The 'Lions led by Donkeys' critique nevertheless carries considerable weight, particularly during the first half of the war when almost everything had to be improvised out of almost nothing, and when both scientists and soldiers were still at an early stage in designing the shape of the new warfare. Beyond these unavoidably crushing problems, moreover, the high command too often wilfully closed its ears to voices of reason from below, too often responded too hastily to imperatives from above and, in Haig's case especially, too often made destabilising last-minute alterations in the perfectly sound plans of subordinates. Not even the most fervent apologists for the generals can ultimately exonerate them from blame for much of what went wrong.

Rather less palatable to patriotic Britons, however, is a widespread second strand of this criticism, whereby the true failure is deemed to have lain less with a few inadequate personalities among the BEF's senior generals than with the entire lower echelon of its officer corps. This was allegedly manned by little more than 'enthusiastic but tactically incompetent schoolboys',[15] who were better at the self-effacing reticences of public school protocol than at the fiery leadership of real men. In both the Cambrai inquiry of January 1918 and in Edmonds's official history of the whole war, the motive for using this slur was clearly to exonerate certain sections of the high command, and on these terms it may be quite easily dismissed. In the daemonology of Lloyd George and his friends, by contrast, the motive was just as much to damn the high command for squandering its trained officers while failing to provide direction and guidance for their successors. This charge may have rather more substance, although we shall see in later chapters that there was in fact a major effort of centralised tactical analysis and training. Finally, it is worth noting that for the very widely quoted Captain Basil Liddell Hart there was an additional — hidden and highly personal — motive for furthering the suggestion that the BEF's junior officers were tactically incompetent. His particular interest lay in propagating the idea that his own rewriting of the infantry manual in 1920 filled a void that had been left entirely unfilled throughout the war,[16] even though such a story was a rather obvious falsehood. To a lesser extent a similar charge may be levelled against General Ivor Maxse, who took control of training and tactics only during the last six months of the war, and who sometimes seemed to suggest that nothing had been done in either field before his arrival.

More recently, a great deal of highly nationalistic historiography has been written by non-Britons, who have condemned almost the entire British effort in the Great War as tactically worse than naïve. Quite apart from the French and Germans, who doubtless have some obvious local axes of their own to grind, there has recently been a veritable storm of protest from most of the colonies and dominions which formed part of the BEF itself. Admittedly we have not yet heard from either Meerut or Lahore, both of which have some very justifiable complaints still to lodge over the events of 1915; but we have certainly heard a very great deal from South Africans about Delville Wood in 1916, and from Australians and Canadians — and even more cheekily from Americans — about the alleged British mishandling of the entire war. Indeed, it would not be too much to claim that a majority of the most interesting scholarly writing about the BEF's tactics during the last decade has come from these 'non-Briton' sources,[17] even though the non-Briton element of the BEF amounted to at best 18 divisions as compared with some 60 divisions from England, Ireland, Scotland and Wales.

The 'colonial' critique is based upon the cultural, political and institutional separation from the mother country which allowed colonial troops to pursue slightly different patterns of organisation and tactics from those advocated by GHQ and Haig. They were a little more free to make innovations than their UK colleagues, which encouraged a certain spirit of criticism and independence. Rightly or wrongly, the colonial contingents liked to think of themselves as an élite, while the 'imperials' were seen as effete stick-in-the-muds whose day had passed. The corollary to this, which has been given a surprising rejuvenation in the US debate since Vietnam, was the idea that all German soldiers were almost always incomparable tacticians, from the lowest 'deeply professional and long service' NCO or stormtrooper, right up to the gifted Colonel Max Bauer, the fiendish Colonel von Lossberg, or the omniscient General Sixt von Armin. Colonel Brüchmuller and Generals Hutier and Ludendorff are naturally elevated to a still higher plane of military insight, such as is accorded to very few mortals indeed....But can we really trust such a critique, in view of the fact that the Germans lost the Great War completely? Is it not far more likely that in tactical affairs the Germans were really stumbling along a track that was every bit as blind and halting as that of the British?

The Germans' inordinate tactical reputation doubtless owes much to the events of 1939–41, combined with their own rewriting of history at around that time. A facile contrast can also be made between the slow British progress against strong defences on the Somme and at Passchendaele, and the much faster German advances against weak defences in the spring of 1918. However, one should add to the equation the devastating casualties suffered by the Germans on the latter occasion, due to their almost total neglect of tactics,[18] as well as the far faster and deeper allied advances which subsequently won the war. As at Appomattox in 1865, the tactical brilliance of the decisive final campaign has often been overshadowed by the varied emotions arising from the war's end.

It is also worth mentioning that much of the German reputation for tactical skill can be traced directly to the promiscuous circulation of translated captured German assault manuals by the British GHQ itself, especially during 1918.[19] The work of G. C. Wynne, one of the official British historians, was also very important in subsequent years, although

Introduction

few of his admirers noticed that he had become so deeply immersed in the German sources that he had badly lost perspective in regard to his own side.[20] Overall, therefore, GHQ may have been indirectly and unconsciously responsible for a long-term denigration of its own progress in attack tactics, merely because many of its more lurid publications about infantry techniques happen to have centred upon German practice, rather than upon the habitual and routine achievements of its own soldiers.

Against all these unfavourable views of BEF tactics, it may be argued that the problems posed by new material and technological conditions were simply too great to be capable of any solution at all, in view of the relatively limited resources available at the time. The defences on the Somme and at Passchendaele really were incredibly strong and difficult to pierce. The dissenting minority answer to the various 'incompetence' theories is therefore that by and large the BEF (whether represented by its high command or by its fresh young subalterns) did as well as could possibly have been expected in novel, strange and unpromising circumstances, and that it was not psychological but material deficiencies which explained the failures.[21] Thus 'Attempts to make scapegoats of Haig or Gough break down on the fact that no one else could do any better',[22] and they failed to find a solution for the simple reason that there was no solution to be found. And besides, by the time of the Hundred Days in the autumn of 1918 the BEF had in fact restored as much mobility to the battlefield as would later be found acceptable during most of the Second World War. It was only the dashing mounted, or mechanised, breakouts that were missing; but even in 1939–45 these would remain much less common than is often alleged.[23]

To a considerable extent the question of competence or incompetence in the BEF's leadership is political and strategic — a matter of whether or not Britain should ever have become involved in the first place, given its initially tiny industrial and military preparations; whether it should then have put as much stress as it did on the Western Front; whether Haig should ever have launched assaults on such unpromising battlefields as the Somme in 1916 or the Ypres Salient in 1917, and having done so, whether he should have thought in terms of 'breakthrough and pursuit' or of 'attrition and limited objectives'. Could there, conversely, be any excuse whatsoever for the debilitating shell shortage of 1915 or for Lloyd George's block on reinforcements towards the end of 1917, which led to something that came perilously close to total defeat in March 1918? Was there any value at all in diverting resources away from the main battlefield in France to exotic 'Eastern' theatres such as Gallipoli, Mesopotamia, Salonika or (most disruptive of all throughout 1917) Italy?

These are all crucial questions for general, socio-political and popular historians of the war, but it is worth remembering that the wider question of competence or incompetence also includes an important specialised stratum of low-level tactics that has perhaps been too often asserted but too little investigated. Certainly some of the early disasters may be directly attributed to faulty tactics; but against this we must remember that the lessons of the first day of the Somme seem to have been learned very quickly — for example in the successful night attack of 14 July 1916. By 1917 the infantry had moved completely to platoon tactics, and many of the problems with artillery cooperation were well on the way to being solved. On the first days at Arras, Messines and even — perhaps surprisingly — in some of the Ypres attacks, it was a finely judged mastery of low-level tactics which led to devastating successes. At Cambrai the innovative use of predicted artillery fire helped the tanks to advance further than they had in earlier battles, while close air support was already commonplace. In technological terms the British were ahead of the Germans in practically every department, in quality as well as quantity, and we shall see just how impressive this overall achievement could be. By 1918 there was definitely an expectation of a smaller proportion of casualties in each attack than there had been in 1916, and a justified expectation that fewer of the attacks would fail.

Nevertheless, there did continue to be some tactical failures until the very end of the war. One of the primary reasons was the reluctance of generals to call off battles which had run out of steam, or to cancel attacks which had been ordered too hurriedly or without proper preparation. There were also undoubtedly too many damaging institutional clashes between arms, especially when novel weapons with specialist operators had to be integrated with the traditional infantry battalions. Nor, around the end of 1917, did the British seem capable of mastering the art of defence. This would be their undoing at Cambrai and in the spring of 1918, although it is scarcely fair to use these lapses as a criticism of British tactics in the offensive. Sound principles had already been worked out for that during the course of 1916, and in some aspects even earlier, and for the remainder of hostilities it was largely a matter of translating the theory into routine practice.

The larger second half of the war

The tactical history of the war may be approached from two opposite ends. The most obvious, perhaps, is to focus especially on prewar visualisations and preparations, and the way they were translated into battlefield action from 4 August 1914 up to the fatal culmination on 1 July 1916. Through an investigation of the foundations of British tactical theory and practice, the general shape of the finished edifice may perhaps be glimpsed. Broadly speaking, this approach was indicated by many of the most famous poetic protesters against the war, who often joined up early

Introduction

and were more or less burned out by the end of 1916.[24] Their opinions were often based more upon their personal experiences of early-war muddle than upon a more reflective analysis of whatever late-war tactical mastery may have actually been achieved.

A somewhat similar early-war bias was also reflected in the British official history, which gives a disproportionate number of its volumes to the relatively small-scale first two years. The same practice has been largely followed by several recent commentators, including Dominick Graham in *Fire-power,* Tim Travers in *The Killing Ground,* and Peter Simkins in *Kitchener's Army*.[25] Its advantage is that it highlights the crucial process of improvisation and learning by which a peacetime BEF of merely ten divisions was able to supervise its own supersession not just by a 'nation in arms', but by nothing less than a 'world empire in arms' with over sixty divisions normally active in France. The first two years of the war represent the essential preliminary charcoal sketch over which the whole war-winning tableau would subsequently be painted in oils.

Unfortunately, however, the first two years of the British war effort also represent the time of greatest amateurism, blundering and fumbling. The retreat from Mons was actually handled far less well than many apologists have asserted, while not even apologists can do much to redeem the generalship displayed even in such comparatively minuscule actions as Neuve Chapelle, Aubers Ridge or Loos. Most of the serious fighting, moreover, was left to our French allies, with whom we were often on terms of such cordial hostility that their absolutely vast contribution has never to this day been fairly laid before the British public.[26] It was the sacrificial but ultimately vital French efforts in 1915 and 1916, when most of the techniques of modern warfare were still in their infancy, that gave Britain the breathing space she needed to assemble and equip a respectable army and make tentative experiments in the methodology for its use. It is not widely remembered in Britain that the BEF was heavily engaged in major battle for little more than thirty days between the Christmas fraternisations of 1914, and 30 June 1916.

Many telling points may certainly be made about the unpreparedness of the prewar army. Despite its enviable regimental cohesion and deep expertise in such diverse — and sometimes apparently mutually exclusive — fields as musketry, aviation, fieldcraft, logistics, camouflage, horse care and motor transport, the BEF of August 1914 was scarcely an instrument for total war.[27] It was in no sense designed to act as the dominant partner in the main battle against the main enemy in the main theatre — nor would its commander even deign to employ an interpreter in his initial dealings with French allies whose language he did not speak and many of whose justified claims he refused to entertain.[28] Behind this initial contingent of 'Old Contemptibles', moreover, lay the need for recruiting and training agencies that had not even been imagined before the war — let alone assembled — and an industrial machine that took well over a year to divert into full war production.[29] The higher staffs and junior *cadres* that the organisation-building process so urgently demanded seemed to be either non-existent or self-consciously suicidal.[30] Many new specialities also had to be improvised almost from nothing. Tanks and aircraft are obvious examples; but perhaps no less significant were signals. In 1914 these depended far more on semaphore flag-wagging than on Morse code, and relatively little use was made of telegraph or telephone at levels lower than divisional HQs. Wireless was deeply distrusted, and even the humble carrier-pigeon was seen as a boldly innovative 'new' technology. The signal service was seriously undermanned, possessed no recognisable chain of command, and came very near to collapse during the winter of 1915–16. Only by the time of the Somme was it starting to adapt itself properly to the demands of positional warfare, and there is a sense in which it can be taken as a barometer for the state of the BEF as a whole.[31]

The generals have often and justly been criticised for all these deficiencies, and it is difficult today to defend very much of their underlying planning, whether before or after hostilities had actually started. The question should not, however, be allowed to rest at that, since it gives us only one part of the story. By basing our view of Great War tactics mainly on the first two years, we are surely restricting our ability to do full justice to whatever tactical mastery was finally achieved during the remainder of hostilities.

It may perhaps now be salutary to look at the whole question from another perspective. By clearing our minds of the many hesitancies, misconceptions and mistakes through which the BEF almost inevitably had to pass before it could take on the full burden of the second half of the war, and by making even undue allowances for early fumbling, we may be better equipped to concentrate properly on the true quality of the final 'finished product' itself. If we focus on the way the British fought their war between 2 July 1916 and 11 November 1918, we shall surely achieve a clearer vision of their eventual tactical achievement — or lack of it — than if we cloud the issue too much with either the heroic passing of the old army in 1914 or the probably necessary — albeit certainly deplorable — initiation rites suffered by the fledgling New Army on the 'first day' of the Somme.[32]

If it turns out that a more or less 'acceptable' or 'best available' methodology for the assault was being put in place by the end of 1916, regardless of whatever tortuous channels may have been necessary to achieve it, then the high command may in a sense be absolved from blame — at least at the tactical level — for the subsequent losses and frustrations. Conversely, if British tactics remained relatively unprofessional and incompetent throughout the second half of the war, then the high command must certainly still shoulder a full weight of obloquy.

Introduction

It is the aim of the present book to investigate precisely how well or badly the mature British imperial war machine conducted tactical attacks during its larger-scale and more high-tempo second half of the war. This was a time when the high command presumably enjoyed the full perspective of hindsight over its earlier mistakes, and ought to have known just what to do and what not to do. How far could the army finally adapt itself to the changed conditions and exploit its new technologies such as mortars, rifle-grenades, Lewis guns, predicted artillery fire or plentiful 'second generation' shell? Conversely, just how far was it dragged down by technological inadequacies or equipment shortages, let alone social rigidities, professional intrigues or the cosy myopias of 'château generalship'? If the army of 1916–18 was ultimately victorious, does that reflect an inherent ability to reform itself in the tactical field, or was it simply a case of the big battalions inevitably winning a war of attrition despite slowness and failure in their self-reform?

As the war entered its second half there was certainly a remorseless process at work by which the more relaxed and discursive officers were replaced by hyperactive, ambitious and hard-hearted 'thrusters'.[33] Poetry may have been intensified among the front-line soldiers, but it successively fled from the high command. This hardening of attitude was partly a matter of tightening the grip held by the top over the bottom of the hierarchy, to prevent the spread of malingering or fraternisation, and partly an expression of the distrust felt by generals — who came from the old regular army — towards the 'temporary gentlemen' and 'New Army' civilians in uniform who were increasingly filling both the ranks and the officer corps. Similar motives also doubtless inspired the multiplication of training schools and tactical manuals, together with the ritualisation and miniaturisation of the offensive spirit into a policy of constant trench-raiding.[34] Yet if seen in a different light, all these measures could equally be attributed to an increased awareness of the need for improved tactics and fighting methods at a time when the war was becoming inexorably more professional. Thus a 'thruster' may legitimately be seen as a clear-headed and efficient tactician as much as a careerist; a training manual or school may legitimately be seen as a means of winning battles more cheaply, and a raiding policy may legitimately be seen as a way to gain information, tactical expertise and a useful routine mastery of No Man's Land.

The degree to which tactical change was really driven by considerations of social control, as opposed to purely practical responses in the face of new battlefield conditions, makes a very fascinating question which has by no means been decided. Despite the 'general' or 'war and society' historian's natural leaning towards the former interpretation, therefore, we should always remember that the purely tactical historian may still be able to put forward a few telling counterarguments in support of the latter.

A further reason for concentrating on the second half of the war is that

there seems to have been far more of it than there was of the first half —
and hence a larger volume of accessible evidence now survives. Whereas
the British offensive battles of 1915 were niggardly in scale and short in
duration, the British and Dominion effort in the last two years of the war
was prodigiously vast. More than six million men came under arms, not
to mention an unprecedented mobilisation of civilian industry, agriculture, shipping and the press. Each big battle could easily involve dozens of
divisions and thousands of guns, and during the Hundred Days alone
there were more than 160 divisional attacks which used tanks. A 'battle'
might well be continued for months, on and off, even though each individual division might be engaged for only a few days at a time, and these
might themselves often each include many hours of inactivity. In its last
two years the BEF certainly presided over some of the most important
episodes in the whole history of the Western Front — which was also the
decisive front. Hence despite the continued muzzling of the press by the
high command and the muting 'long war' effects of exhaustion and routine upon eyewitnesses, we do surely have access to many more details
about British tactics for the second than for the first half of this conflict.[35]

This question of scale is worth careful analysis, since many people
imagine the Western Front to have been one vast uninterrupted battle,
scarcely varying in size or intensity from one month to the next. The
impression is reinforced by published war memoirs from some of the battalions which were particularly heavily engaged, most famously the 2nd
Royal Welch Fusiliers (2/RWF). Thus Robert Graves's *Goodbye to All
That,* a 1929 bestseller, paints a vivid picture of repeated and interminable
combat. When the line was static and 'all quiet' it could still be very dangerous to participants, and even in the 'small first half' of the war Graves
could report that 'casualties remained heavy from trench warfare'.[36]
When it came to launching attacks, furthermore, he contrives to give the
impression that his battalion routinely suffered a new Albuera (an 1811
battle in which the Welch Fusiliers had lost fifty per cent of their strength
in an hour) about once every couple of months through four long years.
He attributes this to the high command's faith in 2/RWF's fighting qualities, since 'No one will mind smashing up over and over again the
divisions that have got used to being smashed up'. And just to finalise the
point, he adds that one of his friends complained on the Somme that 'I've
had five shows [i.e. taken part in five attacks] in just over a fortnight…'.[37]

Graves is perfectly correct to report that 2/RWF, with a strength of
under 800 at any given time, suffered many times that number of casualties in the course of the war.[38] Yet this was undoubtedly an 'élite' unit
with a particularly aggressive tactical posture and, as Graves himself suggests, it was thrown into the cauldron of a 'show' more often than most.
It also arrived in France at the very start of the war and continued
through to the bitter end. But in all of this it was untypical of the BEF as

a whole, as has been very usefully highlighted by two recent analytical works. The first is *Trench Warfare* by Tony Ashworth, who convincingly demonstrates that a 'quiet' sector of the front really could be very quiet indeed, almost entirely devoid of the irony conveyed by the title of E. M. Remarque's book *All Quiet on the Western Front,* and that this happy condition might apply to over two-thirds of the line on any given day. He finds many instances when 'a day of battle' failed to include any actual fighting, and cites at least one battalion which 'fought' pretty continuously in the trenches for a whole year, yet suffered a total officer casualty list of just a single individual.[39] So much for the misleading popular idea that the infantry subaltern's life expectancy in the BEF was no more than a fortnight![40]

In their splendid military biography of Rawlinson, moreover, Robin Prior and Trevor Wilson have inadvertently underlined this same point. One of their major and well-founded criticisms of the high command in the battle of the Somme is that it rarely orchestrated a mass attack along the whole line, with a wide frontage, but allowed many weeks to pass in a series of small, piecemeal attacks that were badly coordinated with flanking formations.[41] The implication is that a so-called 'corps attack' might often actually consist of only a single division, which in turn might involve only one brigade, which itself might have just four companies in the front line. On any given day there might be less than half a dozen such attacks on the entire twenty-seven-mile frontage of the battle, or in other words an army with something like 250,000 infantry might be attacking with no more than perhaps 5,000 combatants, or 2 per cent of its strength. Doubtless the levels of individual stress and danger for these brave men could be every bit as appalling as for participants in a bigger attack; but the point is that even if absolutely all of them were killed, they would still represent a far smaller sacrifice than the 21,000 who died and the 36,000 who were wounded on 1 July. On that occasion there was a total casualty list of something like 33 per cent of all the infantry available for the battle, but the public seems to have run away with the idea that this was a habitual scale of loss in every day of every battle.

It is worth reiterating that 1 July was always seen as exceptional by almost everyone who had anything to do with it. It was the 'Bloody Sunday' or the 'Tiananmen Square' which stood out from the less dramatic, less deadly, but nevertheless far more significant and corrosive structural (or background) violence all around it. It was a bizarre aberration which would never recur, but which has indelibly marked the public imagination.

What tended to happen during most of the war from 2 July onwards was that the 5,000-man attack, incurring perhaps 1–2,000 casualties, became the pattern for the 'normal' day of battle.[42] There were indeed some bigger, more coordinated 'shows' which deserve more prominence in the history books than they have actually received, but they tended to be special and rare. If we take the (very major but oddly forgotten) Vimy–Arras–Bullecourt battle of April 1917 as an example, the action fell on only three out of the five BEF armies, and within them on only about seven corps from a total of something like eighteen in France at the time. Admittedly some thirty-three infantry divisions were eventually involved in one way or another — making for almost half the BEF, and some of these were sent through the mill twice over — but the actual attack frontage was sustained by no more than half that number of divisions at any given time. During the six weeks of the battle, therefore, only something like a quarter of the BEF was actively engaged at any time. Hence it could be said that on average during the whole — otherwise normally 'quiet' — twenty-one-week period between the end of the Somme in November 1916 and the start of Messines in June 1917, approximately only a third of this quarter of the BEF (i.e. just 8.3 per cent of the total) was in fact heavily engaged at any given time. Even then there were really serious assaults in the Arras battle on no more than twenty-six of its days — and never on more than two-thirds of the total frontage — leaving at least fifteen days on which the whole battle front was 'active but without serious engagement', and considerably longer than that for much of it. Admittedly some supposedly 'quiet' sectors well away from the Arras front also got a pretty hot time during the spring of 1917, but many others were almost completely stone cold, and entirely given over to the 'Live and Let Live System'. In any case, in neither 'quiet but hot' sectors nor in 'stone cold' ones did the intensity of action ever approach anything like the level normally experienced on a truly serious attack front.

To take another example, August 1918 is often portrayed as an especially active time for the BEF, but actually it mainly involved just Fourth Army for four days at the battle of Amiens and then Third Army for four days in the battle of Albert — in neither case with very much more than a dozen divisions engaged. The battle of Albert widened out to Fourth and First Armies only during the final seven days of the month, although Second and Fifth Armies still did not take part, since their divisions were not yet considered to have recovered from the fighting in the spring.

Five BEF armies were in the field during the first 243 days of 1918 — making something like 1,100 'Army-Days' of presence in the front line up to 1 September — but of these no more than 100 'Army-Days' (or 9 per cent) were taken up in heavy action. After 4 September, moreover, the onus of the offensive was lifted off the BEF almost entirely until the formal assault on the Hindenburg Line on 27 September. Admittedly many of the remaining forty-six days between 27 September and 11 November were closely engaged by all five armies — making a potential maximum of 230 'Army-Days' — but it still remains true that far from all of those days were actually dedicated to close or fierce action, and certainly not on every front at once.

We may further refine all this by reference to the composition of the BEF itself at various times. By June 1915 it had no more than 23 infantry divisions, but by the end of 1916 it had become standardised for the remainder of the war at something between 50 and 70 divisions. The increase is striking, and we must agree that there were far more British soldiers in the line in 1917 than there had ever been in 1915. The period between June 1915 and the end of 1916 was the main time of expansion, and 1 July 1916 was very much the moment when the British really entered the war in the strength at which they were destined to carry it on to the end.

If we now return to the question of when the main battles were fought, we find that even after we have made all due caveats for quiet times and quiet sectors, the balance of serious action still rests heavily in the second half of the war (see Table 1). It is only on and after 1 July 1916 that the main bulk of the British war effort was able to get fully up to top speed, so that date may be said to represent a very significant 'changing of gear' — however graunchingly the synchromesh may have been strained by the experience. Before it, the sheltered British staff officer must surely sometimes have been tempted to think that the whole 'show' was somehow, at some subconsciously ultimate level, essentially frivolous and unserious — a friendly but relatively small helping hand offered by John Bull to what was essentially a Franco-Belgian war against the Boche, and stemming from little more than altruistic motives. After 1 July, however, he must surely have perceived that the matter had suddenly become bleakly serious: it had turned into a vicious grudge-match in which Britain had somehow, and perhaps even despite herself, become very centrally engaged indeed. From 2 July onwards 'the British staff officer' could surely no longer see the war as essentially a French pantomime, and he must have experienced a correspondingly heavy realisation that his mistakes might alter the whole history of the world, and not merely his personal reputation or his unit's honour.[43] With that in mind, it would seem that he worked harder to reduce the number of mistakes and increase the fighting power of his unit.

The present book will try to assess the eventual achievement of 'the British staff officer' and his subordinates in the tactical field during the last two years of the war, especially as regards the infantry. Its hope is to tap the wealth of available evidence in depth, if not in breadth. It cannot claim to be exhaustive or particularly systematic in its research, since it is not written by a long-term expert in Great War studies. It can, however, at least hope to illuminate some important features of tactical history that have often been overlooked amid the very heated debates about other aspects of the Western Front.

Table 1: Approximate number of 'divisional battles' fought by the BEF

This shows occasions on which a BEF infantry division was committed to a major action, usually lasting several days, as opposed to mere 'line holding'.

Principal Source: appendices and footnotes to Edmonds's 14 vol. OH *Military Operations France and Belgium,* and Becke's 8 vol. OH *Orders of Battle.*

Note that two sets of figures are presented here. The first is my own subjective set, weighted for what I perceive as the true intensity of the fighting, and including the colonial divisions. The second, in parentheses, is the official listing, derived from Becke's OB, which is more exhaustive (or pedantic if you will) for British divisions, but unfortunately excludes the colonials. I have not attempted to assess an 'official' figure for the colonials.

(i) 1914–15
(opening moves, 1st and 2nd Ypres, Neuve Chapelle, Loos etc.)

UK 'Divisional Battles'	56 (or 140)
Non-UK 'Divisional Battles'	9
Total	65

(ii) 1916
(with the overwhelming majority in the Battle of the Somme, July–November, including approx. eighteen to twenty during 1 July and its immediate aftermath)

UK 'Divisional Battles'	59 (or 183)
Non-UK 'Divisional Battles'	10
Total	69

(iii) 1917–18
(Arras, Messines, 3rd Ypres, Cambrai, 1918 March, Lys, 100 Days etc.)

UK 'Divisional Battles'	158 (or 726 — includes 291 in 1917)
Non-UK 'Divisional Battles'	47
Total	205

By either of these criteria, the five-month battle of the Somme, 1916, turns out to have been the biggest single BEF 'battle', and on approximately the same scale as all the BEF action that had taken place in the 23 months preceding it. The 23 months that followed it, however, included roughly three times as much action. Although that may technically add up to rather less action per month than at the Somme, it still makes a significant majority of the BEF's 'active war' as a whole.

Total no. of UK infantry divisions in France at some time during the war	58
Total UK 'Divisional Battles'	273 (or 909)
Hence average no. of battles per division	4.7 (or 15.7)
Total no. of non-UK infantry divisions in BEF in France at some time	17
Total non-UK 'Divisional Battles'	66
Hence average no. of battles per division	3.9

Introduction 19

Note, however, that this includes five Portuguese or US divisions which saw major action at most only once each, and it excludes the South African Brigade which was committed some seven times. ANZACs, Canadians and South Africans — not to mention members of the oft-overlooked 1915 Lahore and Meerut Divisions — might therefore argue that the true figures should look more like 73 battles by 12 divisions, giving an average of 6.1 per division: i.e. realistically more than the Pommy Imperials!

Against this there were six UK infantry divisions that fought in France for only a few weeks, which would give a corrected total of UK 'divisional battles' of 5.3 (or 17.5).

2

The Tactical Dilemma

> The war had become undisguisedly mechanical and inhuman. What in earlier days had been drafts of volunteers were now droves of victims. I was just beginning to be aware of this.
>
> Siegfried Sassoon, *Memoirs of an Infantry Officer*, p.104

The nature of tactics

In approaching the question 'what are tactics?' we immediately run into an important conceptual problem, since the very idea of 'tactics on the Western Front' has often been seen as a logical contradiction in terms, or even a complete nonsense. Trenchlock and attrition were the result precisely of a breakdown in tactics, and their effect was to stifle any further use of tactics for the duration of hostilities. Many participants complained bitterly that even in the relatively fluid 'open warfare' phases there was 'little scope for tactical manoeuvering...as flanks were still, in the big sense, "un-get-at-able"'.[1] Man and beast were bogged down by the mud and pinned down by the enemy's fire. They could do little more than return the fire as best they might, and endure. It was the side that endured longest that won the victory, not the side which manoeuvred most elegantly: 'Of the finer elements of tactics there were none. It was simply hammer and tongs all the time....'[2]

The implications of this view are that 'tactics' are all about finding open flanks and making wide movements at the quick step: about dazzling breakthroughs and the truly 'open warfare' which the cavalrymen of the high command always longed for but which never fully materialised after October 1914. To that extent 'tactics' became a symbol for an ideal war that did not exist, however much commanders like Haig or Gough might set up their battles on the assumption that a breakout was imminent. This expectation represented a very costly form of wishful thinking, and it was left to infantrymen such as Rawlinson or Plumer to plan more realistic and more methodical short advances with limited objectives. That approach, however, brought them under criticism for showing a supine acceptance of all the notorious evils associated with deadlock, attrition and siege. Such generals seemed to offer no end to the slaughter, and no hope of victory apart from mutual exhaustion. Their only common ground with believers in breakthrough, perhaps, was agreement that trench warfare was indeed a very different animal from what was normally known as 'tactics'.

Upon closer examination it nevertheless transpires that even the most static phases of trench warfare did have identifiable 'tactics' of their own. According to one authority, every use of musketry was automatically an act of tactics, since the two things were 'so absolutely interdependent and inseparable that it is impossible to discuss one, as a thing apart from the other'.[3] Even without going to this semantic extreme, one can agree that certain policies or doctrines could be relied upon to create one type of end result — for example, high levels of activity and casualties — whereas other methods produced the opposite effect. Some approaches could win tangible results such as information or security; others might lead to complete inertia and an epidemic of trench feet — which was normally seen as an indicator of poor morale. In one division the snipers might be energetically encouraged; in another there might be an emphasis on the artillery's counter-battery programme, and so on. Despite the wide gulf apparently separating such policies from the classical conception of tactics, therefore, there did in fact remain a very large arena in which the tactician could still exercise his skills.

Strictly speaking, 'minor tactics' are normally construed as any specific ploys or arrangements that are agreed upon for use by a small group of soldiers in battle, in order to achieve their immediate objective. These might depend on questions such as whether or not the Lewis gun will be deployed to a flank, or whether the bombing party can expect covering fire from trench mortars. At the start of the war, when there were no bombs or mortars at all and only two machine guns per battalion, this type of decision often applied to a whole battalion at a time. It would be ordained from on high, by no less a personage than the battalion commander, who would surely be a regular army lieutenant colonel. Later on, however, as bombs and machine guns became very much more plentiful, the decision devolved to companies — which were commanded by a decreasingly predictable social mix of acting captains, 'temporary gentlemen' or hostilities-only lieutenants. Later still the onus of tactics went right down to the platoon, which might perhaps be commanded by the regulation, if inevitably very raw, second lieutenant, but more often by the theoretical second in command — an NCO who would probably have volunteered in 1914 or 1915. There was even a quite widespread suggestion in 1917 that the section should be the tactical unit, and that might well turn out to be commanded by absolutely any type of lance corporal. In the prewar army a million-mile gulf had separated the regular colonel from the amateur NCO — yet by the end of 1918 the latter was often performing many of the combat functions that the former had imagined would forever be his exclusive preserve.

These changes represented an erosion of the powers not merely of the old regular army in general, but also of its more senior officers in particular. In these circumstances it is perhaps scarcely surprising that there was something of a backlash as senior officers became alarmed about what was going on, and attempted to multiply the methods by which they could hope to monitor and direct each local action. Whenever they granted some extra local authority to their juniors, it was clearly in their own interest to win back some additional counterbalancing control, rather than simply assume that their subordinates would automatically try to use their newly granted authority for the general benefit of the whole. The revolutionary decentralisation of tactics to corporals and sergeants was thus in some senses the most unsettling of all the war's innovations, and we should not underestimate its deep psychological effects. Among other things, it lent new meaning to the official historian Edmonds's apparently cynical scapegoating condemnation of 'poor junior leaders' in 1918,[4] and its equally misleading corollary that only the Germans knew about decentralised command. Not only was it horribly true that hundreds of thousands of natural junior leaders had already been lost to the army by March 1918, but it was also true that the objective tactical need for them had become far greater than ever before (see Fig. 1). Yet Edmonds found himself professionally unable to accept that they should ever have been entrusted with the sort of authority that they were actually being given. In an important respect, he was trying to deny the basic reality of the very war that he, above all men, has succeeded in making his own.

At a level higher than this sociotechnical revolution in 'minor tactics', the fighting methods used within the brigade, the division or the army corps are often termed 'grand tactics', representing a level of action which — although considerably higher than the platoon — is still lower than 'operational' or 'strategic' in scope. Whereas the level of 'operational art' in this war usually implied the participation of GHQ (the General Head Quarters of the BEF as a whole)[5] or at least of an Army HQ, grand tactics might embrace such things as the timing of a creeping barrage, the coordination of low-flying ground-attack aircraft squadrons or the provision of engineer detachments to improve the signposting along the roads to the front line. They would normally be the responsibility of corps and

The Tactical Dilemma 23

divisional generals and their command staffs, including BGGSs, BGRAs, Royal Artillery), CREs, GSOs1, DA and QMGs, ADVSs, RAF and often also Allied Army Liaison Officers. Among all these distinguished glitterati, the apparently humble BM stands out as a particularly significant individual, since he was in effect the chief of staff to each brigade and its BGC. As such, he was the essential link between the world of grand tactics and that of minor tactics.[6]

Fig. 1: (Notional) Comparison between the loss of trained tactical leaders and the need for them

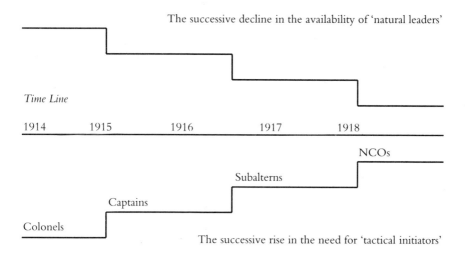

For our present purposes we shall not try to look at the operational and strategic levels of action except perhaps briefly in passing; but we shall concentrate specifically on grand and minor tactics, especially the latter. The relationship between these two tactical levels is also itself a matter of great interest, since in the conditions of the Western Front the higher staffs could no longer interfere directly in minor tactics, as Wellington had been able to do at Waterloo. 'Château Generalship' was a notorious feature of the Great War, whereby the practitioners of grand tactics lived in a totally alien physical environment from the exponents of minor tactics — and often had not even a working telephone to establish contact with them.[7] Higher battle handling therefore became exceptionally difficult unless everything ran exactly according to a very precise timetable; and if anything went wrong the front-line soldier would quickly find himself unsupported from the rear. In a sense this was the most important tactical difficulty of all, although it has often been overshadowded by fearsome descriptions of the shells, bombs and machine guns that actually did the killing. As the Second World War would show, however, these weapons could all be overcome provided good 'intercommunication' could be established between the front line and the rear.

Linked to this problem is the question of intercom between different arms and services, and especially between the infantry and its supporting artillery. That question had already been identified as central in many formations during 1915, and certainly by everyone in July 1916, not merely because many of the artillery's shells were badly made and tended to drop short, but because a realisation was dawning that the barrage now determined almost everything that the infantry was or was not able to achieve. In the past the infantry had usually been able to operate with relative freedom even if it had no artillery — and this was especially true in the British service, where the varied needs of imperial policing had frequently demanded light columns of foot soldiers unencumbered by extensive trains of cannon.[8] But now the 'storm of steel'[9] which was likely to greet any attack seemed so fearsome that some exceptional measures were needed to suppress it. Many such measures were suggested, ranging from tanks or gas to improved trench-raiding techniques that might help to keep the colonial traditions alive. For most tacticians, however, it was a more precise use of artillery that seemed to promise the best way forward. The main difficulty with it was nevertheless a very formidable one indeed, namely that the artillery was almost always frustratingly remote from the infantry it was supposed to be supporting.

The remoteness of the artillery was nobody's fault in particular, since it arose not only from the appalling difficulties of combat signalling in the conditions of the Western Front, but also from the deep-seated cap-badge exclusivity of the British Army. The artillery was, in effect, practically as remote as the high command, and became still more so as soon as it abandoned its traditional rôle of direct fire from the foremost infantry line. It changed its support for the infantry to more indirect, longer-range fireplans such as preplanned and pretimed barrages to help attacks forward, or instant SOS barrages to protect a presurveyed friendly trench that was being assailed. However, the disadvantage was that if the artillery was called upon to do something unplanned at a crucial juncture in a battle, problems were almost certain to arise. The front-line infantry would probably be cut off from communication with its own brigade HQ, which might not be fully efficient in passing information back to the guns. Conversely the gunners themselves might not have an observer far enough forward to read the battle correctly, or might not feel sufficiently confident in aircraft reports to initiate complete new fireplans on the strength of them.[10]

Similar problems pervaded the infantry's relations with absolutely all its other supporting arms. Trench mortar experts were execrated for their independence and aggressiveness of spirit, typically firing a few shots from the infantry's lines, and then disappearing before the inevitable retaliation arrived.[11] Machine gunners were berated for firing apparently futile overhead barrages while obstinately refusing to chance their arm in the front line. Tanks and cavalry often appeared from nowhere — or at least as often failed to appear — without adequate preliminary rehearsals in tactical coordination. In all but very rare instances air power was either something to be watched from afar, like a circus high-wire act, or something to find oneself attacked by at some particularly inconvenient or vulnerable moment. Engineers were known as officers who undermined your trenches, blew gas into your headwind or — most usually — took large parts of your battalion as working parties that had to hurry through the mud to some urgent rendezvous that was then not kept.[12] The infantry's grumbles and complaints were apparently endless, and the increasing sophistication and complexity of supporting arms as the war progressed only served to exacerbate them. The front-line infantry could apparently be satisfied only if it was allowed complete sovereignty over everything that affected it — but of course such a utopia remained permanently beyond reach. Not even the Germans could achieve it, despite their greater readiness than the regimental British to use all-arms battlegroups under a unified and permanent tactical commander. Such a system, in fact, often meant that the infantry was still further subordinated to staff officers and the other arms, rather than liberated from them. It is noticeable, for example, that the famous stormtroop tactics originated with the engineers (or rather 'Pioneers') and not with the infantry at all.[13]

All this means that to many Great War soldiers the concept of 'tactics' must have represented something akin to an acrimonious shouting match that we regularly conduct with staff officers and other arms over bad telephone lines. It was not so much that the execution was especially difficult (even though it might very well be hair-raisingly dangerous), but that the preliminary arranging and haggling certainly was. Six hours was established as the minimum time needed for an order to pass from Corps HQ to Company HQ;[14] but the same time limit also applied to even the smallest alteration to the details of the order, not just to the order itself. A divisional plan of attack might thus take several hours to prepare, most of the day to distribute and then still find itself stymied at the last minute because a few precise details of the barrage had to be changed.[15] The armchair commentator of the 1990s therefore finds himself astonished that anything was ever achieved at all, rather than that it ultimately seemed to achieve so little.

A rather different interpretation of the expression 'tactics' might perhaps be found in the idea of 'drill manuals', 'laid-down principles' or 'training guidelines'. Different types of imposed or top-down tactics affected different parts of the BEF at different times; but attempts were increasingly made throughout the war to extend uniform methods across all of it. Many of the battlefield habits of 1918 were already extant as theoretical precepts at least as early as May 1916; but it naturally took time for them to work their way down — through directives, guidebooks and training courses — into practical everyday application. Such guidance was also very often misunderstood, not least because the front-line recipients of good advice were likely to resent any suggestion that some detached rear-echelon intellectual could possibly possess a wider perspective on tactical needs than they did.[16] An unfortunate but probably unavoidable anti-tactical culture therefore developed — and it must surely also have had its parallel in the German service — whereby ordinary officers reacted with hostility against general directives from above which seemed to limit their freedom of action or even their use of language when talking about it. Sassoon put his finger on this problem when he complained that 'technical talk in the Army always made me feel mutely inefficient',[17] or when he tarted up his written plan for a trench raid with 'a little bit of skite about consolidation and defensive flanks',[18] simply because he thought it would please his superiors' sense of what was 'tactical'. In the event he knew exactly how to plan and execute an effective trench raid, but his problem was more a matter of countering GHQ linguistics than of overcoming the German defences.

At this point we have to confront the eternal philosophical problem of the relationship between theory and practice, or between intention and action. The two are notoriously difficult to match together, since all sorts of frictions can upset even the best-laid plans; and this was perhaps more true on the Western Front than at any time before or since. Tactical historians have often blithely assumed a near-total unity between what was taught and what was actually done; but the whole Western Front experience should serve as a dire warning against any such assumption.

In purely physical terms the Western Front was certainly a highly resistant medium for the application of theoretical 'tactics'. Drillbook formations could not be held together for long in any operation, even when everything was going smoothly. For example, when the 6th battalion of the Royal West Kents (6/RWK) took part in the brilliant five-mile advance to Monchy le Preux on 9 April 1917, they had just finished several weeks' intensive rehearsals on a replica of the terrain, and believed themselves to be highly proficient in keeping to section-strong 'blob' formations until ordered by an officer to deploy into lines. As they crossed No Man's Land under shell fire Captain Alan Thomas, a company commander, noted:

> At present the company was holding its general formation. But the ground was rough and pitted with shell-holes, and as the men scram-

bled in and out of them some of the 'blobs' were beginning to straggle.[19]

They then easily crossed the deserted German front line without any infantry fighting, but already Thomas observed:

> As for our 'blobs', they were hardly recognisable. They had in fact opened out of their own accord. Since we were now making our way over a system of deep trenches it was impossible for the men to keep to any fixed formation. Parties of them were advancing along communication trenches. Others were running forward over the top....
>
> Now our men, taking advantage of such cover as there was, were going forward more or less at random. Mingled with them were men of the flanking battalions, and they were taking orders, all of them, from any officers that were about.[20]

It was only then that the first Germans were sighted, although they were either running away or surrendering. For his part, Thomas completely lost his company and finished the day with soldiers of another battalion.

Even if things went well, as in the case of 6/RWK before Monchy, the success itself would be so disruptive that consolidation immediately became a major undertaking in its own right, and could sometimes continue for several days. But if the enemy offered resistance, the extent of disruption would naturally be immeasurably greater. Not only would any drill formations be destroyed, but the movement of essentials such as food, ammunition or messages would also be greatly hindered. Add to this the fact that key personalities always seemed to be struck down at the most critical moments, not to mention the still more disruptive fact that any battalion committed to an attack tended to suffer heavy casualties. It only took one unsuppressed German machine gun, or one minute's accurate counterbarrage, to blow all tactical plans apart. The 'real' tactics that were actually used on the ground therefore represented something completely different from the theoretical tactics of the drillbook.

As if all this were not enough, throughout the war most battalions, brigades and divisions liked deliberately to throw away much of their received wisdom and top-down drill teaching. They invented their own characteristic tactics and operating procedures, often due to the idiosyncratic whims of an individual commander. The whims of commanders, in fact, were always a very major determinant of how every battle would be fought. Just as Lieutenant General Maxse was free to decide that his XVIII Corps would henceforth conduct 'platoon training' even though the platoons were sometimes only four men strong,[21] so his deeply unpopular Major General Pinney was free unilaterally to decide that the

33rd Division would fight its whole war without the solace of alcohol. Within that division Brigadier Heriot-Maitland regularly chose to use the conferences of his 98th Brigade as merely an occasion for his own monomaniacal monologues; while in due turn his Lieutenant Colonel de Miremont, commanding 2/RWF, remained perfectly free to impose a unilateral ban on the use of tommy cookers in the freezing trenches of January 1917.[22] By the same token, equally, the staff officers of Furse's 9th Scottish Division were free to initiate such a heavy use of smoke shells in their barrage at the battle of Arras that they used up Third Army's entire holding of that particular munition.[23] Within this division Brigadier Maxwell in 27th Brigade was yet another believer in 'talking to the men'; but his formula also included emphasis on contact sports, smart turnout, liberal use of the bayonet — and winning control of No Man's Land by superiority in sniping. Against these he strongly opposed both the use of trench periscopes and acceptance of orders from above which banned the use of blankets in the front line.[24] Within 27th Brigade Lieutenant Colonel Croft, commanding 11th Royal Scots (11/RS), naturally added his own highly personal twist, insofar as he had a particularly idiosyncratic aversion to both dawn attacks and — in the interests of musketry excellence — the issue of grenades to assault troops.[25]

In the face of such an anarchy of personal whims, prejudices, eccentricities, partial solutions and even downright lunacies, the modern reader is tempted to despair of finding any coherent tactical doctrine in the BEF at all. We are reminded very precisely of the tactical anarchy that obtained throughout the hastily improvised armies which had to fight the American Civil War. Nevertheless, it is worth noting that many of the officers involved in taking these diverse decisions were individually celebrated for their strong tactical expertise and success. Pinney, Heriot-Maitland and de Miremont were not perhaps in anything like a high-flying league; but Maxse has recently been held up as the very epitome of the genuinely effective red-tabbed tactician, and Furse, Maxwell and Croft should scarcely be placed very far behind. They in turn can all be trumped by even more colourful — not to say bizarre — characters who still nevertheless somehow managed to deliver the tactical goods. Brigadier F. P. Crozier, for example, was promoted as a 'thruster' and did well in many battles, but the secret of his success seemed to rest to a considerable extent on his willingness to use his revolver on his own men whenever they showed signs of wavering. Major General Bethell was another notorious pirate, who re-created 66th Division from scratch in the autumn of 1918, stealing guns, ammunition trains and even whole infantry brigades from their officially intended formations, to forge them all into a successful vanguard for Fourth Army's October advances.

Emerging from all this, the last definition of 'tactics' that we should consider is the whole question of personnel policy. It has often been said that the most important skill a general must possess is not so much an ability to inspire men or tell them how to fight, but an ability to promote talented subordinates and dismiss weak ones. In this perspective one of Sir John French's worst mistakes was perhaps to get rid of the talented Smith-Dorrien after the Second battle of Ypres, although one of his better decisions was to replace him by the meticulous Plumer. We find many similar stories enacted on a smaller scale at lower levels of command, where senior officers often saw their key decisions as a matter of selecting particular subordinates for particular tasks — or for the sack if they were thought to have failed. The philosophy of the 'thruster', which increasingly became the dominant philosophy of the BEF as a whole, was that a bad CO could spoil a good battalion within three months, and would probably be personally responsible if his attack did not succeed. Any senior officer who appeared to be 'tired' or 'old in body or mind' therefore had to be ruthlessly weeded out.[26] If he in turn could escape retirement by setting up one of his own subordinates as a scapegoat, he might well try to do so.[27]

More ink has probably been spilled over scapegoating and related issues of personnel selection than over any other aspect of Western Front leadership, and by external commentators almost as much as by self-interested autobiographers. Yet the whole question almost inevitably remains irritatingly inconclusive and subjective, and is too specialised for coverage in the present volume. Let us simply agree that the hiring and firing of subordinates remains among the most important aspects of all leadership or generalship, and perhaps even of 'tactics'. Though it has little direct connection with the technicalities of weaponry or the precise direction of manoeuvres, it is one of the principal ways in which a commander can hope to shape the outcome of his battles.

Lines, densities and timescales

In the field of operational art it was discovered in late 1914 that a number of important factors — such as improved firepower, the obstruction of the battlefield by shell craters and barbed wire, and especially the sheer volume of rail-mobile manpower available — had conspired to give the defensive an unprecedented robustness. For the first time since the late Roman Empire an army could spread itself out in a more or less straight line, several hundred miles long but with no very great depth, and yet still retain very great defensive strength. According to eighteenth- and nineteenth-century operational theories such a formation should have been suicidal, since dispersion had always spelled weakness. Conversely, impenetrable strongpoints had always, almost by definition, had to be small and concentrated. Only prepared and flankless positions like Sebastopol or

Port Arthur had been able to hold out for long periods against great odds, preferably when they contained field armies gathered on a frontage of only a few miles. For example, Jackson's line of field fortifications at New Orleans in 1815 had been less than a mile long; Wellington's at Torres Vedras in 1810 had been twenty-nine miles at most, and not even the famous trenchworks of the American Civil War had very greatly exceeded these dimensions, with Petersburg being held by just fifty-three miles of fortification — and even then it had eventually been decisively outflanked. Indeed, a major lesson of that war had been that if you can't take a position frontally, you can always expect to outflank or bypass it. The Japanese had clearly taken this lesson to heart at Mukden in 1905 when they broke the trench deadlock by their mobility. It may be worth noting, incidentally, that this notorious 'greatly prolonged trench battle' lasted merely twenty days, and even then many experts believed that its duration could have been much reduced under European conditions, if the strategic imperative for a quick victory had happened to be more urgent.[28]

None of New Orleans or Torres Vedras, Sebastopol or Petersburg, Port Arthur or Mukden may really be cited as a valid precedent for the Western Front, since in none of those cases was there a flankless position more than a few miles long. In France from October 1914, by contrast, it was found that field armies had at last become big enough to fill the whole length of a national frontier[29] — in this case the 400 or so miles between the Channel and Switzerland. Paradoxically, this was very much a matter of the increased availability of manpower rather than of any thinning out that might have been made possible by improved firepower. At Torres Vedras and Petersburg the garrison's troop density was actually less than on the Western Front, at something like 1,000–2,000 men per mile as contrasted with perhaps 5,000 in the more recent conflict. To put this another way, the 1915 armies of around two million men on each side could presumably have been stretched to hold frontages of between 1,000 and 2,000 miles, if the technological conditions of Torres Vedras or Petersburg had still applied.[30]

This statistic should warn us that the generals of the Great War believed that a single defensive line[31] remained vulnerable to a concentrated or surprise attack, and not even the wonders of the new firepower or the new barbed wire could offer a total guarantee of security. Even if an attacking battalion had to lose ninety per cent of its strength in order to capture the enemy's front trench, it was an all too notorious fact that far too many battalions were indeed very ready and willing to pay that price. The popular view of the Western Front as a smooth, inviolable 'surface' is therefore highly misleading. The front line could always be punctured, at more or less cost, and it very often was. No defender could ever be sure that his front line could be relied upon to beat off an assailant.

This lesson was learned early. Within the BEF the most important revelation was perhaps the battle of Neuve Chapelle on 10 March 1915, when Sir Henry Rawlinson's IV Corps came within an ace of breaking through a single German defence line. This experience helped to encourage the British, especially Haig himself, to mount a series of further offensives — although these would unfortunately usually meet with rather less success until such time as the key tactical lessons had been properly learned some two years later. For the Germans, however, Neuve Chapelle immediately highlighted the lesson that at least two more defensive lines were needed immediately behind the first, so that even if the front line should be lost in an initial surprise, the battle could still continue and time could be gained for reinforcements to be brought in from other sectors.[32] As the war progressed, the same principle would be gradually extended to covering an ever greater depth of terrain with defences, until by the end of the battle of the Somme the Germans would have built no less than seven successive lines covering Bapaume, making a total zone of trench works some ten miles deep.[33]

Ultimately, all this was made possible only by the dense rail grid, which permitted unprecedentedly vast national mobilisations to be supported by unprecedentedly good logistic access. It was the awesome concentration of steam power available in this industrial heartland of Europe which guaranteed that such large numbers of green troops could be deployed straight into the firing line, and then plentifully supplied with food and munitions. In crude industrial terms, there could not possibly have been a better place to fight a war than the very rich border area between Belgium, northeastern France and central western Germany, which was the economic 'golden triangle' from which the European Common Market would eventually emerge. In 1914 it was the next best place to Birmingham, Liverpool and Manchester, in the whole wide world, to fight a high-tech war — in rather the same way as California's 'Silicon Valley' might today be considered the next best place after the five industrious islands of Japan and Taiwan.[34]

By late 1916 the world's munition industries had been mobilised to such an extent that virtually unlimited supplies of ammunition could be reliably delivered to railheads just behind the trenches. The chief restrictions upon their use, apparently, were only the build-up times needed to assemble them in ready-use dumps, and the sheer physical fatigue of the men who had to distribute them to the guns and then fire them. Even with light tramways and pack animals there was always a considerable amount of heavy lifting to be done by the soldiers themselves. Merely assembling and belting the ammunition for a machine gun barrage of a million rounds, for example, might require three or four days' work by a whole company of gunners.[35] The shells for a major bombardment, equally, might take weeks to assemble and camouflage, involving the sweat of thousands of labourers.[36]

Delays of this type meant that offensives could not be mounted quickly or easily, but that a defender could bring in new men from other sectors within perhaps forty-eight hours. His locally pre-positioned reserves, furthermore, could be committed to the battle within a very few hours, or even minutes. This was especially true with the Germans, whose doctrine called for immediate counterattacks, almost without consideration of the local tactical circumstances, whenever one of their front-line positions had been captured. There was hence an imbalance between the slow speed at which an attack could be mounted, which was mainly determined by the build-up of munitions, and the fast speed at which the defending enemy could call in extra manpower, by rail or road, to counterattack or plug a gap. The defender had a natural advantage in this unequal race, although of course in the longer term he, too, would also be forced to limit the pace of his activities in order to gather munitions.

If he was to overcome the defender's advantage of rapid reinforcement, an attacker had to choose between two different types of operation. The first and easiest was to mount a 'bite and hold' attack, in which he hoped to occupy a limited portion of the defender's front line before the latter could react. The attacker could then rapidly revert to a defensive posture on his newly won position, to beat off the inevitable counterattacks. This procedure was always perfectly possible on the Western Front and was repeatedly performed successfully, even though it is fair to comment that it was performed unsuccessfully far more often than it should have been — and it almost always involved a high level of casualties. Even so, and although it was scarcely a startling or decisive approach, it still represented progress of a kind, and it did at least have the virtue of demonstrating to the defending troops that they were being defeated. For all the losses, mud and muddle of the protracted nibbling attacks at the Somme and Passchendaele, it remains true that the BEF did eventually capture the main German positions in both those battles. This may have done little to raise British morale, but it surely helped prevent a collapse, just as it helped to depress the morale of the Germans — and in the case of the Somme it sent them scuttling still further back a couple of months later.

Secondly, a more difficult but potentially much more satisfying option for an attacker was to mount a 'breakthrough' operation. In this case the assault would aim to break right through all the defender's trench systems so rapidly that not even his fastest-moving reserves would have time to come into play. The gap would be made so big so fast that it could not be plugged at all. The cavalry — or the tanks, according to one's fantasies or one's cap-badge loyalty — would accelerate through 'the "G" in Gap', and the enemy's logistic rear would be successfully pillaged, his HQs ransacked and his railways torn asunder. This alluring possibility was always high in Haig's mind whenever he contemplated any battle, whether past, present or future, and it deeply coloured the way he regarded his subordinates. Thus with the cavalryman Allenby at Arras in 1917 Haig believed that he could be certain of a decisive breakout, and was bitterly disappointed when it turned out that not even 'The Bull' could manage the trick. With the infantryman Rawlinson, by contrast, his expections were lower and he saw it as his duty to keep reminding his subordinate of the need for more 'breakthrough' and less 'bite and hold'.[37] On the Somme Haig insultingly superimposed the cavalryman Gough above Rawlinson's head, in the expectation of a triumphant breakout by Gough's Reserve Army, including the Cavalry Corps, on about 2 or 3 July. In the event they were all, alas, disappointed; but this did not prevent Haig from trying again at Ypres the following July, when he superimposed Gough over the infantryman Plumer in a very similar manner. When Gough himself then started to argue for 'bite and hold', yet failed to bite into the enemy line very convincingly, Haig had the good sense to reinstate Plumer pretty rapidly — but still kept nagging at him to go for a breakout after all. There is a somewhat pathetic correspondence from the October battles around Passchendaele, in which Haig is constantly reminding Plumer to be sure to keep the cavalry well towards the front, 'just in case'; but Plumer constantly replies that he will certainly bear it in mind but doesn't think it is very likely to work.[38]

No true breakthrough was ever achieved on the Western Front by either side. In part this may be attributed to the unrealistic optimism (or cap-badge pride) of cavalrymen — and indeed of tankies. In part, however, it may also be attributed to the unrealistic pessimism of infantrymen who failed to look sufficiently deeply into the technique of leapfrogging fresh formations forward through the initial assault troops, or of giving them sufficient mobility. The second bound will always be more difficult to envisage than the first, and the third than the second, and so on. In the conditions obtaining on most of the Western Front after the 'one line' battle of Neuve Chapelle, something like five distinct bounds would usually have been needed to achieve a breakthrough, with the last ones extending well beyond the range of the gun positions that had supported the first, and hence necessitating an extremely ticklish leapfrogging forward of the guns themselves, at the same time as the attack was still in progress.[39]

Ultimately the complexity of such a proposition proved to be too much for even the most imaginative generals to achieve within a forty-eight-hour timescale, and normally the best that could be achieved was just two or three bounds before the attack lost impetus. A 'five phase' attack was found to be impossible within a single surge forward, hence in one form or another the generals were all compelled, sooner or later, to revert to the more modest 'bite and hold' assumptions of the infantry. By the time of the Passchendaele battle it had become generally accepted that there should be a logistic pause of several days between each lunge.

The cavalrymen hoped that these 'several days' would still represent a shorter time than the enemy needed to reorganise his defences, thus still allowing the attacker to move faster than the defender's reaction time; but in practice there were always too many sources of friction to permit this to happen. Whether because it was raining, or because the troops were too tired, or because some commanding strongpoint had not been captured in time, the planned two- or three-day interval between bounds would normally stretch out into a week or a fortnight. The defender almost always won more time than the attackers had estimated, and so he could almost always keep at least one jump ahead of the game. Even in the Hundred Days in the autumn of 1918, when the BEF had achieved an unquestioned moral and material superiority over the Germans, there remained an important feeling that the attack could progress only relatively methodically, one step at a time, and that even the most desperately unsupported enemy rearguards could still pose a significant problem, at least in terms of the time needed to suppress them.

One element that might greatly help or hinder an offensive was the choice of ground. At Ypres in the autumn of 1917 the Germans held the high ground and the outer circle of a salient. They could overlook the British preparations and bring down converging fire from over 500 guns.[40] This greatly complicated the British problem, especially when it came to finding and suppressing the enemy artillery. The fact that the whole area had already been a battlefield for three years had also left it heavily cratered and practically impassable, at the same time as it had alerted the enemy. They had been able to prepare the battlefield with concrete pillboxes in depth. At Cambrai on 20 November 1917, by contrast, the contours were generally downhill from the attacker's start line; the ground was unbroken, and there was no great salient. Admittedly there was a canal to cross, and Bourlon wood topped an awkward hill towards the rear of the German position. That position was itself the formidable Hindenburg Line, with its notoriously deep belts of wire. Nevertheless, it was recognised to be a 'quiet' sector, and initially the Germans had a mere thirty-four guns covering the front.[41] In these circumstances the attack almost inevitably made famous headway.

Regardless of whether an attacker set out with a 'bite and hold' or a 'breakthrough' plan, perhaps the most potent weapon in his armoury would almost always be surprise. The attacker who could successfully keep his preparations secret before his zero hour could usually count on enjoying a relatively free hand for something like the first forty-eight hours of the battle. The British made very significant progress both at Cambrai on 20 November 1917 and at Amiens on 8 August 1918 largely because they achieved surprise, and in particular because there was no preliminary artillery bombardment to forewarn the enemy on either occasion. In both cases, also, the attack had been decided upon and assembled in a relatively short time — weeks instead of months — which helped concealment, even though it implied a shortage of reserves for the subsequent phases of exploitation. At Arras on 9 April 1917, by contrast, there was no real surprise because of extensive logistic preparations which could not be concealed.[42] Allenby's desire for a two-day hurricane bombardment was also overruled by GHQ, which substituted a preliminary bombardment starting over two weeks before zero and culminating in four days of intense fire. Luckily, however, in the event it was found that the Germans had expected the bombardment to continue for a week longer than it actually did, and had positioned their reserves too far in the rear to make an immediate intervention. This meant their defences could be solidified only around 11 April.[43]

If Allenby was lucky at Arras, Rawlinson was less fortunate on the Somme on 1 July 1916, as was Gough at Third Ypres on 31 July 1917. On both these occasions there had been a very obvious build-up followed by a long bombardment, and hence a painful absence of surprise. Nor had the Germans maldeployed their reserves. Their defences were therefore 'solidified' almost as soon as the British attack had been launched, and it would be only in some of the later phases of the two offensives that the Germans would be wrong-footed. On 14 July 1916, for example, they were tactically surprised by the night attack between Bazentin le Petit and Longueval, although operationally they had plentiful reserves close at hand, and the battle soon became a fierce, slugging deadlock. At Passchendaele, by contrast, there may have been few tactical surprises, but the sheer weight of British shelling eventually persuaded the Germans to reposition their reserves further to the rear. Ludendorff would then complain that the BEF's attacks of 20 and 26 September had caught out his front line with its reserves too far away to intervene usefully, just as had happened at Arras.[44] There were obviously some important practical limitations which restricted the depth to which 'defence in depth' could be stretched.

Besides, at the highest operational levels the positioning and timing of an offensive usually depended far less upon the vibrant technical question of how it might be converted into a decisive breakthrough than upon more mundane political questions such as the esteem in which Haig was held by the prime minister. Still more to the point were the military requirements of Britain's allies, which usually meant the French. For example, the Somme battle had originally been designed to relieve the pressure upon them at Verdun; its positioning had been dictated by their desire to attack alongside the British and its timing, almost down to the minute, was determined by the needs of their artillery. Admittedly they would eventually bear a third of its losses, but liaison and coordination at the interarmy boundary made a constant nagging headache for Rawlinson's Fourth Army. Then again, the Arras battle was timed to support the ill-starred Nivelle offensive on the *chemin des dames,* and positioned to effect a grand encirclement, in cooperation with Nivelle, of the Noyon salient. This left GHQ with little room to make its own operational decisions.

In the case of Passchendaele there has always been a suspicion that the site was chosen out of nostalgia for Haig's finest hour in late 1914; but more credibly it arose from the Royal Navy's need to strike at the U-boat bases on the Belgian coast, linked to Admiral Bacon's longstanding but ultimately abortive plans for amphibious landings. These considerations merged into a complex political compromise between the French need to cover up their summer mutinies and Haig's personal need to demonstrate that his forces were performing essential service somewhere — anywhere — on the Western Front, so that they should not be held back in England nor sent off to support the Italians. The shape of the battle was also very deeply moulded by Haig's bare-knuckle conflict with the severely antimilitary 'Easterner' Lloyd George and his paradoxically anti-Army, but pro-technology 'Westerner' protégé, Churchill. By late 1917 these powerful political actors had regained the ascendency over GHQ which they had lost after Gallipoli in 1915, and they were thenceforth able to undermine Haig's strategic plans at almost every turn. Their agility and fluency in politics were especially starkly contrasted with his own semi-inarticulate image, which they were not slow to caricature for the benefit of the wider public.[45]

On 26 March 1918 Foch was appointed supreme allied commander, taking precedence over Haig and Pétain alike. Some such appointment was by this time four years overdue, and especially useful in view of the increased complexity of an alliance which had now grown to include large American and Portuguese formations — with even a few Italians and Russians making occasional cameo appearances in France. From Haig's point of view, however, Foch's new appointment was scarcely a welcome development. It inserted an additional level of authority above his head, in which his hyperactive opponent, Lieutenant General Henry Wilson, chief British representative at Versailles, was free to exercise his Machiavellian politics. If this hierarchical arrangement had been introduced at the very start of the war it would doubtless have run just as smoothly and with as much mutual benefit as the Anglo-American joint staffs would be able to manage in the Second World War; but as it was, Foch's belated appointment led to many recriminations. It helped to focus the stored-up anti-French feeling that had always been widespread in GHQ, and which was perenially fanned by constant French demands for the BEF to take responsibility for additional mileages of front (see Table 2). The French quite naturally objected that they had to hold something like three times as long a frontage as the BEF, and were unconvinced by the latter's arguments that it was all in an 'inactive' sector. Still less were they impressed by Haig's complaints in the winter of 1917–18 that the British Empire had run out of manpower.

Table 2: *Length of the British line in France (in miles)*

(Source: 'Abstract of Statistics', pp. 639ff.; OH 1918, vol. 5, pp. 4–11)

1914	1915	1916	1917	1918
23 Aug: 25	20 April: 36	22 Feb: 42 + 25	25 Feb: 105–110	4 Feb: 123
16 Sept: 20	25 Sept: 40 + 30	30 June: 80–90	25 April: 90	2 April: c.102
20 Nov: 24	(= 2 sectors)	31 Dec: 85–90	20 June: 4 + 86	9 April: c.105
			9 Dec: 95	17 April: c.101
				21 May: c.88★
				22 July: 93
				11 Aug: 101
				28 Aug: 90
				18 Sept: 83
				16 Oct: 93
				11 Nov: 64

(★Excludes 3 British Divisions on the Aisne, but includes some French at Meteren.)

Expressed graphically (with each 'x' representing a month, and each row of 'xs' representing the occupation of about 20 miles), this looks as follows:

```
                                                               xx
                                       x     xxxxxxxxxxx    xxxx  xx  x
                                 xxxxxxx    xxxxxxxxxxx    xxxxxxxxxx
                       xxx     xxxxxxxxxxx  xxxxxxxxxxx    xxxxxxxxxx
              xxxxxxxxxxx      xxxxxxxxxxx  xxxxxxxxxxx    xxxxxxxxxx
xxxxx         xxxxxxxxxxx      xxxxxxxxxxx  xxxxxxxxxxx    xxxxxxxxxx
```

For most of the war the key players at GHQ were probably in the wrong over this particular issue; but they nevertheless perceived it as a perpetual injustice under which they had to labour. An authentic reflection of their spleen certainly comes across in Edmonds's official history, not least in his last two volumes covering the Hundred Days.[46] They remind us that the important operational decisions for siting and timing the battles of 1916–18 emerged out of a fierce political struggle between GHQ, its rivals and its allies, and not out of any calm contemplation of the tactical realities.

The storm of steel

We, however, do now enjoy the luxury of being able to look in from the outside at more strictly tactical concerns, so let us consider the 'storm of steel' encountered by anyone who rose from the trenches to cross No Man's Land. Even to raise one's head above the parapet in daylight was to court a sniper's bullet, and many are the stories of periscope mirrors being drilled through, or of the sudden death of those who forgot to duck at points where the parapet was particularly low. Rifles could be laid on fixed lines during daylight, moreover, so that they could still be fired during the hours of darkness to hit the well-known spots where sentries were liable to be peering out.[47] When it came to resisting an attack, even a relatively few riflemen could develop prodigious volumes of aimed magazine fire. Maybe not everyone could reach the 15–20 rounds per minute claimed by the Old Contemptibles of August 1914, but it seems clear that the rifle could be a very powerful weapon in defence — if only its owners were willing to stand up and fire it.[48]

Many of the same considerations apply with still more intensity to machine guns. Their heavy mountings allowed them to be set on fixed lines very accurately, so that they could beat No Man's Land systematically or skim the whole length of the enemy's parapet at will, in any conditions of weather or light. With a total replenishment and firing crew of less than ten men — and heroically even at times just one man — a single belt-fed Maxim could sweep an area up to 2,500 yards deep and perhaps 500 yards wide. In favourable terrain it could halt a whole battalion dead in its tracks by its insistent hail of shots — a complete 250-round belt each minute was by no means a difficult rate of fire to sustain. Since it was normally crew-served, moreover, its operators could profit by the extra group cohesion associated with such weapons. Certain 1940s studies of combat psychology have shown that members of a team would be more ready to open fire and keep on firing than individuals who felt themselves more isolated.[49]

The machine gun of 1914 could nevertheless still be seen as representing merely a modern and automatic form of what was already a very ancient tactical genre. In some ways it was simply a new shorthand version of 'direct fire musketry', such as had previously been delivered by massed regiments of Cromwell's arquebusiers or highly drilled companies of Frederick's Pomeranian musketeers. It admittedly enjoyed considerably greater predictability and effective tactical range than most of its hand-held predecessors, and it would also soon extend its repertoire to include many complex forms of indirect, plunging and even unobserved area fire. However, for most defensive purposes its paramount tactical value still surely resided mainly in the infantry's traditional flat trajectory fire against troops who presented themselves in a clear line of sight — even if that line might now be obscured by smoke, mist or darkness — without loss of accuracy. For all its modernity, therefore, the machine gun's direct-fire rôle made it a relatively familiar instrument, and easy for the uninitiated to comprehend. The deeply civilian Lloyd George could make it the crux of one of his earliest 'common sense' calls for military modernisation, and in a way his insight has been followed by several generations of Hollywood film-makers. They have enthusiastically embraced the machine gun as an exciting, potent yet relatively straightforward and uncomplicated symbol of war itself.[50]

Even in unfavourable terrain and with only snapshooting opportunities, a machine gun could represent a threat equivalent to many more riflemen than were numbered within its own crew.[51] Indeed, many British commentators admiringly noted how the Germans seemed to exploit this effect systematically, by basing their defences almost entirely around machine guns, with riflemen being used only as supporters who transported and stockpiled the ammunition. This practice was already observable as early as Loos in 1915 and was not, as is sometimes alleged, a distinctively 'late 1917' or even 'Second World War' innovation.[52] It may, however, be worth pointing out that this increasing German reliance on an entirely machine gun-based defence may be explained less by some fiendishly clever or modern tactic than by precisely the same deterioration in musketry skill that the British regulars were deploring among their own men.[53] The same weapon will always be perceived as more dangerous when the enemy is using it than when one's own side is doing so, and in the case of the BEF on the Western Front, which was habitually in an offensive posture, this awe was especially applicable to a supremely defensive weapon like the machine gun. BEF officers were perhaps too quick to rush to the conclusion that the Germans had far more machine guns than their own men, and too reluctant to admit that a few enemy guns could easily look like a lot.[54]

It may also be worth pointing out that even if a machine gun was worth thirty riflemen, the thirty riflemen would certainly be much harder to snuff out than a single gun. For full efficiency under trench warfare conditions each machine gun in any case needed no less than ten servants to carry its water, its ammunition, and the desperately heavy sleds or tripods on which it had to be mounted.[55] As one British manual put it, 'The mobility of a gun depends...to a great extent upon the mobility of its ammunition'.[56] Machine guns were notoriously less mobile than rifles, and one wonders whether the famous German habit of using their infantry entirely as machine gun teams was really as much of an economy in manpower as it often seemed to their opponents. Conversely the late 1916 British impulse to revive interest in the rifle and the rifleman may itself not have been quite so backward-looking after all.

Both machine guns and rifles could be devastating, provided their operators were alert and unsuppressed by the attacker's fire. This applied especially to weapons that were sited just outside the area under attack, since they might escape the preliminary bombardments and still fire in from the flank or from higher ground to the rear. Time after time we hear of assaults which were able to advance little more than ten yards from their trenches, suffering appalling losses. On 1 July 1916, indeed, some battalions were at first mistakenly thought not to have made an assault at all, because their progress forward had been so trivial before they were shot down. In spite of all that, however, the surprising fact is that large bodies of men could and did habitually move around 'on top' with impunity, whenever it was dark enough or misty enough to blind the enemy's watching snipers or machine gun teams. Wiring parties, advanced outposts and reconnaissance patrols went out every night, and usually came back unscathed once their task had been completed. On many occasions this was made possible only by the action of a deliberate mutual 'live and let live system' between the two sides; but in many other cases it can be attributed directly to the protective cloak of darkness. This might be temporarily dispelled by a parachute flare, which would cause anyone in No Man's Land to freeze until its light had fizzled out. Everyone was also painfully aware that, in even the darkest night, an entirely random burst of fire from a nervous sentry could have fatal consequences at absolutely any time. Nevertheless, the paradoxical fact remains that activity above ground was far from always incompatible with survival in, and even all the way across, No Man's Land. Small-scale raids could very often reach the enemy's trenches at night, although they were certainly risky. Big attacks at night were even riskier still, since they were far harder to control than in daylight, and little easier to conceal. Once the enemy had realised that he was under attack, whether by a small raid or a major offensive, he could lay down as much fire by night as he could by day. Yet big night attacks did sometimes succeed, and they continued to be practised.[57]

When we turn from small arms to artillery, we find the defensive 'storm of steel' was worked by similar but slightly different principles from those of rifles and machine guns. Unlike the infantry, the gunners were not physically present in the front line, so they could not react quite as directly to targets as they appeared. They might fire in response to one of three types of stimulus. Most of the time they fired planned bombardments, as laid out in orders issued by some higher HQ. These might include a very elaborate and precisely timed fireplan, which set up walls of shells at key points on the battlefield, maintained them for a predetermined number of minutes, then moved them round to another part of the battlefield — rather as a builder of card-houses might move around his 'walls' on a card-table. One particularly telling example of this came in the final part of Alan Thomas's attack towards Monchy le Preux on 11 April 1917, when the cavalry had passed through to exploit. Once inside Monchy itself, the horsemen had been caught in a box barrage with four 'walls'. There was no way out of the trap as the German gunners gradually walked these walls inwards to meet each other in the centre, leaving a mass of dead horses and men to block the road.[58]

Secondly, the gunners might depart from a preplanned bombardment in response to new intelligence, which might be collected from air observation, from messages sent in by the infantry, or possibly from an observation post (OP) in the front line. As the war progressed, the gunners' means of intelligence collection did gradually improve, but the simple OP nevertheless tended to remain something of a weak link. Such an OP would probably be manned by a (more or less) senior officer from his own battery or brigade, working (more or less) closely with the local infantry HQ, (more or less) continuously, and (more or less) usefully close to the epicentre of the battle. Ideally each battery's commanding officer would man the OP for twenty-four hours a day, in spiritual harmony with — although physically well forward from — the infantry battalion's command post. He would live up to the 2/RWF's model: 'Our gunner observer is knowledgeable and helpful, unlike the general run of forward observers these days who disclaim being in any sense liaison....'[59] Alas, for very many excellent reasons this lofty ideal was scarcely ever fully achieved in practice. OPs were often entirely non-existent, or manned only intermittently by relatively junior officers, who could see little of what was going on and who enjoyed only very flimsy communications with their parent battery.[60] Because they were junior they could not automatically insist that their fireplans were followed; and because they were somewhat out of touch with the main action, they could not always call fire for maximum tactical efficacy. Many memoirs by OP officers contain laments at juicy targets that were not engaged in time.[61]

Finally, the artillery was committed to respond as soon as it possibly could, whenever the infantry fired signal lights to summon SOS fire. This consisted of an intense preplanned squall of shells designed to catch an attacker in No Man's Land, before he could reach the defender's front trench. A good SOS response could smother him with such an impenetrable wall of bursting shrapnel or high explosive (HE) that he would be entirely obliterated, in every horrific sense of that particular term.

Obviously an SOS bombardment depended for its effect upon two main elements. In the first place the front-line infantry had to perceive that it was really under such a serious threat that an artillery response was appropriate, and there were a number of factors which might militate against that. The 'live and let live system', for example, could encourage a defending sentry to minimise the fire he called down, for fear of the likely retaliation. He might also misinterpret exceptional activity in the enemy line for the normal routine of trench repairs or the arrival of fresh rations. Much would depend on the personal aggressiveness of the sentry, the institutional aggressiveness of his unit, and on the higher intelligence expectations of his divisional or corps HQ. Of course the precise level of alertness of front-line sentries at any given moment has always made a perennially interesting subject of study, and no infallible key to it has ever been found, even today.

Against this, it may be urged that everything really depends less upon the alertness of the sentry than upon the clumsiness of the attacking infantry as it formed up for its assault. In unlucky times an attack might allow itself to be detected several hours, or even days, before it was launched. The defenders' artillery fire could then be concentrated with entirely crippling effect, and might scour the attackers' forming-up and communication trenches long before anyone had risen above the parapet at all. Especially in 1915 the BEF tended to attribute some uncanny powers of divination to the Germans, who always seemed to know what they were doing in advance. This in turn led to the shooting of many French and Belgian peasants suspected, probably incorrectly, of spying, together with a rather more relevant tightening of telephone security.[62]

If the attacker did succeed in keeping his intentions secret, by contrast, he might find a chance to worm his way silently forward into shell holes only a few yards short of the enemy's line. At zero hour the defender would thus be left only a few seconds to realise what was happening, and hence still fewer seconds to fire off his SOS flares. During the 'mature' second half of the war, the BEF's infantry had fully realised that a good attack could be defined as one in which the main German SOS fires landed harmlessly *behind* the attacking waves. Particular stress was therefore laid on precautions of secrecy before zero.

Finally, the effectiveness of an SOS appeal naturally depended very heavily on the ability of the gunners themselves to be alert and ready to fire as soon as possible after they had seen the infantry's signal flares. Towards the end of 1917 this was being recognised as an uncertain and wasteful use of artillery, with too many false alarms, and there was a move to abandon SOS fires altogether.[63] There was also a serious problem if the gunners found themselves shrouded in a blanket of ground mist, when they would be blinded and unable to respond at all. On a misty morning a defending rifleman or machine gunner might eventually still be able to detect an attack upon him — albeit perhaps by sound as much as by sight, and doubtless at lesser ranges than were possible on a brightly shining dawn; but a gunner would probably not be able to see a distant SOS flare, and hence would remain entirely in ignorance that anything was amiss. This single factor, more than any other, was often cited as the true reason for the British collapse at Cambrai on 30 November 1917 and especially on the Fifth Army front on 21 March 1918.[64] No doubt it was also important in the German collapse on 20 November 1917, as well as on many occasions during the misty moisty autumn of 1918.[65] By the end of 1917 it was also becoming a key point of tactics to supplement artificially the natural effect of the weather in blinding enemy gunners. It was thus that the gunners almost inevitably became the choice and ideal recipients of gas shells.

After making all these caveats for the many difficulties of bringing artillery to bear, we must still recognise that modern shells represented a mightily more powerful weapon than did even the machine gun. Both of these weapons had been developed during the 1880s; but by its very name, any 'machine' gun seems to suggest a complex and intricate piece of man-made ingenuity, whereas the expression 'high explosive' suggests just mindless destruction and uncontrolled blowings-up. Yet in fact the various types of shell, and the guns which fired them, made a far more complex weapon system than the humble machine gun, at the same time as they harnessed far more explosive power. This made them the more significant innovation in the art of warfare, and the cause of something like sixty per cent of the casualties. To take just one vivid example of what this could mean in practice, at Bernafay Wood on 3 and 4 July 1916 the 6th King's Own Scottish Borderers had only trifling losses as they quickly cleared the position of German infantry, but then suffered 150 casualties in the terrific counterbombardment which followed.[66] Expressed in very rough statistical averages, the balance of profit and loss for the whole war would appear to have been something like this:

— The Infantry lost one casualty (i.e. all types, not just dead) for every 0.5 it caused, with an average loss during the war of 200 per cent of each division's starting strength, including perhaps 600 per cent in 'élite' battalions (assuming losses were always replaced).

— The Artillery lost one casualty for every 10 it caused, with an average loss of forty per cent of each unit's starting strength.

— The Special Gas Brigade lost one casualty for every 40 it caused, with a turnover of 100 per cent of its starting strength during the war.[67]

Quite apart from its rôle in causing casualties, the delayed action HE shell — as opposed to instantly fused HE or shrapnel — could gouge and crater the ground, transforming a flat field into an uneven and treacherous moonscape that would seriously impede an attacker's progress across it. The craters could also be used as improvised trenches by a defender, who might find them the safest place to hide, once his formal trench networks had been accurately registered by the enemy's guns. If field drainage systems were important, as they were most disastrously in the Ypres Salient, the shells would destroy them and leave deep pools of water, slime and mud. If telephone communications were important — as they inevitably were in almost any battle before radios were widespread — most types of artillery fire would quickly cut the wires unless they were buried at least six feet deep.

It cannot be stressed too strongly that shells represented a far greater technical change than the machine gun. Not only could an SOS wall of shells prevent movement in No Man's Land more completely than machine guns, but their new indirect fire capability allowed them to reach deeply into the enemy's trenches in a way that machine guns usually could not. Their unique combination of very great range, accuracy and hitting power also allowed them a far greater flexibility than machine guns. If a friendly trench was captured by the enemy, he could be pounded to oblivion with numbing precision; but if a suitable target far behind enemy lines was identified, for example by aerial observation, it too could be pulverised. The range of the guns gave them one further advantage of inestimable value, in that it allowed them to be removed from the front line altogether. They were entirely beyond the risk of interference by the enemy's infantry, and were vulnerable only to a complex and often unreliable series of counterbattery measures. Only after the end of the Somme battle did the BEF even begin to master the art and science of suppressing the German artillery, and even then its successes often remained patchy.

If this 'storm of steel' posed a large part of the tactical problem faced by the armies on the Western Front, we are now in a position to look at some of the tactics and weapons which could be used to find a solution. Let us begin with the infantry.

Part Two
Infantry

3

Infantry during the First Two Years of the War

> The capture of a system of hostile trenches is an easy matter compared with the difficulty of retaining it.
> Fourth Army 'Red Book', May 1916;
> OH 1916, Appendix 18, p.138

Among the most deeply cherished traditions of the British Army is the idea, or perhaps 'myth' would be a better word, of the tactically self-sufficient infantryman. This may be traced back to King Harold's heroic shield wall at Hastings, unfairly overwhelmed by overcomplex Norman technology but eventually avenged by Prince Hal's triumphant archers and dismounted knights at Agincourt. Then there was the 'thin red line' holding out steadfastly, from Albuera and Waterloo to Balaklava and Rorke's Drift. For the 1914 generation, Kipling's tales of the great game on the Khyber Pass held a particular resonance, as did Baden Powell's isolated but cheeky stand at Mafeking. From wherever the inspiration might be drawn, the British regimental system seems to have enshrined a notion that any given regiment could always fight its own battles more or less entirely on its own — and indeed many of them were forced to do just that, owing to the demands of an undermanned but exceptionally far-flung Empire. Since a great majority of the army was always infantry, moreover, the notion of self-sufficiency has always tended to be most associated with that specific arm.

In the conditions of the Great War, however, the infantry's independence was for the first time called very seriously into question by the onward march of technology, especially artillery and signals. Tacticians were forced either to include other arms far more closely in their plans than they had before, or to devise some radically new techniques for allowing the infantry to fight successfully on its own. In the event they succeeded in doing both, and the next few chapters will try to look at some of the ways in which they did so. On the whole, however, the picture seems to be that attacks tended to succeed only when the other arms were fully coordinated in adequate strength, and to fail when they were not.[1]

Tactics of the Old Contemptibles

The experience of the Boer War, soon followed by reports from the Russo-Japanese War in Manchuria, had powerfully concentrated the minds of British officers upon the whole question of infantry tactics and increased volumes of fire. Although many still believed in the continuing possibility of an assault supported only by its own rifles, there was at least a new spirit of caution and a widespread awareness that the world was changing. The adoption of khaki uniforms and a stress on 'fieldcraft' was only one symptom; there was also an increased study of firepower in such places as the Staff College, the Royal United Services Institution or the School of Musketry at Hythe.[2]

Probably the most influential of all the tactical works of the period, albeit at first sight also one of the least serious, was entitled *The Defence of Duffer's Drift*. This was published as a magazine article in 1903 by 'Backsight Forethought', the first in a string of pen names that would be used by Ernest Swinton, at that time a captain of the Royal Engineers who had served in South Africa.[3] The book is attractively presented as a series of illustrated 'dreams', but it is actually a reasoned step-by-step analysis of a particular problem in minor infantry tactics. A river-crossing has to be defended against surprise attack by superior numbers of Boers, and it soon becomes clear that merely to mass the defenders at the threatened point, in a 'traditional' way, will lead to abject failure. The solution turns out to be a matter of laying out interlocking fields of fire from a number of concealed positions arranged to the flanks of the objective, in a manner reminiscent of Vauban's classic fortress geometry — Swinton would later become an instructor in fortification at Woolwich. It was also heavily suggestive of the crossfires that would soon become a doctrinal hallmark of the Machine Gun Corps itself.

Although 'Backsight Forethought' was concerned with defence rather than attack, his analytical method appears to have made a profound impression upon many of the upcoming tacticians who would have to fight the Great War. It was based not only on fieldcraft and military common sense, but also on a subtle paradoxical quality which warned the student that it was always dangerous to do the obvious or traditional thing. It also perpetuated the suggestion that a few British infantry, unsupported by artillery, could defeat a large number of enemy. All they needed was sufficient cleverness to exploit the jujitsu principle — or what Orde Wingate would later call the 'Gideon' method — and make the enemy defeat himself by his own weight. One final noteworthy feature of *Duffer's Drift* was that its story was explained in part by perspective sketches as much as by words, appearing at a time when much of the press had become profusely pictorial and 'popular'. The techniques for illustrating text would be relatively little used in the training manuals which appeared before 1914, but thereafter they would gradually worm their way slowly but surely into the technical literature. By the end of the war the infantry manuals were still falling far short of the coloured cartoon strips we have seen in that context since the Vietnam era; but some of them at least were hailed as artistic masterpieces.[4]

In its official doctrine for the attack, the prewar army believed in a cautious approach with 'fire and movement'. One would try to suppress the enemy's fire as much as possible by one's own fire, while the advancing riflemen used fieldcraft, short rushes and extended formations to protect themselves as much as possible. Gradually a firing line would be established at around 200 yards from the enemy, preferably helped by artillery firing over open sights. Then, once the firefight had been won and the enemy's fire abated, the attackers would charge forward and finish the job with the bayonet.[5] Opinions varied widely about just how much stress should be placed on the fire and how much on the bayonet, as also upon the precise interval between men at each successive stage of the operation. Some authorities envisaged a very tightly packed line charging an enemy who had not been fully suppressed at all, on the basis that a battle which consisted only of stand-off fire could never reach a decision. Others believed in depth and a succession of assault waves. Brigadier-General Ivor Maxse, for example, wanted numerous lines or 'waves' one behind the other, and subscribed to the maxim — both before the war and later — that 'a single line will fail; two lines will usually fail; three lines will sometimes fail, but four lines will usually succeed'.[6] This idea was not based on a directly physical theory of weight and impetus, as it is now fashionable to allege, but rather on a realistic understanding that each man could be expected to contribute only so much, after which he would need a relief. In the Fourth Army's *Tactical Notes* of May 1916, for example, we find an admirably modern acceptance:

> Under existing conditions only one definite offensive blow can be expected from one body of infantry in one operation....
>
> There is a limit to human endurance in battle, and once that limit is reached the reaction is severe.[7]

Admittedly the massing of men implied by these tactics was somewhat traditional, running against the *Duffer's Drift* style of thinking to be found elsewhere in Maxse's work. Yet as a theory it had managed to maintain its credibility by virtue of some of the undoubtedly successful frontal attacks seen in the American Civil War, at Plevna in 1877, and even in the Russo-Japanese War itself.[8]

Within the prewar British Army there were other authorities who disliked closely packed waves and followed the spirit of 'Backsight Forethought' more directly, by arguing for a far more empty battlefield. They insisted that massed modern fire was simply too powerful to be overcome quickly or by any sort of massed column, and demanded a battle in which a few dispersed snipers took perhaps several days to work their way carefully forward, or even sapped their way forward with formal siegeworks. Such a solution was clearly distasteful to anyone who wanted a rapid resolution of the battle, so it was not wholeheartedly embraced in the manuals; but the real nub of the problem was that the manuals were ultimately left ambiguous. The whole officer corps therefore remained uncertain about just what assault tactics it was supposed to follow, and there was never any final resolution of the debate between proponents of the 'full' and the 'empty' battlefield. It all made for a central ambiguity which has rightly been blamed for many of the disasters of the war.[9] Although it was far from the case that everyone embraced a reckless cult of the offensive, it remained true that too little prohibition was imposed upon those who might be leaning in that direction. However, it is also probably true to say that at least some of the commanders in 1914 believed the manuals gave them full freedom to use 'fire and movement' in its most sensible and economical form, and that this could be developed into a respectable ideal for the remainder of the war — and also for most other wars ever since.

Perhaps the main difficulty for the BEF in the development of its assault tactics during 1914 was that it was rarely called upon to attack at all. Mons and Le Cateau were almost entirely defensive battles, where 'fieldcraft' meant digging in and keeping your head down, 'winning fire superiority' meant blazing away madly whenever a German came into sight, and 'manoeuvre' often meant skedaddling to the rear as soon as the coast was clear. Admittedly on the Marne there were some quite serious clashes of outposts as the BEF moved forward, but it would not be until the battle of the Aisne in mid-September that a reasonably solid defence would be tackled head-on — although on that occasion the defence was found to be only too solid. There were certainly some interesting technical innovations, such as air observation for the guns, but unfortunately too little communication between the guns and the infantry they were supporting. The Old Contemptibles[10] were exhausted after two months' marching and countermarching; they were attacking up a steep hill, and lacked adequate time to plan their battle properly. The upshot was that things did not go well at all. This point is important because it was often wilfully forgotten later in the war by Old Army generals who blamed the new Kitchener armies for lacking the skills in 'fire and movement' that had supposedly been demonstrated so brilliantly by the original BEF.[11] In reality, these supposed skills were never given a proper chance to be displayed in serious offensive action, even if they existed at all. After the Aisne the BEF was whisked away to desperate defensive battles further north, and before very long its prewar character had been forever lost as a result of its heavy casualties.

A closely related misconception is that the Boer War lessons of 'fieldcraft' or 'fire and movement' implied a rejection of linear tactics. This was not the case, at least during the final phases of an attack when a firing line had to be established which could then be transformed into an assault line, for which the much-loved 1904 manual apparently recommended two paces between each man.[12] Admittedly small platoon or section columns in 'artillery formation' were recommended for the long approach march before that, in order to minimise the effect of long-range shelling; but that was not quite the same thing as actually fighting in small groups. Instead, it looked more like a modernised version of Napoleonic practice, whereby the column had usually been preferred for manoeuvres out of contact with the enemy while the line was usually preferred for close action.

Nor did the Old Contemptibles' doctrine exclude the use of hand-grenades, if ever they had a chance to get hold of any. It is true that they had not envisaged their use in open warfare; but they certainly tried to improvise them the moment trench warfare set in, and carried as much responsibility as anyone else for the 'cult of the bomb' which soon developed. Subsequent attempts to suggest that the prewar regulars would always prefer the rifle and bayonet to the grenade, or that the grenade somehow negated fieldcraft, would seem to be rather wide of the mark.[13]

What the Old Contemptibles did have to offer was a uniquely high volume of musketry, in their 'mad minute' of concentrated fire at relatively close range. This was designed to wreak some five times the damage upon the enemy that was normal with either the musketry of continental armies or the deliberate sniping of trained shots;[14] and it does indeed appear to have been very effective in defence.[15] Still more to the point, however, was the Old Contemptibles' enormous depth of military culture and cohesion: an irreplaceable moral asset born of years living together, often in exotic foreign parts, under a long-service regime of undoubted professionalism. No European organisation of comparable size could match this standard, apart perhaps from the French Coloniale.[16] From the British point of view it was therefore doubly painful that the prewar BEF had been all but chewed to pieces as early as Christmas 1914.

The exceptional prewar pride in this army led to anguished wails of lament at its loss which can still be heard today — mixed in with an unfairly vindictive dismissal of the massive successor armies which followed on so eagerly.

What most continental European armies possessed, but which the British patently lacked, was a peacetime arrangement for training far larger numbers of conscripts for a relatively short time, then returning them to civilian life as reservists, whence they could be plucked back to fill the ranks whenever a new war broke out. This surely gave continental soldiers a 'medium' standard of tactical efficiency and military culture which might not have been as high as that of the best British Old Contemptibles, but which nevertheless enabled them to fight on long after the contemptibly small British regular force had been expended. The French and Germans could field many millions of such men, not just 100,000; and like most soldiers on the Western Front they continued to improve in tactical awareness the longer the war went on.

Besides, not even the British prewar regulars really paid very much attention to 'tactics', since it is a notoriously difficult subject to teach unless one is in a 'genuinely tactical situation' — that is, someone is shooting at you with live ammunition.[17] For most of the prewar BEF, for most of the time before mobilisation, this condition simply did not apply. The idea of 'tactics' was thus accorded considerably less importance than more basic skills such as route marching, musketry or drill. Discipline and cohesion were seen as a very much more potent key to success in battle than any particular set of tactics, which would in any case doubtless have to be reshaped on the spot to meet any given situation.[18] The Old Contemptibles of 1914 took pride in the slogan 'We'll do it — what is it?' even though such an attitude was scarcely conducive to deep study of the finer points. That was left to enthusiasts in specialist institutions such as Camberley or Hythe.

The New Armies arrive

By 1915 the prewar regulars were starting to be replaced by the new Kitchener armies and territorials, amid conditions of gigantic administrative muddle, equipment shortages and universal unpreparedness.[19] Among other things the general standard of musketry training rapidly declined,[20] although this did perhaps help the high command to accept the idea of Lewis guns as a necessary firepower supplement. The threatened disappearance of old tactical skills also boosted the rapid spread of training schools and the distribution of manuals and other tactical ideas, almost regardless of their origin. By 1915 the BEF had worked out few new formal tactics of its own, and would often repeat that 'classic principles' need only be adapted to new situations; but it was notable that a large number of translated French pamphlets were also distributed,[21] as representing the most modern ideas available, in rather the same way that French manuals had been standard issue during the American Civil War. Unlike the earlier American conflict, however, the Great War would soon see a proliferation of radically new home-grown manuals, mixed with a profusion of captured enemy tactics and analyses.

The BEF of 1915 certainly started to make its own serious tactical experiments for the offensive. In the course of the small but fierce battles of Neuve Chapelle, Festubert, Aubers Ridge and especially Loos there grew up at least a rough outline understanding of what ought to be done. Artillery fireplans rose to major prominence among the infantry's concerns, especially the question of their precise timing and density, for which effective formulae were glimpsed but not yet grasped.[22] The first creeping barrage and the first machine gun barrage were fired. Novelties such as aircraft, gas, smoke, trench mortars, Lewis guns and signals — even wireless — began to be absorbed into the general picture, to the extent that it was fair to say that British tactics would continue to be based on the model of Loos until late 1917. They would be characterised mainly by short objectives, caution and ever larger quantities of artillery.[23]

One point that is often missed about these small and unsuccessful battles of 1915 is that several of them could show at least a few moments when the infantry actually came very close to complete victory. At both Neuve Chapelle and Loos some of the assaulting divisions experienced the heady sensation of walking calmly across No Man's Land in their regulation waves, without undue interference from the enemy. In the case of the 15th Scottish Division at Loos, 'The scene resembled nothing so much as a cross-country race with a full field. Men ran as if for a prize...', and continued to do so for some four miles into the very heart of the German defences. Their historian saw it as a major use of 'infiltration tactics' some thirty months before Ludendorff would demonstrate his own more famous version of that genre.[24] Admittedly such successes had a lot to do with the state of the German defences, the cleverness of BEF staffwork and the intensity of the artillery preparation; but to many infantry commanders it must have seemed as though there had been no real break from the nineteenth-century experience. In the past it had generally been possible for massed column attacks to roll over even the most formidable defences, given the right sort of preparation: and now in 1915 exactly the same phenomenon seemed to be repeated.

This consideration did much to perpetuate the infantry's primary assault formation as a series of more or less successive linear waves, moving forward by alternate rushes covered by fire where possible,[25] or simply moving forward directly if the enemy had already been suppressed by artillery. Rearward units would either follow on to consolidate, or might hope to leapfrog through the more forward units at preordained moments to continue the attack. Within the 9th Scottish Division, for example, the attack at Loos was made by four battalions in line on a frontage of 1,600 yards, with each battalion split into three waves, one behind the other. Behind each battalion followed a second one in the same formation, ready to leapfrog through; and behind that came a whole brigade in general reserve. The depth of the attack at any given place was therefore at least six lines of men, at intervals of perhaps two yards between each man.[26] Some variant of this layout, albeit thinner and shallower in later years, would remain standard practice for most of the war, particularly in the battle of the Somme. At Longueval on 14 July 1916, for example, the same division attacked again with four battalions in the front line, but this time with each company in a column of platoons, making four successive platoon waves with seventy yards between them. In 26th Brigade each battalion had a frontage of two companies, with the remaining two in second line, and a second battalion following behind, to give a total depth of sixteen platoon waves attacking any given section of front. In 27th Brigade, by contrast, each battalion had all four companies in the front line, making just four waves, and only one battalion backing up the two in front. Overall the division's foremost front line, with six platoons advancing side by side on a frontage of perhaps 1,000 yards, represented a density of about one man to every five and a half yards — a considerable lightening since Loos and a very far cry from the popular stereotype of a 'mass' attack.[27] What was in theory a 'divisional' attack therefore turned out, in the first instance, to be an attack by just six platoons, or one and a half companies (see Fig. 2).

Much of the thinking behind all this had been home-grown in the British Army before the war; but it now won an important endorsement from the work of the French Captain Laffargue, who has often been hailed as no less than the 'father of infiltration tactics'. He had written an extremely influential pamphlet based on his own front-line experience at Neuville St Vaast on 9 May 1915, in which he pointed out that 'Infantry units burn away in this furnace like bundles of straw'.[28] His recommended solution was to multiply the number of successive waves, usually closely packed together and firing their rifles into the enemy's faces as they rushed forward as deeply into his defensive system as they could possibly sustain. Because they were trying to go deep they were also 'infiltrating'; and because they were using their own rifle fire, supplemented by mobile machine guns, light trench cannons and other organic weapons to suppress enemy machine guns, they could be seen as self-supporting small groups which had all but broken free from artillery support. They hoped to overwhelm the enemy frontally by their own efforts, just as much as they hoped to bypass recalcitrant strongpoints and go deeper. In this they were no different from their German confrères, who also saw 'stormtroops' as primarily a frontal assault weapon.[29]

Fig. 2: Variants of the wave attack by the 9th Division, 1915–16

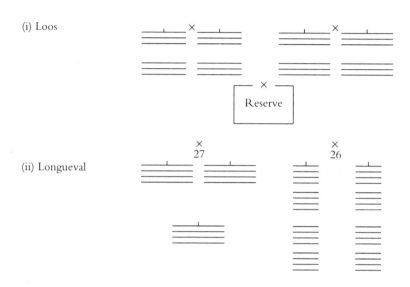

In much modern literature Laffargue's work is correctly held up as a very forward-looking analysis; but the point is usually missed that its recommendations for infiltration were possible only because they incorporated the supposedly outdated doctrine of successive waves. Far from being the antithesis of infiltration, in fact, waves permitted the infiltrators to be systematically backed up and sustained within a short timescale, wherever they found an opening. Yet it remains true that the whole system was a fragile and subtle one — just as it would be for the Germans in 1917–18 — and it contained almost as many ambiguities as the prewar British manuals. Some of its elements could be taken in isolation by those who looked backwards to massed tactics, just as others could be taken in isolation by those who looked to a future of self-sufficient small groups. One possible point of confusion was that Laffargue cited the Boer tactician De Wet as one of his inspirations for massed tactics, whereas at least as early as October 1914 the Germans had already taught 'Boer' tactics as a more dispersed system.[30] Such things only helped to deepen the central ambiguity, or subtlety, of Laffargue's message; although perhaps it was precisely this ambiguity which explained his pamphlet's undoubted popularity. In the event it appealed to a very broad church indeed, and so was distributed to the French armies in August 1915, translated twice over for the BEF around Christmas 1915, and given to both the Americans and Germans later during the course of 1916.[31] It was still being cited in British tactical pamphlets in 1918[32] — which of course does not prevent some modern commentators from repeating the ridiculous notion that the British had never even heard of it.[33]

A part of the 'infiltration' methods implied by the official BEF manual, issued on 8 May 1916 for the *Training of Divisions for Offensive Action* (SS 109), was the general idea of a succession of lines for 'adding fresh impetus and carrying the whole forward to the objective' and keeping enough 'driving power' to get there, plus enough 'residual energy' to consolidate upon it.[34] Unfortunately this was also a highly linear conception which encouraged Edmonds, for example, to misunderstand it. He gave an unfair cue to future generations of Somme-mockers when he took it to task for failing to understand the value of 'infiltration',[35] by which he seemed to mean an attack in small groups or 'blob' formation. In the overcharged perspective of 1932, apparently, 'lines' or 'waves' represented high command rigidity and authoritarianism, whereas more fluidly female words such as 'stalking', 'infiltrations' or 'blobs' were used to represent something altogether more acceptable and humane. This prejudice unfortunately appears to have survived unchallenged into the current debate even though, as we have seen, there was never in fact any true conflict between the two formations. In practice both ideas continued to permeate tactics, hand in hand, in all armies for the remainder of the war.

In Fourth Army's *Tactical Notes* ('The Red Book')[36] issued in May 1916, the option was left open for each battalion to attack with a frontage of between two and four platoons, making a depth of either eight or four successive waves, each no more than 100 yards apart, but all rising from their trenches to start moving forward at the same time 'as one man: this is of the highest importance'.[37] Considerable emphasis was laid on the need to practise the tricky operation of passing one line through another. This was designed to reduce the objective of each individual element to manageable proportions, avoiding wearing it out by demanding too much. A good passage of lines was also designed to avoid bunching one line with another, which had always been a major problem throughout the nineteenth century. If the front line ran into difficulties, therefore, the second line was supposed to stop where it was, and not keep advancing into a traffic jam before the situation could be clarified. The theory of physical weight of numbers was specifically eschewed. In essence, therefore, it was the platoon in the front line which had to fight its own way forward as an independent unit, perhaps shouldering as much as half the responsibility for an entire brigade's offensive effort during the few minutes before it was relieved by a subsequent wave. The platoon was thus already being seen as 'the tactical unit' in a way that would continue to be true for at least the next fifty years.

An important improvement was that it was now quite a widespread practice[38] to divide the attacking waves into several different functions, as between 'fighting platoons', 'mopping up platoons', 'support platoons' and 'carrying platoons'. The fighting platoons were intended to press forward as fast as possible — or 'infiltrate' the enemy's line, if you will — while the moppers up had the task of securing the ground won. Experience had shown that a warren of enemy trenches could often be crossed relatively easily by a formal attack; but stray Germans were liable to keep popping up out of deep shelters for several hours thereafter, disrupting the whole battle directly in the rear of the front-line troops and threatening to cut them off. These strongpoints had to be suppressed by the moppers up. While this action was going on, the support and carrying platoons would be moving forward 'in small groups. Supports in Platoon, Section or Squad Columns can pick their way through a hostile barrage with fewer casualties than in extended lines, and can be more easily controlled.'[39] Thus the traditional concept of 'lines for fighting, columns for movement' was preserved. Regardless of their formation, however, the task of these supports was to consolidate the ground won and, with the help of engineers and machine gunners, to build and munition their own 'cruciform strongpoints' within the newly captured line. They would then be able to resist the almost inevitable counterattack, and organise the ground ready for the next bound forward.

The BEF always saw 'consolidation' as particularly important, not only because of the likelihood of an immediate German counterattack, but also because every objective captured would inevitably be the start-line for the next bound forward, and this needed to be properly prepared as quickly as possible. The process of consolidating one line before sending the next wave forward from it was perhaps the most difficult trick of all in maintaining the momentum of an offensive. It was still more difficult to pull off in time against an actively counterattacking enemy like the Germans of 1916–17, at least before the 1917 certainty that one could call down overwhelmingly responsive and accurate protective fires from artillery.

As with Laffargue's teaching, the BEF's 1916 doctrine of an aggressive front line to push forward wherever possible, and rearward lines to mop up and consolidate, already amounts to something very similar to the German 'stormtroop' tactics which would be hailed as such a wonderful innovation towards the end of 1917. We should add that at this stage of the war the British, in common with the German stormtroops later, were already laying particular emphasis upon the hand-grenade as the essential weapon for trench clearing. They also understood the need to push machine guns (by now increasingly Lewis guns) well forward, even before zero hour,[40] and many of them already believed in 'bags of smoke'. At Loos they had used gas to supplement the thinness of their artillery barrage, and they had then supplemented the gaps in their gas coverage with smoke as a deceptive ruse. They also used smoke to screen open flanks. By the time of the Somme they had moved forward to a massive programme of smoke as part of the 'Chinese' attacks on sectors which were not to be attacked seriously; and they soon developed it as a vital flank screening to many of the more genuine attacks. The British had, in effect, already seized upon the *Nacht und Nebel* principle for which the Germans would later become more notorious.

There was even an understanding, at least in theory, of the much-vaunted 'mission orders' or 'directive command'. This was a system, usually attributed to the Germans, whereby minutely detailed rigid orders were avoided in favour of general guidelines which left full discretion to the man on the ground to interpret as he saw fit. Admittedly some of the British operation orders issued for 1 July can appear very daunting: that for XIII Corps, for example, runs to some thirty pages and covers apparently everything from pigeons to flares and from Russian saps to metal identity discs on each man's back.[41] Yet this was essentially an administrative order for a complete army corps, and it did not purport to determine the tactical action of each platoon. More relevant to that was the following passage from a lecture given by G. M. Lindsay, the creator of the Machine Gun Corps, on 10 August 1915:

> If everyone who gets into a scrap knows what the legitimate object to be achieved is, and what the object of the people on the right and on the left is, then, when the unforeseen circumstances arise and things go wrong, they will be able to alter their plans to meet those circumstances, but they cannot possibly continue to fall in with the object and arrangements made, and fit in with the movements of those on their right and left if they do not understand what the job is, and what we are trying to achieve. (Therefore they should be given definite instructions)...
>
> Now, 'definite instructions' do not mean do this, and do that and the other and nothing else in the world. You give a man definite jobs to do, and let him do them his own way, and if everyone knows his job, and knows what other people are trying to do, then you will have gone a long way towards preventing regrettable incidents of chaos arising.[42]

A similar point was repeated at the very start of the Fourth Army's 'Red Book' of May 1916. After complaining that the troops have become too accustomed to 'deliberate action based on precise and detailed orders', it calls for them to rely constantly upon the exercise of initiative by their officers and NCOs or, if none of those are at hand, by themselves:

> With the above objects in view special exercises should be held during the period of training... to consider the action to be taken by subordinate commanders in local unexpected situations, such as a portion of a line being held up, impassable obstacles encountered, or local counter-attacks by themselves or the enemy.[43]

This theme was continued throughout the war in training schemes, with constant encouragement for junior officers and senior NCOs to think for themselves.[44]

Of course everyone in the infantry knew that flexibility and initiative were essential — indeed unavoidable — within the confusion of an attack. Provided everyone knew the general plan, they had to be trusted to set about it in their own way. It did not take the Teutonic formulae of 1917 to point this out: it was common sense at least as early as 1915. One may indeed object that practice all too often strayed far away from precept, since many of the attacks were either unnecessary or badly prepared from above, and often failed. On 3 August Gough himself complained of inadequate energy in preparation, and of a tendency to use platoons for company operations, or companies for battalion attacks.[45] Prior and Wilson's scholarly analysis has recently confirmed a very similar point at the corps and divisional level, by demonstrating that most of the attacks launched on the Somme were too small and too lightly supported to stand much chance of success.[46] However, this is not quite the same thing as saying that the assault troops were too rigidly controlled in their minor tactics, which they were not. In many units there was a ready acceptance of flexibility and innovation. The 19th Division at La Boisselle, for example, was making important changes in official assault tactics as early as 2 July.

If we take the case of Charles Carrington of 1/5th Warwickshires, in his attack against Ovillers at 1 am on 16 July, we find that his troops were exhausted before they reached their start-line, and were suffering from the effects of lachrymatory gas shells. The attack was unpromising, with the men too closely bunched shoulder to shoulder, and with the first two waves only fifteen yards apart (front to rear). It had no artillery support and was on ground which had previously been held intact by a ferocious German defence. Yet in the dark, Carrington's command was lucky enough to hit a specific point where no enemy machine gun was firing, and without losing a man it captured a trench well behind the main German position. In stormtrooper language it 'infiltrated through a weak spot'. Carrington then exerted his personal initiative to persuade his superiors not to order him to withdraw. During the following morning his men beat off counterattacks and prepared for a long — and relatively undersupplied — siege. Eventually it paid off completely and the main German position surrendered because he had cut it off from the rear.[47]

An important point to remember is that although the Old Contemptibles of 1914 had been unable to develop assault tactics because of their generally defensive tasks, exactly the same could be said of the Germans in the West during most of 1915, 1916 and 1917. Throughout this time they were busily perfecting their defensive techniques — complete with deep shelters in suitable soil, or concrete pillboxes where there was a high water-table, and multiple lines with exceptionally thick belts of wire. The Germans also worked out effective command and control arrangements for a long haul, even if they could not match their opponents' technological lead in many other areas. Conversely it was the Western allies who launched most of the attacks apart from Second Ypres, the first few months at Verdun, and a relatively few raids and *divertissements* elsewhere. It was therefore the Western allies, not the Germans, who most learned by trial and error just what was necessary to an attacker, whereas the Germans learned most about the defensive. Much of the fashionable adulation of the latter's 'stormtroop' methods towards the end of the war therefore misses the fact that the Western allies had already reached a very similar theoretical understanding almost two years earlier — although in practice they had usually encountered far more robust defensive systems than the Germans would ever encounter on 30 November 1917 at Cambrai, or on 21 March 1918 against Gough's Fifth Army.[48]

No objective student of the Great War can avoid the impression that although the Germans undoubtedly did have some very efficient storm units and raiding parties, from Sturmabteilung Rohr in 1915 onwards, they also habitually used disastrously massed and unsubtle tactics in many of their attacks. At Mons in 1914, one observer reported that German tactics 'seemed to me to be very much like the ordinary opening advance we ourselves had always practised in peacetime',[49] whereas in March 1918 British gunners described the Germans as shambling around in 'cricket match crowds', and making succulent targets as a result.[50] In between these two dates the British infantry, and especially its artillery, had established a very marked ascendancy over the habitually massed and predictable German counterattacks.

In the German army a major source of inspiration for the development of new stormtroop tactics had come from the experience of planning and executing trench raids.[51] In the British army the same was also true, except that the raids tended to be more numerous yet every bit as fiendish and deadly. The tactical benefits of British raids have nevertheless been buried in the literature under a mountain of ill feeling, since the troops disliked them as being expensive in casualties yet unproductive in ground gained — and with some justification they suspected the high command of regarding raids as merely a means of building abstract moral qualities such as 'offensive spirit'. This was particularly true since the New Armies were regarded with great suspicion by many senior regular officers: for example, Haig had used the 21st and 24th 'K' divisions as his convenient scapegoats for Loos, and among other things had accused them of inadequate training with hand-grenades.[52] The result was not only a multiplication of bombing schools, but also a stress on the more 'practical' use of bombs in trench raids. Over the winter of 1915–1916 each fresh division which came into the line was met by an incessant demand for raids and *actions d'éclat* which could show that it had been 'blooded' and was unafraid of the storm of steel.[53] This was not popular among the men who had to risk their lives in the attempt. However, it must be said that there is nothing in military history to suggest that it was in any way unreasonable to 'blood' these New Armies in small operations before they were committed to a big one. On the contrary, the experience of almost all other wars is nearly unanimous in recommending such a policy. Green formations have to be given their battle inoculation gradually, otherwise they will be too fragile when it comes to a really severe test. And in the specific case of junior tactical leaders, there was every argument in favour of giving them experience: 'It is most important that young officers should be trained in patrol work and should accustom themselves to moving around in No Man's Land.'[54]

Trench raids also offered many background technical benefits, which extended far further than familiarity with bombs. A routine became established by which the terrain would be studied in every possible way, including air photographs. Then a cloth, plasticine or clay model would be built for briefing, and a full-scale replica of the terrain taped out behind the line for practical rehearsals.[55] The plan would be minutely worked out and everyone fully practised in his rôle, including such specialised arts as wire-crossing or trench-blocking, often with the use of hand-picked assault troops who had been rested for several days out of the front line. Artillery box barrages, trench-mortar and machine gun support would be stitched into the plan, as would the dumping of munitions or the distribution of sacks of bombs. Flanking units would be warned what to expect, complete with passwords, synchronised watches, flare signals and precise timings.[56] Personal camouflage, including blackened faces, noise security and foliage tastefully arranged in steel helmets, or white coats in the event of snow — all of which has often wrongly been associated only with the infantry of the Second World War — would be routinely applied by trench raiders as early as 1915.[57] Every trench raid therefore amounted to a dress rehearsal in miniature for the large-scale attacks that would only too soon be demanded on the Somme. As a means of familiarising all ranks with the details of assault technique, this must surely have played an important rôle.

Trench raids were admittedly not always successful, and many were unnecessarily costly; but by the end of the war they had evolved into almost a complete alternative to the set-piece attack by a large body of men. It was not only the Australians who seem to have converted such raids from merely a hit and run technique into a totally decisive method (wittily known as 'peaceful penetration') for winning and holding significant slices of enemy terrain. Other, non-colonial, outfits were also engaged in very much the same game.[58] In their larger manifestations such as Hamel on 4 July 1918, these 'raids' might admittedly be effectively indistinguishable from a formal all-arms assault, complete with tanks,[59] artillery and machine gun barrages. But on many smaller occasions they must surely have been difficult to distinguish from the good old Northwest Frontier ideal of 'infantry fighting forward without outside support'.

A less well-known variation on the general theme of the trench raid was the surprise and even apparently 'random' gas cloud attack, not necessarily connected with any infantry action, in which the British appear to have excelled above all other nations.[60] Very deliberately and deviously orchestrated by a small handful of highly educated and well-motivated self-styled 'Pioneers' — who were surely every bit as innovatively lethal as Rohr's German pioneer assault sections — these attacks each habitually claimed several hundred German casualties at the cost of merely single figures of British killed or wounded. Their impact was distinctly lessened both by problems of procurement within the industrial base and by the scepticism of senior commanders; but it must be recorded that time after time they do seem to have shattered absolutely all resistance in the trenches in front of them, sometimes up to a depth of fifteen kilometres. It is only a pity that these repeated successes were never properly followed up by infantry.

Really it must be admitted that the BEF had already discovered most of the key points of modern assault tactics even before the first day of the Somme. It seems to be a perfectly respectable and supportable proposition to state that it had by then already achieved a theoretical understanding of what would later be termed 'stormtroop tactics' — including infiltration, machine guns well forward, directive command and decentralised initiative, sacks full of grenades carried by the leading echelons of stormers, 'bags of smoke', and even a somewhat imperfect acceptance of all-arms support. Admittedly the full orchestration of the artillery's creepers and 'depth battle' would have to wait until 1917, but even then it seems to have been both quantitatively and technically ahead of Bruchmüller in most departments apart from the particular types of gas to be used as a shell filling. In several important fields such as machine gun barrages, gas clouds, tanks, nebelwerfers — and perhaps partially even in aircraft — the British in mid-1916 probably enjoyed a lead of about a year over their opponents. In a few auxiliary fields such as flamethrowers and front-line assault cannon the Germans were ahead, and in their quaint 'pipe-pusher' the British really do seem to have backed a major technological loser,[61] but overall it normally seems to have been the Germans who were struggling to keep up.

There were, nevertheless, many practical problems facing the BEF. The first was that neither its artillery techniques nor its shell supply would reach maturity until 1917, thereby often exposing attacks to withering fire from unsuppressed enemy positions within the first few yards of the advance. Even a potentially very successful deep penetration would remain vulnerable to flanking fire and second-echelon enemy reserves, from the moment its own flanking formations chanced to run into serious difficulties. When the 36th Ulster Division quickly overran the Schwaben Redoubt on 1 July, for example, the Orangemen discovered that even their most beautiful 'infiltration' was doomed to ultimate failure when no equally beautiful infiltrations materialised on either side to carry it forward. Not even the best German infiltrators of March 1918 could have coped with that type of situation, if the defence had been as well layered and as well dug in as that of the defending Germans in July 1916.

Because the defences on the Somme proved to be so well prepared, moreover, the British quickly came to the conclusion that they should place additional emphasis on formal massed artillery in systematic barrages. This was often an exceptionally effective tactic that should not be airily dismissed as 'unimaginative', even though one of its inevitable side-effects was that it seriously limited the scope of infiltration tactics. Obviously one could not expect to press boldly forward with daring machine gun teams deep into enemy territory if one was simultaneously regulating the pace of one's own main advance by an impenetrable wall of advancing shells. Provided it was properly administered, the line of shells would surely prove to be just as impenetrable to one's own intrepid and bright-eyed machine gunners as it was intended to be psychologically devastating to the defender's lacklustre trench troglodytes. Yet against a strong enemy with interlocking defences there could be no hope of infiltration at all, however 'imaginative' it might seem, without the devastating wall of shells. The Germans would lose mountains of casualties in the spring of 1918 whenever they encountered such a defence, since they never properly mastered the creeping barrage.

There were many other problems facing the British adoption of advanced tactics. By virtue of their rapid and ramshackle method of formation, the New Armies did indeed often find it difficult to absorb the tactics handed down from above, especially when these demanded 'do it yourself' initiatives from below which broke free from firm guidelines.[62] The same, however, could also be said of the regular formations which had already suffered heavy casualties too often for true continuity to be maintained. In the event, too great a proportion of too many units proved to have been inadequately 'blooded' by 1 July — and then, when they were all too quickly and too radically blooded, they tended to find themselves left with too many new drafts and too few veterans to handle the full subtlety of 'tactics'. Inexperience in one battle ensured the perpetuation of inexperience in the next, and so on in a vicious spiral. It often took a long time, and many many casualties, before units could hope to build up the significant hard cores of veterans that they needed for full combat efficiency. The result was that although the high command had actually grasped much of the theory of tactics even before the battle of the Somme, and was encouraged by that fact to maintain its optimism for the offensive, it still lacked an army that was fully capable of putting the tactics into practice. It realised that it had a problem with the New Army infantry, but failed to see that it also had just as much of a problem with the regulars, with the staff and with the artillery. It kept on trying to launch attacks with a machine that looked magnificent on the drawing board, but which in reality was still unready in almost all of its parts.

4

The Lessons of the Somme

> The British Army learned its lesson the hard way, during the middle part of the Somme battle and, for the rest of the war, was the best army in the field.
>
> Charles Carrington, *Soldier From the Wars Returning*, p.120

Rifles, bayonets and the cult of the bomb

The five-month Somme battle taught the BEF many lessons and transformed it from a largely inexperienced mass army into a largely experienced one. Perhaps the most painful lesson, and the one most difficult to put into practice, was that very close cooperation was necessary between infantry and artillery. In particular, this meant that creeping barrages should normally be used, and the attacking troops should hug their barrage as closely as humanly possible: 'Experience has shown that it is far better to risk a few casualties from an occasional short round from our own artillery than to suffer the many casualties which occur when the bombardment is not closely followed up.'[1] This official acceptance that dropshorts could in fact occur contrasts refreshingly with General Haking's condemnatory disbelief at Loos, when he had blamed the 21st Division for using such incidents as an excuse for their own poor performance. On that occasion he had said:

> Troops that have failed in an attack are very apt to believe that they have suffered from their own shell fire when really it was the enemy's, and many cases of this nature have been investigated and it has been shown that there was no real foundation for the allegation.[2]

Artillery cooperation had of course already been identified as important from the very first moments of the war, and it was already very much a 'respectable subject'. For example, Captain Wyn Griffith of 17/RWF (38th Welsh Division) had spent four days in April 1916 attending a conference on the subject at Aire. However, at that time he had found the issue irrelevant and boring. He regarded the conference only as a reasonably pleasant way to get out of front-line duty and so, since most of the participants were senior to him, he remained content to sit back and let them talk it out among themselves. Griffith's attitude would change dramatically in his attack on Mametz Wood on 7 July when in the HQ of 115th Brigade as a staff captain — later acting brigade major — he found all telephone links to supporting batteries cut within a few minutes of zero. Confusion resulted, and it was only by good luck that the brigade managed to retain enough strength to fight again on 11 July. But then the artillery fired too early and in the wrong place, negating the brigade's attack. Sufficient communication from the rear nevertheless remained for higher authorities to insist on a renewal of the offensive; but this was an entirely unrealistic order.[3] Captain Griffith's own brother was killed while delivering a message, and all the brigade signallers were annihilated by the enemy shelling. No forward progress was possible on the ground. After the battle he was left in a state of dazed shock and grief; but above all he was certain that the main problem had been the failure to coordinate the artillery with the infantry. He believed there had been nothing inevitable in the 'constant failure to link up the loose threads of intention scattered over the battlefield, an omission that was made catastrophically evident to us in the gap that divorced artillery support from infantry action'.[4]

Another pertinent testimony came from Lieutenant Colonel Frank Maxwell, VC, commanding the 12th Middlesex, who was thrilled by his chance to orchestrate the fire of large groups of guns, but still remained very suspicious. On 29 June 1916 he reflected that 'Much too often... something goes wrong, and guns or infantry get tangled up in their plans, with the result that our infantry run into the fire of our guns, and have to come back for fear of being killed.'[5] His conclusion, as an infantryman of the old school, was that it would be better to do without guns altogether. When this course was disallowed by higher authority, he used some (unspecified) devious means to make sure that his supporting artillery fired as far away as possible from his men.

The lesson that good artillery cooperation was vital nevertheless seems to have been learned quite well throughout the BEF by the later stages of the Somme, and by 6 October we find an aerial observer's report being circulated throughout Fourth Army, telling of 'a most perfect wall of fire in which it was inconceivable that anything could live'. The 50th Division's infantry followed it within fifty yards; it kept a dead straight line, and the village of Le Sars was taken without opposition. As in some phases of both Neuve Chapelle and Loos, in other words, the foot soldiers found they were able to walk through the German positions, in line and with smiles on their faces.⁶

Such successes certainly represented an advance in technique, and pointed a way forward to the future; but in 1916 the barrages were still dogged by too many problems. Defective shells continued to be supplied, and the techniques of accurate gunlaying remained to be mastered gradually during the course of 1917. Towards the end of the Somme battle, moreover, the barrages were often too thin because deep mud imposed excessive delays upon the guns' deployment and resupply. When seen from the viewpoint of traditional infantry culture, moreover, the very idea of a creeping barrage in itself carried some deeply troubling implications. For example, one highly influential battalion commander came out of the fighting with a belief that

> The two fundamental facts which govern the modern assault are these, viz:-
> (a) The assault no longer depends upon rifle fire supported by artillery fire, but upon the artillery solely with very slight support from selected snipers and company sharp-shooters....
> (b) The decisive factor in every attack is the bayonet.⁷

At a stroke this abolished the teaching of centuries that infantry could fight forward without outside support. Still more astonishing, perhaps, was its almost total dismissal of the infantry's own firepower of bombs and bullets as irrelevant, leaving only the cold steel as a means of killing Germans.

From its very first days on the Somme the BEF learned that the 'bomb' or hand-grenade could easily bring as many problems as it solved. According to this critique the infantry might stand a better chance of fighting its own way forward if only it could manage to liberate itself from the bomb. As early as July the official position had become that

> It must be realised by all ranks that the rifle and the bayonet is the main infantry weapon. Grenades are useful for clearing small lengths of trench and for close fighting after a trench has been rushed; but no great or rapid progress will ever be made by bombing, and an assault across the open after adequate preparation will usually be a quicker and in the long run less costly operation than bombing attacks on a large scale.⁸

This formula began to be a standard orthodoxy. Grenades were good for rolling into dug-outs after a trench had been entered, but less good for capturing trenches in the first place.

An apparently modernist bomber's (or 'stormtrooper's', if you wish) approach had been creeping up on the BEF throughout 1915 and the first half of 1916; but there was now an important 'post-modernist' reaction against it which would continue for all the remainder of the war. It seemed to many that the cult of the bomb could only too easily be used as an excuse for timid soldiers who wanted quietly to forsake the offensive and revert to the 'Live and Let Live System'. The bomber's contract seemed to include too much emphasis on 'not exposing oneself', and too little on such things as 'capturing the enemy's trench'. Far from storming forward, in fact, such troopers were now perceived as preferring to cower behind a traverse. When the 9th Division entered the line near Armentières in May 1915, for example, it had been given considerable training and indoctrination in bombing. But it gradually came to see that 'the trouble was that when troops became stale with months of underground warfare, the bomb fight tended to result in a stationary conflict, no serious effort being made to gain any ground'.[9]

The tank expert Major W. H. Watson agreed with the opinion that 'When attacking troops are reduced to bombing down a trench, the attack is as good as over....'[10] Charles Carrington, the hero of Ovillers, also describes how bombing tactics tended to bog down as soon as the enemy started throwing bombs back.[11] Towards the end of 1916 many similar cries could be heard in other parts of the BEF, and they were doubtless well founded. If so, they may also be seen as a valid criticism of the German stormtroopers' own emphasis on grenades, which continued until the very end of the war. This was sometimes even made specific, for example in a Fifth Army circular of 16 September:

> The Germans counter-attacked in parties of 20–40 men in shirt sleeves with no equipment only bombs: they should have been dealt with by rifle fire. The rifle is the infantry soldier's best weapon and he must use it.[12]

On this occasion the defenders had tried to retaliate with grenades, but were ineffective and attracted the wrath of General Gough himself.

The BEF's commanders did nevertheless face a serious problem of dissemination when they tried to cut back the bomb culture, since it was already very deeply rooted. Attempts to banish the bomb completely led to immediate objections from the ranks,[13] and in fact the official manuals were themselves continuing to recommend plentiful bombers.[14] Yet they were simultaneously now starting to echo the pleas for a return to emphasis on the rifle and the bayonet instead of the bomb.[15] It was all a little confusing; but perhaps the lyrical palm should go to GHQ for its instructional war poem of 5 June 1917:

> 'Bombs,' says Alf, 'are good for a change
> But it's the rifle will pull you through....'[16]

Alf's emphasis on the rifle certainly reflected the thinking of many senior officers throughout the second half of the war, who rarely seem to have paused to reflect very deeply about the many ways in which their own official training manuals during the first half had already, and almost irreversibly, fostered the cult of the bomb. Obviously they were looking for a spur to keep their men moving forward in the assault without going to ground; but this cut clean across the natural prudence which encourages almost everyone in a battle to seek cover whenever they can.

It may well be easier, and appear more aggressive, to attack over the top with a rifle against a suppressed enemy than to weave one's way below ground through a maze of trenches with a bucketful of bombs. However, in both cases the key question is surely whether or not the enemy has first been truly suppressed. If he has been, the frontal assault with the rifle is doubtless the best approach; but if not — as was all too often the case in practice — then any frontal approach may very easily turn out to be suicidal. However uncertain his progress, the bomber working his way forward from one shell hole to another, or from one traverse to another, may well have the better part of the bargain, as the New Armies frequently found when they came into action during 1915 and the first half of 1916. This factor, rather than any particular neglect of musketry, surely explains their original emphasis on grenades.

The official 'anti-bomb, but pro-rifle and bayonet' polemic was certainly one of the most striking and widespread changes in infantry doctrine to emerge from the Somme fighting; but perhaps little less important was the concurrent debate between the rifle and the bayonet. As we have already seen, at least one battalion commander dismissed rifle fire in the attack almost completely, in favour of the bayonet. For him, rifle fire slowed down the attack and encouraged it to fall behind the all-important creeping barrage. He wanted to stop anyone firing or lying down in the assault — 'otherwise all control will be lost'[17] — and he wanted the attack echelons to bound forward 'on top' rather than becoming bogged down in narrow communication trenches.[18] Speed in getting forward was everything, in order to follow the barrage and reach the enemy's parapet before the enemy could man it and open fire. This emphasis on speed and momentum absolutely resonated with the teachings of Laffargue, although it ran counter to the latter's distrust of artillery, and his belief that the infantry could shoot its own way forward. Whereas our British battalion commander had placed all his faith in the gunners and had told his men not to fire, Laffargue had personally emptied his rifle magazine into the faces of the Germans during the final rush at

Neuville St Vaast, wounding one and — more importantly — intimidating the rest. He believed it was psychologically impossible to stop troops firing in the assault, unless by the personal example of picked leaders and skirmishers who led from several paces in front and did the firing for all the rest.[19]

A slightly different, but related, conclusion came from Brigadier General Kellett, temporarily commanding 2nd Division in November 1916, when he said:

> With men whose period of training is necessarily short, and habits of discipline in consequence not sufficiently inculcated, there is no doubt a great temptation for men under heavy fire to drop into shell holes or trenches instead of going forward, and thereby escape observation....[20]

In this case the prescribed remedy was to keep the soldiers in close groups to prevent shirking, and thereby to encourage the time-honoured moral qualities of mutual surveillance, emulation and cohesion.

In Napoleonic times it had been commonplace, especially in siege warfare, to maximise the forward impetus of attacking troops by keeping them in columns and ordering them not to fire. This had never been a stupid or 'Luddite' response from remote senior officers, as is often claimed by ignorant modern commentators, but was a reasoned analysis of what was needed, by some of the finest tacticians of the day.[21] Admittedly there is little evidence that much true bayonet fighting ever took place in the Napoleonic or any other era, or that the percentage of casualties attributable to bayonets was ever anything but negligible. Nevertheless, an appeal to the bayonet could often be a highly effective spur for troops who might otherwise hesitate to press forward to occupy the enemy's position. When seen in this light, the new creeping barrage represented merely a more powerful and more technological adjunct to the bayonet, since it served the same function of intimidating the enemy at the same time as it accompanied, accelerated and paced the attacker's charge.

The perfectly simple logic behind appeals to the bayonet has nevertheless repeatedly been obscured and denied by self-interested parties. Certainly the gunners could never come to think of it as in any way analogous to their own awesome power, even when they were firing in merely a 'neutralising' mode; and Tank Corps champions such as J. F. C. Fuller gradually came to doubt that infantry still had any significant rôle in modern battle at all. He himself was originally an infantryman, and his pamphlet on infantry and tank cooperation issued on 27 January 1918 fairly stressed that the infantry would normally enjoy primacy, and that 'bullets and shells must be employed to facilitate or ward off the use of the bayonet';[22] but throughout the remainder of his long influential life he would become successively more sarcastic and extreme about that particular arm. Even more effective as propaganda against the humble bayonet, however, has been the humanitarian disgust universally evoked in sensitive poetic souls by the blood-curdling lectures of Major (later Lieutenant Colonel) Ronald B. Campbell, DSO, of the Gordon Highlanders, GHQ's Assistant (later Deputy) Inspector of Physical and Bayonet Training.

Few individuals on the Western Front managed to evoke such universal hostility among the literary élite as did Ronnie Campbell. Haig's very talented Chief of Staff Kiggell may be a close contender with his apocryphal exclamation upon belatedly seeing the Passchendaele battlefield — 'My God! Did we really send men into that?'[23] A majority of the corps commanders also seem to have impressed most of the men they inspected and reviewed with their evident personal pettiness and almost criminal remoteness. But in both these cases the offence was at least in some sense accidental and unintended. With Campbell, by contrast, it was planned with very deliberate forethought and skill and, still worse, it seems to have been a 'star turn' which made a deep positive impact upon most of the lumpen rank and file at whom it was really directed.

It was Campbell's job to tour around the whole of the BEF with a team of instructors and accomplices, giving a highly rehearsed lecture on the importance of the bayonet and the 'spirit' of the bayonet. Apparently he and his team possessed a unique flair for the work, using many pyrotechnic effects, stomach-turning visual aids and anatomical specificities. After the lecture students went through an assault course to practise bayoneting straw men while screaming their atavistic hatred for the Boche, and adopting the correct 'killing face'.[24] Blunden called the show 'more disgusting than inspiring', while 'Mark VII' (Max Plowman) was reminded that he had a weak stomach.[25] R. F. Callaway admired it as 'extraordinarily good' at the same time as he was personally revolted by it, whereas Colonel Jack thought it 'cold blooded & rather horrid' and Colonel F. Mitchell found it 'at the same time impressive and repellent'.[26] Liddell Hart, however, regarded it as 'comic relief', while Graham H. Greenwell was more impressed and less revolted, accepting the Campbell thesis that the Somme fighting meant there would be more hand-to-hand work and that British superiority in this department would soon prove decisive.[27] There is an irony that the arch-primitivist Campbell here seems to have been echoing the thoughts of the arch-technologist Winston Churchill, who in December 1915 had looked forward to a great 'Attack by the Spade' with 300 saps over a 30,000-yard front. His idea had been simply to dig forward into the German trenches, as in classical siege warfare, and then conduct a ten-day hand-to-hand fight 'flat out' and with rested troops. Churchill had seemed quite happy to envisage that 'a long series of bloody local struggles will develop along the whole front of the selected sector. The more this goes on, the better for us....'[28] In later years, however, he may perhaps have come to regret this only too prescient enthusiasm for unmitigated slaughter at close quarters.

Campbell did at least seem to speak for an important strand in British tactical thinking, and his ringing phrases, or something close to them, kept resurfacing in the works of both official and unofficial tactical spokesmen. In his 26 September 1916 attack on Thiepval, for example, Colonel Frank Maxwell, VC, was not taking any prisoners, on the grounds that all Germans should be exterminated,[29] and if anything he seemed to think the 'Bayonet and Physical' instructors themselves were too moderate and prim.[30] In his II Corps notes 'from recent fighting' of 17 August 1916, General Jacob urged that no prisoners should be taken, since they would only hinder mopping up.[31] Even the apparently more humane Christopher Stone insisted that 'a live Boche is of no use to us or to the world in general'.[32] Such ferocious attitudes remained widespread throughout the war, and in a letter of 20 August 1918 the peppery Ivor Maxse was still referring to front-line combatants as the men who 'put bullet or bayonet into Hun belly',[33] rather than by any more delicate cognomen. In the same letter, incidentally, Maxse made it clear that he had abandoned his earlier tendency[34] to see the bullet as distinctly superior to the bayonet. Campbell himself had always apparently been consistent in urging rifle fire as an equal partner to his own beloved bayonet — and by 20 August 1918 Maxse had come round to a similar view. He now wanted to combine the two weapons as a single subject in the training curriculum.

Others, however, took an opposite opinion, and believed that the rifle was the only true infantry weapon. Such officers referred back specifically to the Old Contemptibles' lost skills in accurate musketry, and deplored the poor training of the New Armies. Thus Lieutenant Colonel Croft's disgust knew no bounds at Longueval on 14 July 1916, when he found that his 11/RS rifle fire was unable to hit enemy soldiers fully visible at a range of only 200 yards. This, he said, fell 100 yards short of longbow range at the battle of Crécy.[35] Taking his cue from *Duffer's Drift*,[36] he therefore sat down to analyse and reflect on the 'bad dream' of Longueval, in which his battalion had lost 624 casualties before the enemy finally broke and fled. Apart from his regret at the lack of cavalry close enough at hand to exploit the victory, his main conclusion was that an intensive programme of musketry training was required, and as much under realistic battle conditions as possible.[37] Quite undeterred by an unfortunate experience in October at the 'Pimple' — when the rifles were so clogged with mud that the enemy had to be repulsed with grenades alone — he set his face firmly away from the latter and in favour of the former. 'If you met a tiger coming down the street,' he asked, 'would you shoot him or chuck a bomb at him?...Think of it! Chucking a bomb like a blighted anarchist....'.[38] There was no mention of the bayonet in any of this, but only an ever-growing obsession with rifle fire.

One somewhat specialised application of rifle fire was sniping, which grew up in 1915 mainly for static warfare rather than for the assault. It soon become an élite technical activity within each battalion, in a similar way to bombing or machine gunnery, although unlike bombing it never gravitated back into the ranks as a universal skill, and unlike machine guns it never levitated to a completely separate higher status. Both Robert Graves and 'Mark VII' complained at the lack of telescopic sights which prevented them from applying the lessons they had learned in sniping courses, and the latter further complained that official attempts to turn it into a 'sport' made him feel queasy.[39] Nevertheless, sniping did become a reality in a relatively small way within many battalions,[40] and more or less of a platoon of specialist snipers would be commanded by the intelligence officer at each battalion HQ, and tasked in pairs to watch the whole enemy line opposite.[41] At least in theory a group of less expert marksmen was also maintained within each company, although as the war progressed these men tended to become confused with the 'scouts'.[42] Probably they rarely amounted to anything very different from ordinary riflemen or sentries, since they are so rarely mentioned. Indeed, all snipers appear to have been so successful in their undercover activities that they have left very few traces in the general literature of the war. Apart from GHQ's specialist sniping school,[43] they did not have any organisation higher than a single platoon in each battalion, and so they caused no institutional waves of the sort that tended to give machine guns and mortars a bad name. Within the battalion, furthermore, this platoon enjoyed a relatively sheltered life as part of the vast semi-visible, semi-combatant 'HQ' echelon which normally contained some 200 men, and occasionally rose to as many as 350.

Snipers did nevertheless have at least a theoretical rôle in the offensive, even if their practical utility remains to be documented. In the anonymous battalion commander's *Memorandum on Trench to Trench Attack* of 31 October 1916,[44] the regimental snipers are intended to position themselves before dawn within fifty to sixty yards of the enemy front line, to snuff out any troublesome machine guns during the crucial moments of the attack itself. Others would occupy positions overwatching more distant flanks, with the same purpose. The company marksmen, by contrast, were to accompany the attack but not join in the trench-clearing. Instead, they would lie out watching for any enemy response, in effect serving as the initial phase of consolidation. In all this the rôles of snipers were intended to be very similar to those intended for Lewis guns in the May 1916 'Red Book', and in both cases they represented a serious attempt to deal with the potentially disastrous German machine guns which might survive the preliminary bombardment. This was the 1916 version of 'pillbox busting'; or the first generation of an endeavour which would soon be formally extended to include the Stokes mortar, the rifle-grenade, the tank and even, by late 1918, a reversion to direct fire from field guns[45] or aircraft.

Alongside the interest in various forms of musketry and sniping, there was also a widespread nostalgia for 'fieldcraft', although it was never made entirely clear just what, in the conditions of trench warfare, this was supposed to mean. One would have thought it meant precisely the skills of the night patroller, the sniper and the camouflage expert, all of which took root and flourished, in varying degrees, in every New Army division almost as soon as it arrived in France. Yet what was usually implied was something rather different, more closely related to 'offensive spirit', 'morale', or even 'regular army pedigree'. Thus outside Fontaine les Croisilles near Arras, on 23 April 1917, the veteran Captain J. C. Dunn, RAMC and 2/RWF, felt that the 98th brigade's attack on the Hindenburg Line had collapsed because of a failure to use the 'minor infantry tactics, learned from the Boer riflemen in South Africa, which were routine instruction before the war'.[46] By the start of 1918, however, as he surveyed callow recruits dispersed in section posts 100–150 yards apart, even the reactionary Dr Dunn seemed to be dimly aware that the prewar tactics had not really worked for the Old Contemptibles, and had included dangerous overcrowding and bunching in the trenches. His poignantly unhelpful comment was that 'In early 1915 men worth ten of these — were something like one to the yard'.[47]

The importance of careful preparation

The Old Army had certainly wanted its troops to be aggressive and enterprising in the attack, but it often failed to see that those troops, in turn, needed a much higher level of support from the rear than they generally received. In 1916 this lesson was gradually being learned and, although it would not be fully mastered even in 1917, no fair commentary on the Western Front can deny that reform did eventually occur. A major part of it was a matter of improving artillery support, which was largely a technical matter; but in many other departments it was more a matter of changing attitudes among the high command. In particular, it gradually dawned upon the better commanders that there was really no substitute for careful preparation.

No one who reads the monumental staffwork and planning for 1 July on the Somme can doubt that the preparation had been very thorough indeed. When it led to failure, however, the high command did not sit back quietly to analyse the technical reasons, in order to make another careful assault later; but instead it called for an immediately renewed

The Lessons of the Somme

effort, in the hope of snatching victory from the jaws of defeat. Impatient generals cast aside their admirable earlier acceptance of such things as logistic buildup, rest and breakfast for the assault troops, reconnaissance, rehearsal, detailed coordination, and briefing of all participants... and simply kept on demanding instant results. As we have seen in Wyn Griffith's attack on Mametz Wood on 11 July, such 'decision-making on the hoof' was almost always disastrous. Yet it was a bad habit which persisted well into 1917,[48] leading repeatedly to attacks that were called at too short notice and on too narrow frontages, before essential preparations could possibly have been made.[49]

Unlike the battle of Salamanca in 1812, where Wellington had overthrown the French army by snapping out a single unexpected but decisive order in the space of perhaps five seconds, the Great War offered no opportunities for such inspired instantaneous generalship. On the contrary, it gradually became very clear that the best generals were those who made haste slowly and carefully, trying to leave absolutely nothing to chance, and giving their men plenty of time to rest and recover from each exposure to the cauldron of battle.[50]

A part of this process was systematic tactical analysis. As the Somme fighting progressed, many parts of the May planning assumptions were gradually reexamined and refined. Thus the precise intervals between echelons and waves were reconsidered and one corps commander, Jacob, even went so far as to contemplate a single-line attack. He argued that most casualties occurred in the support and reserve lines, regardless of whether the initial assault line succeeded or failed. This was because No Man's Land became the area of greatest shelling as soon as the German artillery had responded to an SOS call — probably six to eight minutes after the British zero[51] — so any follow-up lines were bound to be caught too closely bunched within that zone. The front line should therefore be left to sort out its own destiny; but if its creeping barrage had been good, Jacob believed the attack would always succeed. If not, then nothing could save it. He did, however, recommend winning additional surprise by attacking in the afternoon rather than at dawn, and showed that all six afternoon attacks in his corps had succeeded whereas both the dawn attacks had failed.[52]

Gough, Jacob's superior, was not converted to the 'single echelon' attack, although he did stress the need for every echelon to move forward simultaneously and pass the danger area quickly.[53] He also, perhaps surprisingly, drew some tactical lessons from one very modern piece of technology: the cinematograph. A film of a real attack had persuaded him by 18 September that the use of small groups bunching together, rather than advancing in regular lines, caused a whole clutch of problems.[54] Not only were the bunches themselves vulnerable to artillery, but they did not attack the enemy trenches and shell holes in the intervals between groups, thereby allowing the Germans to continue a prolonged resistance from flank and rear.

Many other details were brought up to date during the later Somme fighting, such as the need to signpost and tape out approach routes and the start-line, to avoid confusion as the assault troops formed up. Sapping across No Man's Land to hasten subsequent consolidation was another major preoccupation, as were signalling, timing and, above all, direction keeping during the advance. In the weeks before the autumnal rains finally closed down operations, analyses of this type did in fact lead to reforms which helped win a fair number of local successes.

One particularly interesting subject for debate was the immediate German counterattack, which was generally found to be as ineffective as it would continue to be throughout the remainder of the war[55] — and also through most of the next war. Jacob believed that it could easily be beaten off by infantry alone, but that this did not happen often because it was usually dispersed by artillery first:

> The German infantry counter attack is talked a lot about in theory but, in practice, is, unfortunately, rare. We have had on the II Corps front, no authenticated case of an immediate infantry counter attack, other than by bombers when trenches have been left undestroyed by which the latter might approach.
>
> We have had reports of German infantry forming up to counter-attack but overland [i.e. 'on top' rather than via trenches] attacks, if attempted, have never been allowed to materialise. The artillery see to that.[56]

Finally, the Somme battle underlined the importance of Lewis guns, rifle-grenades and Stokes mortars as close supports for attacking infantry — not only in helping consolidation, but especially in 'pillbox busting', or in other words completing the work of the artillery and thus enabling the infantry to fight its own way forward. One report on the Reserve Army's experience noted with satisfaction that all three of these weapons were now becoming more plentiful, and it was particularly fulsome in praise of the help all three could offer to an attacker.[57] This opinion had its detractors, such as the rifle or Vickers purists, or the anti-Stokes bitter-enders; but by and large it was sufficiently widespread to lay the foundation for what would be the decisive 'third generation' of tactical reappraisal in the BEF.

If 1915 had seen the first such reappraisal and May 1916 the second, then the winter of 1916–17 brought the whole movement to fruition with the production of two vitally important manuals which would survive, in essence, for the remainder of the war. These were the sixty-page SS 135, *Instructions for the Training of Divisions for Offensive Action,* issued in

December; and the fifteen-page SS 143, *Instructions for the Training of Platoons for Offensive Action*, issued on 14 February. SS 135 was deeply influenced by the July 1916 *Preliminary Notes on the Tactical Lessons of the Recent Operations* (SS 119); also by Jacob's September analyses, and by the October *Memorandum on Trench To Trench Attacks* written by our anonymous battalion commander. SS 143 seems to have drawn upon some very different inspirations, probably including Laffargue, although it does still attempt to crystallise the general thinking of the day, not least within Fifth Army.[58]

With SS 135, *Instructions for the Training of Divisions for Offensive Action*, we have moved a long way forward from the early-war shibboleth that tactics were merely a matter of finding specific applications for 'eternal principles' (which usually meant Haig's 1909 FSR). We have even moved beyond the May 1916 'self-supporting stormtroop' conceptions. Instead, we have an acceptance that the creeping barrage is the key thing; that it must be followed as closely and as quickly as possible by the assault troops. Even so, there is still plenty of room for infantry to fight without help from artillery, if enemy machine guns are still active after the creeping barrage has passed. In this case snipers, Lewis guns, Stokes guns, smoke barrages — and everyone else — must try to work together to deal with the threat. Perhaps this reduces the rôle of infantry to little more than mopping up after the artillery has carved out the general shape of the battle; but at least it is a more realistic conception than the earlier, more heroic, formulae. It certainly includes a growing understanding of the need for careful preparation, as well as a devilish awareness of what can be achieved by surprise, or by the types of paradoxical indirect approaches that had originally been suggested by *Duffer's Drift*. The battlefield portrayed here is no place for the inspirational prima donna type of general; but it is a welcoming home for the shrewdly methodical plodder who is fully conversant with the whole range of modern technology. It is an ultimately refined battlefield which includes all weapons — even tanks — and which is at long last starting to find ways of coordinating them all into a single whole.

In SS 143, *Instructions for the Training of Platoons for Offensive Action*, we find a radical departure from the earlier assumption that there should be a solid line of riflemen accompanied by a few specialists detached from brigade, battalion or company HQs. Instead, we now have a self-contained unit which is divided into a small platoon HQ plus four fighting sections, each with its own speciality. The first section has two expert bombthrowers and three accomplices, the second has a Lewis gun with thirty ammunition drums and nine servants, the third has nine riflemen including a sniper and a scout, while the fourth section has a battery of four rifle-grenades — called 'the infantry's howitzers' — manned by a further nine men. The platoon is thus a complete and independent 'tactical unit', in a way that was implied but never fully elaborated in May 1916. Some authorities went as far as to interpret the new arrangements as elevating the section into the new 'tactical unit',[59] but really this cannot be sustained. Each section was intended to be centred around one particular type of weapon, and therefore it could not be seen as properly self-sufficient. The platoon as a whole, however, amounted to almost a complete army in miniature — apart from its lack of cavalry.

The platoon attack was envisaged in two waves. The first wave would consist of the two 'manoeuvre' sections in line abreast — the rifles and advanced scouts to the left and the hand-bombers to the right. Some fifteen to twenty-five yards behind them would come the 'firepower' elements — the rifle-grenades to the left and the Lewis gun to the right. The platoon HQ would be in the centre somewhere between the two lines. Upon contact with an unsuppressed enemy the 'firepower' line would attempt to neutralise him while the 'manoeuvre' line tried to move forward, especially to a flank. Thus the enemy would be suppressed by fire from the front while he was overrun by bayonets and bombs from a flank. While they were thus overrunning the Germans, the attacking riflemen were urged to recall Major Campbell's precept that our aim 'is to be *bloodthirsty* and for ever think how to kill the enemy...'.[60] (See Fig. 3.)

Fig. 3: Platoon tactics, February 1917

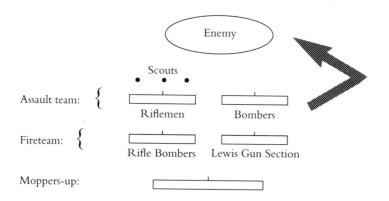

SS 143 may be seen as a vital milestone in tactics, marking a changeover from the Victorian era of riflemen in lines to the twentieth-century era of flexible small groups built around a variety of high-firepower weapons. Once each platoon had its own Lewis gun, as was starting to be true in the winter of 1916–17, combined tactics had begun to be possible even at this lowly level. Admittedly the Lewis gun itself was still being classed merely as a useful auxiliary — 'a weapon of opportunity' — while the rifle and bayonet were placed on a higher pedestal.[61] Nevertheless, in reality the Lewis had become the true centre of the platoon's firepower and the pivot of its manoeuvre. As Maxse would exclaim in February 1918, 'A platoon without a Lewis gun is not a platoon at all!'[62] It had not at all suppressed the idea of linear tactics, since SS 143 was still essentially linear; but it had certainly reformed and reshaped the inner spirit.

In common with all other prescriptive manuals, both SS 135 and SS 143 were followed as much in the breach as in the observance. Some attacks continued to be ordered at excessively short notice and without a full 'combination of all arms'. The cries of château generals for 'more platoon training' often seemed laughably unrealistic to battalion commanders whose men were exhausted by labouring duties or pitifully reduced in number. The schools to teach the new tactics themselves often appeared very rigid and unimaginative to troops who had by now often seen a great deal of real warfare.[63] Charles Carrington, for example, felt particularly proud that he and his friends had understood the general concept of all-arm cooperation within the platoon long before it surfaced in any training course or in SS 143.[64] For all that, however, the new system was in many senses fully suited to its purpose, and gave the junior tactician plenty of appropriate guidelines. It made an excellent basis upon which the BEF could move forward to victories in 1917.

The assault spearhead of the BEF

By the start of 1917 the BEF's infantry units had come of age and had won distinctive characters above and beyond their simple original classifications as either 'Regular' or 'New Army'. Some battalions, such as Dunn's regular 2/RWF[65] or Croft's New Army 11/RS, had maintained a reputation for good discipline, cohesion and an aggressive desire to dominate No Man's Land at every opportunity. They have sometimes been seen as the British equivalent of the specialised German stormtroops, and indeed often established a supremacy over them. Other BEF battalions, by contrast, were known for their easygoing ways, their epidemics of trench foot and their general failure to make an impact. Such corporate identities were often attributed to the specific personalities of individual commanding officers; but in fairness they doubtless had as much to do with the unit's longer-term traditions and self-images. Each battalion might have as many as half a dozen different COs during the course of the war but, short of some catastrophic rupture, its inner 'essence' might change only once or twice, if at all.

As with individual battalions, so with whole brigades and divisions. The longer they lived together as distinct formations, the more they would accumulate a corporate identity. In one of his most pregnant passages for the military technician, Robert Graves — by then fully indoctrinated in the regular army's ethos — recounts how among the instructors at the Harfleur bull ring in early 1916

> It seemed to be agreed that about a third of the troops forming the BEF were dependable on all occasions: those always called on for important tasks. About a third were variable: divisions that contained one or two weak battalions but could usually be trusted. The remainder were more or less untrustworthy: being put in places of comparative safety, they lost about a quarter of the men that the best troops did.[66]

He went on to list the divisions thought to be 'dependable on all occasions' which, apart from the 1st Canadians, all turned out to be British regulars.[67] This, however, was still in the days before the New Armies had arrived in force, and we are fully entitled to speculate that if the same list had been drawn up a year later, it would have included a significant number of New Army entrants.

The 9th (Scottish) Division was the first New Army division to take the field in France, and for much of the war it incorporated the lively South African brigade. It produced more genuine tactical innovations than any other division; more of its commanders went on to command army corps than would be true of any other division and, as the most senior, it was the only New Army formation allowed to join the occupation forces in Germany after the armistice. In the long view of history it probably deserves to be seen as the most 'élite' of all the British divisions — not least because it managed to create a potential breakthrough on three separate occasions,[68] and all without any help from tanks whatsoever.[69]

Other early Kitchener divisions with a claim to 'élite' status include 11th (Northern), 14th (Light), 15th ('Thistle' Scottish), 18th (Eastern — which was very successfully commanded on the Somme by Ivor Maxse), 19th (Western), Jacob's 21st (despite blotting its copybook at Loos), 30th (Lancashire), Pinney's 33rd and the 36th (Ulster) Division. The 63rd (Royal Naval) Division also has a very valid claim, as do the territorial 47th and 56th (London) divisions, and 51st (Highland — known as 'Harper's Duds' after its somewhat erratic commander). Other, perhaps more questionable, claims to élite status may be filed by such formations as the 46th (North Midland territorial) Division, which had widely been consided 'dud' until it crossed the Hindenburg Line in a trice in the autumn of 1918. Then again there was the 55th ('Red Rose' Lancashire territorial) Division, which did well until it bolted on 30 November 1917 at Cambrai — although even after that it has generally won a good press, since it chanced to have been commanded by the official historian Edmonds's good friend General Jeudwine.

The BEF can certainly boast more than a dozen 'élite' divisions which originated in the British Isles; but to these we must also add the ten ANZAC and Canadian divisions which have always clamoured very stridently and insistently for a still higher recognition. These colonial formations enjoyed many organisational advantages over their UK colleagues, not least their independent political status which allowed their commanders to question GHQ policy at almost every turn. Still more to the point, perhaps, was the fact that both the Australian and the Canadian corps eventually managed to establish themselves as permanently self-contained formations. They cut themselves free from the normal BEF (and German) practice whereby a corps might take in any division almost at random, only to spit it out again a few weeks later.[70] Within both the Australian and Canadian corps, by contrast, the order of battle very soon became permanently fixed at the same five and four divisions, which gave them an enviable continuity of leadership downwards from their respective corps HQs.[71] Only the New Zealanders and South Africans were cast loose to take their chance within the more complicated rough-and-tumble of the 'normal' BEF arrangements.

The frequency with which a particular division would be sent into action naturally depended in part upon the length of time it was in France; but it also depended to a very great extent upon the approval or favouritism of the high command. Thus all five of the original infantry divisions of August 1914 went into serious action more than thirty times, although the 2nd Division, with 44 occasions, seems to have been specially favoured over the 4th Division with only 31. Regular divisions unlucky enough to have missed Mons even by a few weeks seem to have been cut back savagely to around 20 occasions. Equally the first tranche of the New Armies had almost as high an experience of front-line action as the Old Contemptibles, with 9th Division going in 26 times, 25th Division 27 times, 21st Division 29 times, and Maxse's fighting 18th Division no less than 32 times. Yet against these, some of the other long-service New Army divisions went in only 12 or 14 times (16th and 14th Divisions respectively), and four of them escaped quickly to other theatres. All twelve of the second tranche of Kitchener divisions managed to stay in France for most of the war, with their average number of times in action standing at 15.3. Yet among them, the ill-fated 'Bantam' 35th Division went in only 11 times, and was ignominiously de-bantamised in the spring of 1917, while the partially Bantam 40th 'Mongrel' Division seems to have been so little trusted that it was committed only 9 times.[72] When we come to the territorial divisions that were actually sent to

France we usually find a similarly low level of trust, with their average number of times in action standing at around 16. These generally seem to have been the divisions that were least well viewed by the high command, as is particularly reflected in the high frequency of their despatch to Egypt, Salonika or Italy. Far more of the TF divisions were relegated to secondary theatres, it seems, than were the regulars or first-line 'K' divisions. Ironically, the Western Front appears to have been a place where the harder you fought, the less easily you could be excused from the fighting thereafter.

The unavoidable conclusion is that the high command observed an informal 'pecking order' of divisions, by which the most reputedly trustworthy formations would often be chosen for the dirtiest jobs. This pecking order was surely based upon little more than prejudice, hearsay and the cut of the division commander's jaw when he turned up at GHQ; but it nevertheless appears to have been a reality in the minds of the higher staff. To some extent it even fed upon itself, insofar as good divisions would improve by the very experience of being placed alongside other good divisions, in rather the same way as middle-class schoolchildren are supposed to improve merely by being placed alongside other middle-class schoolchildren in our great public schools. The élite divisions' HQ and command echelons would thus absorb technique and sophistication by a painless process of osmosis, just by working together with their peers from other élite divisions. For example, Congreve's very successful XIII Corps on 14 July 1916 included the 3rd, 9th, 18th and 55th divisions — all leading players in the 'premier league'. In late September at Thiepval Jacob's equally successful II Corps included the 18th and 11th divisions, flanked by the Canadians themselves. In November at Beaumont Hamel E. A. Fanshawe's V Corps included the 2nd, 3rd, 51st (Highland) and 63rd (Naval) divisions, and they in turn were flanked by Jacob's II Corps. It all made for a dense matrix of cameraderie and expertise among the leading tacticians of the BEF, who were given plentiful opportunities for sharing ideas and attitudes, although it also perhaps implied an equal but opposite paucity of such stimulation on other fronts.

This system of grouping 'assault' or 'élite' divisions together for grand occasions seems to have persisted throughout the remainder of the war. At Broodseinde and Passchendaele in early October 1917, for example, two ANZAC corps rubbed shoulders with three of the regular divisions plus 11th, 21st and 48th divisions, which were later relieved by 9th, 18th and 17th divisions. Subsequently the Canadians took up the baton for the final sprint into Passchendaele itself. It was surely no accident that out of the total of seventeen divisions involved in this notoriously tricky attack, all but two were drawn from our list of 'élite' formations. The Cambrai attack which followed soon afterwards, by contrast, had the character of a last-

The Lessons of the Somme 83

minute improvisation, and therefore had to make do with only two 'élite' divisions[73] out of its total of eight committed during the first two days.

Perhaps the most celebrated example of deliberate 'élitism' in the BEF came nine months later at the battle of Amiens, on 8 August 1918. On that occasion great logistic and deception efforts were made to ensure the surprise concentration of both the Australian and Canadian corps at the spearhead of the attack; and almost equal efforts have subsequently been made to contrast their superlative performance with the poor showing of their neighbouring British III Corps. The only potentially 'star' performer in the latter formation was 18th Division, but that was now Maxse-less, devastated by Maxse's own March defeat, and it did not do particularly well.

Beyond the quality of particular formations, by the end of the Somme battle we can also start to glimpse some of the truly effective individual commanders within the BEF. This was not merely a matter of the weak being weeded out by higher authority, but often also a matter of experience kindling a creative interest in the subject among personally active men who were ready to learn. Among the army commanders, Plumer and Horne stand out as 'tacticians' in a way that was scarcely true of their contemporaries, apart from the possible exceptions of Birdwood and Byng. At lower command levels there were such men as the careful Jacob from the Indian Army; the asthmatic Congreve (a VC from Colenso); the loquacious Maxse, who has left us the best of all the tactical archives; the controversial Monash of the 3rd ('Neutral') Australian Division and then of the whole Australian Corps; the gunner Du Cane of XV Corps; the aristocratic cavalryman De Lisle of 29th (regular) Division and then of XV Corps; the pushy Brigadier Solly Flood of X Corps and soon of the GHQ 'training directorate'; and the scientific Captain Lindsay of Hythe, who created the Machine Gun Corps. There were eccentrics such as the near-bankrupt (albeit militarily highly competent) Currie of the Canadians, and the restless spiritualist Fuller of the tank staff. There was Jeudwine of 55th Division and, from the gunners, such men as Furse, Tudor (both of 9th Scottish Division), Uniacke and Birch.

Taken together, this cohort of officers formed an undoubted[74] 'élite', and was responsible for actualising many of the reforms which pushed forward the art of war between 1916 and 1918. Perhaps they sometimes suffered from a lack of formal recognition either as 'tacticians' or even as an 'élite'; but it remains true that many of them were responsible for writing manuals which enjoyed wide circulation, and by their personal example they often exerted an important influence upon their contemporaries. As we have seen, élite formations tended to be deliberately grouped together by the good offices of the GHQ staff.

5

The Final Eighteen Months

> Rapidity, surprise and smoke.
> Key motto from *Suggestions For Attack*,
> in Maxse papers, Box 69–53–11, File 53

The formal battles of 1917 and the chaotic battles of 1918

Although the start of 1917 represented the fourth year of the war and the fourth rekindling of hope for victory — by which time there might have been a certain scepticism — it did not yet actually see the onset of disillusionment and weariness. Much of the BEF had arrived in France only in 1916 and had been heavily engaged for only a relatively few weeks on the Somme. Despite the deep attrition that was unquestionably gnawing its destructive path through units which had been in the line since 1914–15, the Somme battle actually left the majority of its healthy survivors in surprisingly good shape. Morale was further lifted when the Germans obligingly agreed that they had lost the battle after all, and pulled back to 'previously prepared positions' in the Hindenburg Line during February and March. The excessive optimism of Haig's intelligence chief, Brigadier John Charteris, has long been a point of controversy;[1] yet whatever else one may say about him, there does at least appear to be a good foundation for his view that the Germans did in fact feel themselves defeated on the Somme, albeit not routed. Having striven so hard and for so long to maintain their positions, they now retreated entirely and left their enemies to 'take possession of the place of slaughter'.[2]

For the British this meant a welcome reversion to the mobile warfare which had been so long awaited, but it came so suddenly and unexpectedly that they could not really take advantage of it. There were a few successful cavalry charges and a few infantry attacks which encountered surprisingly little resistance; but there were still some ill-prepared attacks which met disaster, and a great deal of deliberate destruction and obstruction by the retreating enemy. For the most part, however, the BEF was simply happy to rise from its muddy trenches and shake itself down. The serious business of taking the next swing at the Germans was postponed until the big push in April.

The battle of Arras would be the first in a series of deliberate attacks designed around the technical lessons learned in 1916, especially as regards artillery. The numbers, range, accuracy and reliability of the guns had all improved greatly during the winter — as they would continue to do throughout the remainder of the year — despite an alarming but temporary surge in German air successes in the spring. The result was that counterbattery fire and 'the deep battle' could now greatly enhance the preliminary bombardment, and could start to tackle such ambitious objectives as the total destruction of the German artillery, or even the physical starvation of his front-line infantry through the interdiction of communication trenches.[3]

Alongside the deep battle, the creeping barrage for the main assault had also now come of age, and would typically include five or six successive lines of shells, all lifting in unison to cover the whole area to at least 2,000 yards ahead of the advancing infantry. This was further reinforced by massed machine gun barrages and a number of 'devices of frightfulness' such as mines, Livens projectors firing gas or Thermit, and a gradually increasing supply of tanks. Smoke shell was beginning to appear in quantity, as were back-barrages to redouble the intensity of fire and false or camouflaged barrages to encourage the enemy to betray his positions. To read the detailed barrage arrangements at Arras, Messines and Third Ypres is to stand in awe and trembling before the sheer scale and power of the aggression which they implied.[4] It would certainly be a great mistake to suppose that just because many attacks supported by barrages came to grief, or that because many shells did undoubtedly fail to hit anyone, or to inconvenience anyone, or even to detonate at all, the artillery could therefore be dismissed as some sort of paper tiger. On the contrary, the British artillery from 1917 onwards could make even the most hardened German stormtrooper turn pale.[5]

At Vimy-Arras on 9 April there were some eighteen infantry divisions supported by 2,817 guns, 2,340 Livens projectors and sixty tanks. This made it fully comparable in its general scale to 1 July 1916; but in gunner terms it was actually twice or three times as big and complex. The preliminary bombardment lasted for three weeks, and built up for five days before zero to what has fairly been described as 'the greatest barrage ever seen'.[6] To add a final point of luxury to the arrangements, many of the attacking troops were further protected in their approach march by a warren of underground tunnels running right up to the front;[7] and they were lucky enough to have enjoyed an obscuration of rain turning to snow on zero day. In these conditions the attack was a spectacular success for both the Canadians and the Third Army, who seized most of their objectives with relatively light, and in places almost minimal, losses.[8] The 9th and 4th Divisions made a leapfrog advance of over 6,000 yards which was described as 'more like a Salisbury Plain ceremonial manoeuvre',[9] and which represented a record penetration for the whole war. Here, indeed, was the happy consummation of all the tactical reforms which had taken place during the winter.

Then again, at Messines on 7 June an array of twenty big mines was supplemented by a monumental two-week artillery fireplan for 2,266 guns, but on a shorter frontage and with only half the infantry of Arras. There was also a major effort by the RFC in close support of both infantry and artillery, which was designed to wrest back the command of the air that had for a time been claimed by Baron von Richthofen. Overall it amounted to an an almighty concentration of fighting power, and it suffered only from the penalties of an unexpectedly excessive triumph, since too many of the enthusiastically victorious troops crowded too far forwards at the end.[10]

Some of the 1917 battles were planned as single-action 'bite and hold' operations, to be closed down almost as soon as they had been successful. Vimy Ridge and Messines had this character, which in retrospect has given their commanders the credit for not undertaking more than they could achieve. Other operations, however, were supposed to open the way for rapid exploitation by mobile forces. The main assault at Arras was intended to open a breach through which the cavalry might charge forward all the way to Cambrai. Yet in the event the spectacular initial advance could not be continued on favourable terms until artillery had been moved forward to support the newly won front at Monchy and Roeux.[11] Around Bullecourt shortly afterwards Gough's Fifth Army launched an even more ill-fated offensive against the newly manned Hindenburg Line. The tanks were particularly disappointing, and so for a more deliberate and careful version of the tank breakthrough the world would have to await the famous surprise attack at Cambrai in November — although that too eventually turned sour, and the Tank Corps proved to be rather too happy to shift the odium once again on to the horsed cavalry.

Among all these 'mature' 1917 battles the case of Third Ypres is particularly interesting, not only because it was the biggest, but also because it probably represented the most sophisticated of all the BEF's attack plans in the entire war. Unlike any others until the hasty improvisations of the

Hundred Days, it was eventually planned as neither a limited 'bite and hold' attack, nor as a single-action breakthrough. Instead, it envisaged a high-tempo succession of many 'bite and holds' spread over many days, with breakout and exploitation still being the final goal, but a goal which was envisaged only as an eventual final phase — almost an ideal vista which mere mortals should not even try to imagine until they had first made the necessarily laborious climb up to the viewpoint on the high mountain summit.

This concept had been glimpsed on the Somme but had usually been overtaken by events too quickly to become a central tenet of deliberate planning. In an important GHQ analysis of 22 June 1917, however, Brigadier J. H. Davidson (BGGS/Operations a) concluded that 'It has been proved beyond doubt that with sufficient and efficient artillery preparation we can push our infantry through to the depth of a mile without undue losses or disorganisation....'[12] If this could be repeated every two or three days, he believed, then the German reserves could be destroyed without too much being demanded of the attacking infantry. Provided there was good organisation, only partial reliefs and relatively minor artillery displacements would be needed to maintain momentum, and the cumulative pressure of the sustained succession of blows would soon be decisive.

There was, however, a preliminary delay of more than six weeks, as the fine summer weather gradually slipped away. Quite apart from the meteorological implications of this, the Germans were given the chance of launching a model spoiling attack at Nieuport on 10 July, which effectively knocked out the intended coastal prong of the British offensive. Nor was Haig's choice of commander for Third Ypres a happy one, since he entrusted the operation to the 'harooshing'[13] cavalier Gough rather than to either Plumer or Rawlinson, both of whom had been given grounds for expecting the command and both of whom were more methodical and careful with their infantry. When Walter Guinness, a brigade major in 25th Division, moved from Plumer's command to Gough's after Messines, his heart sank as he noted that Fifth Army seemed 'very haphazard in its methods' after Second Army.[14]

Following three weeks of artillery preparation with some 3,106 guns, 31 July was the 'first day' of the Third Ypres offensive. The attack was very comparable in frontage and infantry numbers to both 1 July 1916 on the Somme and 9 April 1917 at Arras; but the weight of gunfire was greater than the second, and the territorial gain was greater than the first.[15] In effect the mile of ground surrendered in the spring of 1915 was retaken in a single bound with relatively few casualties, although it was admittedly being held only lightly by the line of German outposts. More serious strongpoints soon started to be encountered further to the rear, coinciding with the ominous start of the autumn rains. The battlefield quickly degenerated into a quagmire which could not be wished away even by Edmonds's cheery reassurance that 'mud in war is not unusual'.[16]

Fifth Army hammered away for a month; but its 'succession of blows' became ever less coherent and it had petered out by 24 August. To his credit, and perhaps uncharacteristically, Haig then accepted that the attempt had failed. He finally listened to his advisers who wanted a more deliberate series of limited attacks, which he characterised as 'siege' as opposed to Gough's vision of 'semi-open warfare'.[17] He reduced Gough to a secondary rôle and now entrusted the major assault on the Gheluvelt-Passchendaele plateau to the methodical Plumer and his Second Army. This change came belatedly, however, since it required a further pause of three weeks to reorganise the guns and communications. It was doubly infuriating that the sun reappeared and the mud dried to dust during the inactive period in September, but gave way to rain again on 4 October, just after the offensive had been resumed.

The pause did at least ensure a convincing resumption of the assault on 20 September, after a particularly intense week-long bombardment.[18] A new spirit of enterprise and deliberation was abroad among the infantry, who were again able to advance up to 1,500 yards in each attack. The RFC had now regained command of the air, while the gunners were orchestrating some devastating shoots against both the Germans' artillery and their massed infantry counterattacks. These changes were reflected in a reduced rate of BEF casualties, with a total of 36,000 in a whole week of hard but successful fighting as contrasted with 32,000 for just the three days starting on 31 July.[19] During this phase of the battle Davidson's ideal game-plan appeared to have been realised at last, and the scale of success seemed almost comparable with that of Arras or Messines.

The good progress was maintained on both 26 September and 4 October, bringing the line forward from Langemarck to Poelcappelle, from Frezenberg to Broodseinde, and from 'Clapham Junction' through Polygon Wood to Reutel. 4 October at Broodseinde was particularly notable as one of a growing number of 'black days of the German army', in which their entire front-line system was overrun with relatively light casualties to the attacker.[20] As it had at the same time on the Somme the year before, however, the weather now closed down definitively for the winter, and the mud took over decisively. The mud of August had been bad enough; but it had been but a pale foretaste of the new conditions which stymied practically every type of military undertaking. Gough's HQ, for example, soon discovered that absolutely no class of weapon could any longer be considered effective, owing to the mud. HE shells became buried and smothered in it. If they could fly at all, aircraft could not tell friend from foe since the uniforms of both had been equally discoloured by it. Mortar base plates sank into it every time a bomb was fired, and tanks bogged down into it as soon as they left the roads. Small arms became hopelessly clogged by it within an hour of opening fire — and the wounded often drowned in it.[21]

Haig and his two army staffs really had no excuse for persisting with their preplanned programme of assaults once these awful conditions had fully set in, yet this is precisely what they did do on 9, 12, 26 and 30 October, as well as 6 November — when the Canadians took the rubble of Passchendaele village by a superhuman effort — and 10 November, after which the battle was at long last finally closed down. Little more than a mile of ground was won in a month, on a frontage that narrowed down successively to that of a single corps. The cost in casualties was around 100,000, or about forty per cent of the total of about 250,000 for the whole battle since 31 July. Admittedly this latter figure was significantly lower than the total losses suffered on the Somme the previous year,[22] but the fact remains that the final month at Passchendaele had been largely fruitless. An encouraging attempt to relieve the gloom was made by Third Army's surprise attack at Cambrai on 20 November, but hope was once more dashed by the German counterstroke ten days later, which recaptured most of the ground won. When the BEF turned to a defensive posture in December it was generally recognised not only that the year's early promise had been seriously betrayed, but that the newly won extension to the Ypres salient would itself in effect be indefensible against a major assault — and this proved to be precisely the case in April 1918.

By the fourth Christmas of the war the BEF had fallen into a painfully acute crisis. Passchendaele, and the bitter disappointment of the German counterstroke on 30 November, had exhausted not only the immediate participants but also the political credibility at court of most believers in a 'Western' strategy. Lloyd George was finding the tide running strongly in favour of his 'Eastern' ideas, and so was encouraged to undermine Haig in every way he could think of, short of actually dismissing him. He replaced many of the key GHQ staff officers, notably Kiggell and Charteris; he diverted important sections of the BEF to the Italian front, not excluding General Plumer himself for a time. He held back the flow of replacements in England and insisted that infantry strength be reduced by almost a quarter; yet he also allowed the French to palm off an additional twenty-eight miles of front on to their reluctant ally.

Rarely has the incompetence of politicians in military affairs been as clearly demonstrated as in this episode; nor, for that matter, the incompetence of soldiers in politics. With all the benefits of hindsight we can now see that Lloyd George should have either sacked Haig outright or backed him to the hilt: there should have been no middle way. From Haig's point of view, by contrast, the correct course should have been to make a clear, coherent and comprehensible explanation of the true military situation, as a resignation issue. There should, in short, have been a productive dialogue between the military and political branches commensurate with the seriousness of the case. But alas, on neither side does the sensible policy appear to have been followed, and on neither side was there a proper understanding of the underlying realities. The result was a highly unsatisfactory fudge, which led to catastrophe when the Germans launched their offensive on 21 March 1918.

The fudge was based on an assumption that the Americans would satisfactorily fill out the defensive line, while the enemy assault would be held to a stalemate similar to that in Haig's attacks on the Somme and at Passchendaele; yet in reality neither prop of this analysis stood up for even a moment. Partly because of deep institutional unpreparedness, and partly because of Pershing's personal insistence on the status of a 'cobelligerent' rather than of an 'ally', the Americans failed to play any truly effective part in the war before September — a full six months too late for Ludendorff's March offensive.[23] On the German side, by contrast, there was a far greater aggressive drive than either Lloyd George or Haig seems to have imagined — or at least it might be more reasonable to say that there was a far weaker defensive response within the BEF than either of them had been led to believe was possible.

In January and February of 1918 the British armies had desperately tried to reorganise their defences in both depth and breadth but, with greatly reduced manpower spread over a considerably increased frontage, they often lacked the reserves necessary to man the crucial rearward lines. The result was a near-satisfactory defence on the northern part of the front, but a dangerously overstretched sector further south, particularly in Gough's Fifth Army between Bapaume and the Oise. Naturally it was here that Ludendorff chose to strike, and although he lacked true surprise, his blow fell on a morning when there were again the '30 November 1917' advantages of deep ground mist combined with deep BEF unreadiness and lack of reserves.

The German machine gun teams often managed to work their way between and behind the British 'blob' defences, which were themselves prevented — by the mist which negated visual signals, and by shelling which destroyed telephone cables — from calling up artillery support on preplanned SOS fire missions. Battalion and brigade HQs thus found themselves sitting silent and inactive, encased in a numbing fog which was only intensified by gas shell and HE detonations.[24] They were unable to make contact with friendly forces either to the front or the rear, and their outnumbered artillery was condemned to stand powerless while the enemy picked off the forward British positions one by one. A mixture of panic, bewilderment and fear for the flanks then loosened up the cement holding together the depth defences. Even though it was found that the unsophisticated German frontal assaults could be held in check very easily, there always seemed to be at least one undefended avenue through which their neighbouring units could push forward and outflank the defenders. At first the situation was similar to the many German frontline collapses that had been seen on the first days of Loos, Arras, Messines or Cambrai; but after about the first thirty-six hours it became clear that there was in fact a difference. The main British defence zone had been destroyed; but because of their inexperience in receiving such blows there were inadequate depth reserves to form a solid new front towards the rear. Each division therefore had to make a fragmented and divergent retreat in which coherence was in effect lost.[25] (See Fig. 4.)

Fig. 4: Successive fragmentation of XVIII Corps in March 1918

(Source: Brigadier Hollond's *Lessons,* no. 1, 28 April, in Maxse papers, Box 69–53–10, File 45)

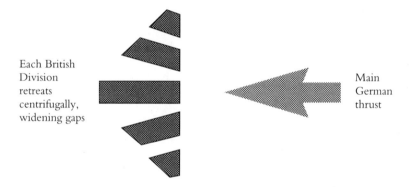

One cannot help believing that Fifth Army really ought to have made a stiffer resistance than it did, especially in the second phase when its main positions came under attack. If that battle had been handled better, the Germans might well have been held to a break-in of only a few thousand yards, as they were on much of Third Army's front not only on 21 March, but still more conclusively when they tried another major attack on 28 March. As it was, however, their maximum penetration against Fifth Army was forty miles, giving them not only all the ground they had lost since 1916, but also most of what had been the British rest areas during the battle of the Somme. Despite the pleading of his many apologists, therefore, the dismissal of Gough seems to have been eminently justified, especially since he had already patently mishandled both Bullecourt in 1917 and the Passchendaele battle in its early stages. Still more appropriate, perhaps, might have been the dismissal of both Haig and Lloyd George, although in the event they avoided such an outcome by managing to stabilise the line in front of Amiens. Marshal Foch was at last belatedly appointed allied supreme commander, as a response to the crisis.[26]

As soon as Ludendorff had been held in front of Amiens, he struck a new blow on the Lys in early April. At first it produced scenes similar to those of 21 March, as BEF units leapfrogged past each other towards the rear in bewildered confusion. However, in this case there were more reserves at hand and the Germans were halted after advancing only ten miles, around Meteren and Béthune. But at the end of May they tried again on the Aisne, where they once more successfully exploited a criminal allied maldeployment and disregard of intelligence.[27] They advanced twenty-seven miles before finally being halted.

Despite these successes, by midsummer Ludendorff was starting to see that he had effectively shot his bolt. Brilliant staffwork and effective initial assaults were not enough, since the mass of his army possessed none of the tactics or training that might have allowed them to sustain the offensive with an acceptable level of casualties. They proved to be just as unreliable in sustaining a chaotic forward movement as the BEF seemed to be in sustaining a rearward one. Command and control broke down too quickly on both sides, once fully open warfare had replaced the static certainties of trench life. The artillery often failed to keep in touch with its infantry; lateral communications between formations tended to break down every time there was a move,[28] and headquarters were often — for the first time — as much concerned with personal survival as with directing the battle. These factors all provided invaluable experience which allowed the BEF to fight its equally mobile battles of the autumn much more professionally; but in the spring everything moved too fast for the lessons to be learned in time. At least there seemed to be a paradoxical balance between the two sides which ultimately saved the BEF from ruin and deprived the Germans of the decisive victory that surely lay within their grasp.

As the German attacks were still spluttering to their close, Foch launched his first counterstroke at Soissons on 18 July, with a largely French array.[29] It was a grandiose success and recaptured most of the area lost in May. Being fought over uncratered and unfortified ground, moreover, it was one of the very few moments in the war when large masses of tanks could be used effectively. Most of them were Renault light tanks, which despite their many limitations and low speed did at least physically resemble the tanks of 1939–41, and hence offered some inspiration to later German theorists.[30]

A month after Soissons it was the BEF's turn to attack, this time at Amiens on 8 August. Once again, the result was a shattering success which helped to persuade Ludendorff that his army had suffered yet another 'black day', and was no longer as reliable as it had been a year before.[31] Armoured cars and even cavalry found opportunities deep behind the German front line, although neither tanks nor aircraft fully lived up to their promise. There was no genuine breakthrough, and the attack was called off after three days following an advance of around six miles. However, the speedy cancellation in itself represented conceptual progress, insofar as Haig had persisted with earlier breakthrough attempts long after they had run out of steam, and it had only been limited 'bite and hold' attacks like Messines which he had been willing to shut down quickly.

After Amiens the BEF's advance gradually picked up impetus and extended its frontage. The Third and Fourth Armies gradually closed up to the Hindenburg Line south of Arras, and were joined by First Army. Then they mounted a formal assault upon that imposing position at the end of September. This operation — like the Gulf War of 1991 — had been widely dreaded as likely to be a bloody repulse, but fortunately it turned out to be surprisingly successful.[32] Whereas a predicted hurricane bombardment starting at zero had sufficed at Amiens, the Hindenburg line was thought to require a return to a prolonged preliminary bombardment of the 1917 type. Especially careful arrangements were also made in every other department, and they paid off in a sustained advance[33] beyond Douai and Cambrai in early October, which was shared by a corresponding Second Army advance to recapture all of the old Ypres and Lys battlefields.

By mid-October German resistance had been reduced to a very pale shadow of its former solidity, and the pace accelerated markedly during the final two weeks of the war. For the very first time for either side on the Western Front, a deep penetration was able to feed on itself and continue progressively ever deeper, rather than wasting away in direct proportion as it moved further from its start-line. In technical terms there was never a true 'breakout'; but almost daily there were spectacular infiltrations and big bounds forward, including many assault river crossings and even street fighting in towns (see Fig. 5). Only the armistice of 11 November prevented the complete destruction of the German army, since the allies had now achieved such a total supremacy over it. The problems of mobile warfare, which had caused such difficulties in both March 1917 and March 1918, had now been very largely overcome.

Flexible formations for mobile war

As far as British infantry tactics are concerned, the intense combat of 1917 and 1918 led to little revolution but much evolution. Apart from the final mobile operations of the Hundred Days, when both time and artillery were in short supply, the key thing was still a matter of charging

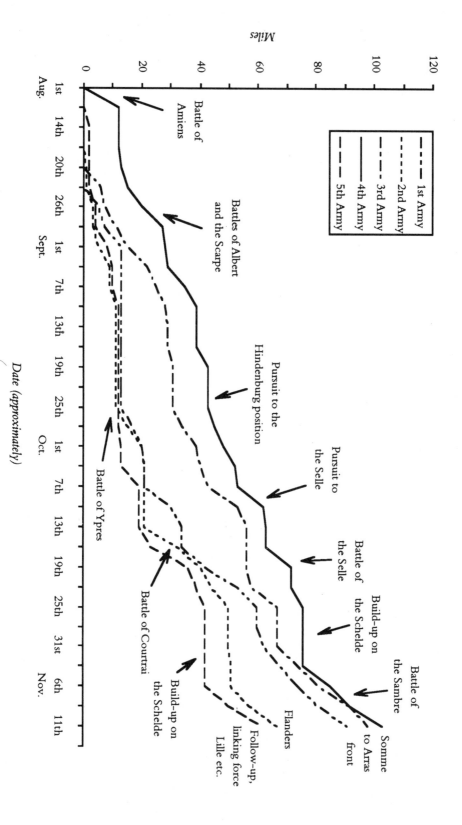

Fig. 5: Advances of the BEF in the Hundred Days 1918

forward as closely as possible behind a creeping barrage. If all went well the barrage would do all the 'fighting'; but if not, the infantry could still fight its own way forward using an inspirational combination of rifles, bombs, rifle-grenades, Lewis guns, Stokes guns and, if by any chance they were capable of arriving on time in running order, tanks. Regardless of whether the essential covering fire was to be provided by the barrage or by the infantry's own weapons, the key motto was that 'To advance is to win'.[34]

Within this general framework there was nevertheless considerable debate about the details, and many local reinterpretations and variants of the early 1917 orthodoxy as laid down in SS 143 for platoon tactics. The first question was that of platoon organisation, a subject which General Maxse made distinctively his own. He was particularly anxious that each platoon commander should train his own men, that he should know them well, and vice versa, that therefore every platoon in the army should be fixed in a standardised layout and organisation, and 'The section as a fighting unit will consist of 1 NCO and 6 men, no more, no less'.[35] SS 143 had called for four sections each with a different specialisation; but it appears that this was far from universally applied during 1917. Some platoons had fewer than four sections, and their specialisations were far from standardised in a regulation manner, which was a situation that the general was determined to reform.

However, in February 1917 Maxse had somewhat undermined his own argument by pursuing the highly non-regulation theory that every section should contain rifles, bombs and rifle-grenades, rather than having each of these weapons ring-fenced into separate sections; and he wanted the Lewis guns to be removed as a platoon specialisation altogether, to be held at company level.[36] As more Lewis guns became available, however, he soon had to accept that they should indeed be admitted into the platoon.[37] Some commanders were eventually even using two of the four sections as Lewis groups,[38] including at least one revolutionary suggestion that the crews of between six and eight men per gun might be cut to just three.[39] The final official abolition of the specialist bombing and rifle-grenade sections nevertheless came only on 23 July 1918 — presumably some eighteen months too late for Maxse.[40]

Overall, the gradual disappearance of specialisations represented an increase in the 'mechanical power' available to ordinary infantrymen, since everyone now had access to bombs, rifle-grenades and even Lewis guns. The platoon was very well armed, even before the attachment from outside of heavier weapons such as mortars or — in the Hunded Days — eighteen-pounder field guns for direct fire. This was not therefore an army which sent its soldiers naked into battle, and despite his unfortunate tactical failure in March 1918 we can certainly applaud Maxse's sentiments when he said,

I believe in husbanding manpower. I believe in making up for our shortage in manpower by mechanical power and training and not by rushing a lot of people into a fight merely to stand up against artillery without firing a shot themselves.[41]

Next there was the question of formations.[42] Since at least the spring of 1916 everyone had agreed that the rearward lines of an advance should be arranged in 'artillery formation' of sections and platoons in single file, for easier movement across the broken battlefield. The most forward lines, however, should be in extended 'waves' of skirmishers with intervals which gradually expanded to five or six yards between men. As with platoon organisation, General Maxse may perhaps be seen as the senior officer most interested in codifying these issues — especially after he became Inspector General of Training in mid-1918 — and he certainly seems to have held to the system of 'waves followed by worms' with little variance throughout the war, at least when there was a creeper to be followed.[43] The system was also often used in practice, for example it was reported by the French Lieutenant Colonel Duffour in October 1917 at Ypres,[44] although he complained that the British lines tended to bunch together into small groups so they were no longer true lines at all, and failed to use short rushes or proper fire and movement. This, however, was a fault of application rather than of principle, and lines continued to be widely used throughout 1918.[45]

There were, nevertheless, always some tacticians who wondered whether the front line should not also be deliberately placed in a formation of 'snakes' or 'worms', just like the more rearward lines. When it evaluated the special qualities of the Passchendaele battlefield in September 1917, for example, the 9th Division had come round to a belief that in certain circumstances a line of section columns in single file was the only practical way to move forward at all, even though this placed more individual initiative on the shoulders of section commanders.[46] In a pamphlet issued by the 4th Seaforth Highlanders in late 1917 there was a suggestion that counterattack forces should be built up 'dribbling' forward in section columns 'on the German model' — rather than advancing in waves — and then form a firing line to advance by 'controlled irregularity'. This idea seems to have won Maxse's approval for his XVIII Corps defensive arrangements in early 1918,[47] although one might perhaps accuse him of clinging too closely to the habitual practice of an enemy who knew more about defence than about attack. A variant of the same system was the 1st Canadian Division's practice of advancing in worms until contact was made with the enemy, then deploying into waves to fight.[48]

For others the original idea of 'platoon training', as encapsulated in SS 143, was itself an acceptance of fighting in informal section groups, or

The Final Eighteen Months

'blobs', rather than in lines. This idea has often been misrepresented as a panacea which could at a stroke prevent the heavy casualties of the 'bad old' massed waves; but in reality it was no such thing, since a blob could offer just as much of a massed target as a wave. In both cases it all depended how closely bunched the attackers were, and how well the enemy had been suppressed. Indeed, it was found by experience in the spring of 1918 that in defence, at least, a blob should be more rather than less closely bunched, in order to maintain the morale of the defenders. Thus in February Maxse had called for blobs in his XVIII Corps defensive scheme[49] — once again 'on the German model' — but in the March fighting these blobs were found to crumble too easily when they were composed only of sections rather than of complete platoons.[50]

In the attack, the advantage of section blobs was that they could move more easily in unexpected directions than could either columns or lines. Blobs were therefore especially appropriate when there was no creeper to set the pace and direction; and it is noticeable that they came to the fore particularly in the open warfare of the Hundred Days. At the level of the platoon, the blob system implied either a 'square' or 'diamond' layout of the four sections, so that any of the four could take the lead in any unexpected change of direction (see Fig. 6).

Fig. 6: The flexibility of diamonds: the BEF's 1918 concept of a fluid 'infiltration' attack

(Adapted from IT training leaflet 13, *The Soft Spot,* January 1919)

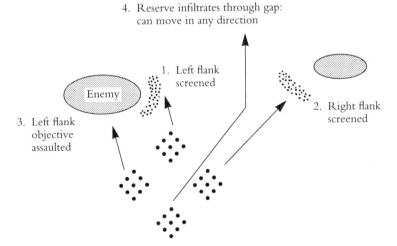

The diamond layout had apparently already been suggested in some editions of SS 143 before the end of 1917, although it was not made compulsory.[51] It was picked up by Maxse in his mid-1917 pamphlet *Hints*

on *Training issued by XVIII Corps,* which he circulated more widely in August 1918 as 'The Brown Book'.[52] In the training recommendations for his XV Corps, De Lisle also seemed to favour the flexibility of diamonds as an alternative to 'lines then worms', even though he retained the latter as an option.[53] Finally, in the official Inspectorate of Training (IT) pamphlet of October 1918 infantry commanders were offered a free choice between squares, diamonds, worms or blobs, according to local circumstances.[54]

Whatever the formation, the idea of flexibility and local initiative was certainly accepted as essential in all assault tactics. For example, in Maxse's *Notes on Training* for infantry companies of June 1917, there is an acceptance that each section need not be aligned on its neighbour, whatever their respective formations.[55] This view is reflected in an 18th Division circular of 28 September which calls for careful alignments during the forming up phase:

> ...but once the barrage opens all ideas of rigid line must be abolished. The main idea must be to keep up to within 70 yards of the barrage and to fight quickly under cover of the barrage, the front line commencing to fire directly enemy rifle fire is detected.[56]

A favourite saying was 'When in doubt go ahead. When uncertain, do that which will kill most Germans. Don't fear an exposed flank.'[57] The idea that flanks are relatively unimportant to an aggressive attacker is of course a central tenet of 'infiltration' tactics, and it recurs persistently in much of the BEF's tactical writing of 1917–18.[58] For example, the Canadians had reported that the system was 'sound in every way' at Amiens on 8 August 1918,[59] while a XVII Corps note entitled *Lessons Learnt During the Operations 21 Aug–7 Sept 1918* complains that 'The expression "I cannot push on because my flanks are in the air" is far too common'.[60]

Linked to all this was an idea current in the BEF towards the end of 1918, to the effect that scouts should move ahead of an attack to identify 'soft spots' or 'gaps' in the enemy defences which could then be exploited by the attacker's main body. Successes should be reinforced, not failures. Just as the Germans had briefly enjoyed the ability to make spectacular infiltrations against Gough's porous Fifth Army in March, so the British now seized similar chances in the autumn. Such a system could never have worked against solidly fortified fronts, even though it had been long discussed and was fully understood in theory. But in the Hundred Days the opportunity presented itself and was often used. Unfortunately for the reputation of the BEF's tacticians, however, the fully developed 'Soft Spot' manual appeared only rather belatedly, in January 1919,[61] although in many ways it remained close to the recommendations of SS 143 for pinning an enemy machine gun or pillbox in front by fire while manoeuvring around it to flank or rear. It was certainly a commonplace in the BEF that 'Turning a flank is not being able to fire through the side windows, but into the back door'.[62] This method was used, for example, by the 9th Division at Frezenburg in September 1917.[63] Another commonplace was that the enemy 'invariably chucked it when he felt that people were working round his flank',[64] and 'It was found that the best way to take on the Boche machine guns was to turn Lewis guns on to them and manoeuvre the riflemen'.[65] Still more to the point for infiltration, perhaps, were Godley's orders to his XXII Corps in July 1918 to engage simultaneously, and hence distract, the overwatching enemy strongpoints *further* in rear, during the time when the more accessible strongpoints in the front line were being systematically knocked out.[66]

All this stress on flexibility and local initiative seems to run directly counter to the 'old army' emphasis on rigid drills and top-down discipline. The conservative guardsman Maxse was himself as keen as anyone for his men to salute smartly, practise close-order drill at every opportunity, and listen to cautionary lectures on 'what has happened in Russia due to indiscipline'.[67] One may indeed raise an eyebrow at his fanatical emphasis on regularity in route marches or in 'intensive digging';[68] but one should not, perhaps, scorn his extension of the same drill principle into such combat functions as platoon organisation, deployment into tactical formations, or the advance to contact. Whatever one may think of parade-ground bull — or indeed of Bolshevism — there seems to be little doubt that most troops in most circumstances will fight best if they have practised at least the outlines of their fighting organisation and techniques beforehand. When Maxse wanted to apply the principle of drill to tactical training for modern warfare, therefore, he was merely harnessing contemporary common sense to a tried and tested tradition. Close-order parade-ground drill had itself always been synonymous with battlefield tactics until at least as late as the American Civil War,[69] and even when close-order tactics had finally been abandoned after that conflict there still remained an obvious utility in rehearsing assault troops in the skirmish lines (or perhaps 'worms') in which they were intended to fight. As Maxse explained, 'all details such as "forming up" lines, formations, objective, etc., must be prepared *beforehand* and rehearsed *beforehand*. It won't be "all right on the night"'.[70] His very sensible idea was that tactics would be 'so ingrained on the training ground as to become second nature in battle',[71] and it was his constant cry that platoon commanders should seize every possible moment for training, even in the front line.

Despite his many rigidities, Maxse actually displayed a deep inner consistency in his approach to the things that mattered most. His constant thrust towards regularity and system, combined with repeated exhortations for lively training methods and individual initiative in battle, was all designed to make sense to the ordinary soldier and guide him through the shock of real combat. The soldier would then be trained up — or 'indoctrinated' — to the point where he could be coaxed into giving of his very best in an unforeseen crisis. This approach was indeed practically identical to the supposedly 'Prussian' tradition of severity combined with decentralisation which went all the way back to Frederick the Great, and then reappeared in Ludendorff's stormtroopers. It was an outlook which simply came naturally to all intelligent soldiers, of whatever nationality, who fully understood the inward workings of their chosen profession. It should not therefore be criticised as mere 'rigidity' or 'authoritarianism' by armchair critics.

Particularly irritating, perhaps, are the claims by Captain Basil Liddell Hart that it was he who had originally invented the whole modern idea of Battle Drill, in his cosy billets at Stroud and Cambridge,[72] rather than the grizzled commanders in France who had had to fight their divisions and army corps all the way through the Somme, Arras and Passchendaele. Perhaps it is indeed true that Hart[73] did originally coin the specific phrase 'Battle Drill'; but he certainly did not invent the inner concept, particularly not as it was habitually applied to rehearsing and preparing for attacks on the Western Front. Within that specific context many BEF figures had already spelled it all out very clearly by the summer of 1917, right down to such details as hand signals or concern about 'the load of the soldier'. The only thing they lacked, perhaps, was a tame journalist who could trumpet their achievements to an otherwise ignorant posterity.

Maxse himself was keen on drilling for battle, as we have seen, and to that extent he may have given inspiration to Hart. He also certainly shared the latter's penchant for making scintillating denunciations of his predecessors and precursors. However, still more similarities with Hart's writings may be found in the tactical work of General De Lisle.[74] Hart cannot, therefore, be seen as creating any new edifice of his own, since he made merely a personal interpretation of the manuals circulating in France at around that time.[75] Maxse, conversely, appears to have encountered Hart only in 1920, and does not seem to have drawn any particular inspiration from him. Just what, then, was Hart trying to prove? We must conclude that he was ruthlessly attempting to establish his own name as 'The Father of Modern Tactics', even though many others had gone before him far more painfully and creatively, and in the process had innocently fed him with all 'his' best ideas. Unfashionable though it may seem, we should today surely identify the true fathers of modern tactics among the staff officers of GHQ and Fourth Army around the spring of 1916, or at the very least among the numerous army and corps HQs who knew how to win so many victories during 1917 and in the Hundred Days.

Part Three
Heavier Weapons

6

The Search for New Weapons

> All night we lie ready for action with our gas masks on. The wounded and gas cases are carried off in batches. There are many killed by gas. We are quite powerless against the English.
> German letter from Wytschaete, 6 June 1917,
> quoted in Foulkes, *Gas!*, p.192

Among the very many specialised weapons and contraptions which were brought forward by the Great War, there can be no doubt that the new generation of artillery exerted by far the greatest influence upon the battlefield. One may perhaps debate whether this was because the gunners already enjoyed a commanding institutional position within the army, or whether it was entirely due to their new technologies. Perhaps it was simply that the new generation of guns had been largely perfected before the war, and was therefore already a part of tacticians' mental universe, whereas such novelties as bombs, gas, tanks and coordinated machine gun tactics appeared only after hostilities had commenced. It is at least certain that gunners such as Horne could hope to rise to the command of armies, whereas most of the leading technicians in the other arms finished the war as colonels or, at best, brigadiers. This was especially unfair to the Royal Engineers, who actually took a key rôle in most of the new technologies, from tanks to survey and from signals to gas — not to mention official historiography! — but their political impact was seriously reduced by the very diversity of their contribution. In the present section we shall attempt to review this plethora of new weaponry, and penetrate some of its technical and institutional complexity.

The mobilisation of invention

The Western Front had become a zone of static trenchlock by Christmas 1914, and the infantry who manned it had already fully realised that these new conditions demanded new weapons to supplement their time-honoured rifles and bayonets. It did not take many weeks before they were improvising bombs from jam tins, trench catapults from large rubber bands or springs, and periscopes from shaving mirrors. The leading London department stores were soon competing avidly with each other to supply these multifarious new requirements, and the Army and Navy Stores, for example, opened a 'weapon department' which unconsciously won immortality on the day when it chanced to supply Siegfried Sassoon with his pistol and wire-cutters.[1] Gamages were not far behind with their patent trench catapults,[2] and many must have been the newspaper advertisements for body shields which persuaded anxious parents to press these cumbersome and unpromising items of personal impedimenta upon their reluctant warrior sons.[3]

The war rapidly became a festival of unlikely ingenuity and crackpot invention — a benefit night for whole battalions of wild-eyed mechanics, whether amateur or professional, civilian or military. The government was almost instantly swamped by a flood of well-meaning suggestions from ingenious men with resonant names like Begbie, Bickford, Leeming, Norton-Griffiths or Newton-Pippin...and it would find itself condemned to spending the remainder of hostilities wearily attempting to distinguish the valuable from the improbable and the murderous from the downright impossible.

There is something laughably quaint in all this, leading one to sympathise with Dr Dunn's comment from 2/RWF that Heath Robinson's famous cartoon caricatures of desperately unlikely inventions were often more 'grimly to the point' than one liked to think.[4] Yet some of the inventors were quite highly placed. For example there was Admiral Bacon, with his notorious pontoons for landing tanks on Belgian beaches;[5] or no less a personage than Winston Churchill, who has been acclaimed as effectively the father of the land battleship. Indeed, if the inventors themselves were not highly placed, they were very likely to attract influential lobbyists who were. The correspondence of G. M. Lindsay is littered with attempts to win over members of Parliament to his cause, while his fellow light infantryman J. F. C. Fuller became a prodigious seeker after extrahierarchical 'friends'. These men knew that if they could succeed in obtaining patrons in high places, the bees in their own particular bonnets could be ignored far less readily by the military authorities than if they were being sold merely on their bald intrinsic merits.

During the first year of the war the impetus for new technology tended to come mainly from below, as individuals hit on bright ideas and enthusiastically built private prototypes to demonstrate to the authorities. The official procurement agencies, by contrast, seemed to approach these with extreme caution and conservatism,[6] not to mention a deeply incoherent institutional competition as between one agency and another.[7] Especially while the munitions factories were overstretched to meet the shell shortage, there was a natural reluctance to embark on new projects which might only stretch resources still further. The officers who conducted acceptance tests also quite rightly distrusted any mechanism that might jam when exposed to mud, or explode prematurely if incorrectly handled. They had presided over too many abortive practical demonstrations to be impressed merely by a good theoretical idea; and they sensibly held to the view that a new weapon should be adopted only if it had already been thoroughly combat-proven. Only occasionally did they seem to realise the fallacy which lay at the heart of this last proposition, namely that a weapon could not be combat-proven at all unless it had first been adopted....

Many inventors found they had to conduct a veritable war of attrition before they could make any progress against official resistance. Both of the two mortar designers named Stokes, for example, had to engage in a long fight to get their weapons accepted. Frederick William Stokes, managing director of Ransomes and Rapier engineering works, was eventually by far the more successful of the two, and it was his name which would forever afterwards be associated with the classic 3-inch (later '81 mm') smoothbore infantry mortar.[8] Nevertheless, his original design, conceived in December 1914, 'failed in both accuracy and range' in its first test at Shoeburyness on 30 January 1915. After a second test in February the Director of Artillery turned it down flat. Stokes persisted, however, and made improvements[9] at his own expense which won him another series of tests at Woolwich, leading to another rejection in April. One more test in April led to a third rejection in June, although by this time there were signs that the purely technical objections were starting to give way to more 'political' arguments, such as the fact that there were already too many competing designs of trench mortar either in service or under trial. A political change was certainly needed for the climate to improve; but fortunately for Stokes this chanced to come very soon afterwards, as a result of the cabinet reshuffle and inauguration of the Ministry of Munitions. This organisation reinvestigated the Stokes gun at its test range on Clapham Common, where it was seized upon by Lieutenant F. A. Sutton RE — an officer who had lost a hand at Gallipoli and been sent back by General Hunter-Weston specifically to look for an improved pattern of mortar. He arranged for the Stokes gun to be demonstrated to Lloyd George and Churchill at Wormwood Scrubs on 30 June, and they in turn urged production of 100 pieces, finally leading to the placing of a firm order for 200 on 21 September.[10] Meanwhile Stokes had managed to infiltrate an example of his product into France itself, where Lieutenant General M. F. Rimington of the Indian Cavalry Corps actively championed it against GHQ, supported by Foulkes of the gas brigade, only to have it rejected in August. It would finally be issued to the troops only from January 1916 onwards.

If Mr F. W. Stokes needed five conflicting series of tests and almost a year of intense lobbying before his revolutionary mortar was fully accepted by the authorities, his namesake Captain S. F. Stokes, RE, would encounter more protracted opposition even than that, culminating finally in abject failure.[11] Captain Stokes's gun was considerably more complex and expensive to build than Mr Stokes's, since it had a rifled bore and was designed to fire an automatic stream of between fifty and sixty rounds per minute, rather like a machine gun. By the autumn of 1915 he had started building a prototype with his own money, and had even succeeded in having it taken up by the Munitions Inventions Department with the encouragement of Lloyd George. He himself was seconded to this department in order to supervise development — but unfortunately the first test firings, towards the end of the year, were unimpressive. Despite his appeals for more time to perfect his gun, Captain Stokes was posted to India — or 'deported', as he resentfully put it — on 4 January 1916. To add injury to insult, he wounded himself in the hand soon afterwards, during an experimental firing of the gun. This did not deter him, however, and he eventually managed to obtain further official tests in May 1917 at Claremont Park, Esher, and on Epsom racecourse. He continued to make technical improvements, but all to no avail. At the final test in January 1918, also at Claremont, the verdict remained unswerving that his mortar offered no advantage over the single-shot smoothbore version — which could, after all, keep fifteen bombs in the air at once. The S. F. Stokes gun was furthermore deemed to be too unreliable, too heavy, too complicated, and extremely dangerous to its operator.

Most of the parks and racecourses of southeast England seem to have been converted into free-fire zones for the tactful frustration of hopeful inventors; but the same was still more true of the test ranges which had already existed before the war. The School of Musketry at Hythe, for example, became the scene of a large number of trials for a weird and wonderful diversity of patent devices.[12] Taking reports from fifteen months in the middle of the war as representative, we find that Major H. W. Todhunter, the school's 'Experimental Officer', made a series of decisions which usually turned out to be almost retrogressively negative. Thus King's rifle rest for night sniping was rejected in favour of the less cumbersome improvisations and 'rifle batteries' that might be constructed in the trenches themselves.[13] Austin's retractable trench turret, although portable by four men, would not extend high enough to stick out of the top of a particularly deep trench, yet might expose its user's stomach if he was in a particularly shallow one. Longfield's muzzle pivot machine gun mounting allowed the gun to kick up too violently when it was fired, and 'the pattern with bicycle tubing is quite unsuitable'. Lieutenant Branson's rifle stock, as modified for bayonet fencing, was not thought to be any good — but was referred to the gymnastic staff just in case. A portable shield tested in April 1916 was found to be as big as a field gun and too heavy to merit even the expense of returning to its inventor, whose name has unfortunately not come down to us. Sergeant John Ross's loopholed shield for machine guns was rejected as not bullet-proof, while a smooth-bore Maxim gun specially designed for wire-cutting was less effective in that task than an ordinary Vickers. Major Atkinson's 'flash hider' for the Vickers suffered from the difficulty that it produced bright sparks visible at 400 yards' range, while Paterson's spotlight attachment for rifles was thought to be militarily useless. Paper-tipped bullets, however, were at least found to possess greater penetrative powers than ordinary ones; and although 'Sparklet' tracer bullets were admittedly effective to 400 yards, 'Aerators' Tracing Composition' was effective to no less than 600.

Apart from its 'bread and butter' work in testing and retesting all aspects of the service rifle, the other trials at Hythe between September 1915 and November 1916 included assessments of the Colt, Beardmore and Vilar Perosa Revelli machine guns; of pocket range-finders and position-finders; a prismatic compass-cum-periscope (!); Lieutenant Bellingham's elevating bracket connector (?) for artillery; and several portable shields for infantry. There were also many new designs for machine gun belts, tripods, sights and carrying slings; in fact almost every conceivable aspect of the infantryman's armament was systematically considered. Even if one may censure them for conservatism or scepticism, therefore, one can scarcely accuse the authorities of any lack of diligence in analysing what was submitted by their ingenious supplicants. Despite all his rather negative reports, it must be accepted that Major Todhunter tackled his work very thoroughly and well. One is tempted to the conclusion, furthermore, that many of the inventions themselves were every bit as 'militarily useless' as he usually believed them to be.

On the other hand these authorities were almost always operating in a 'reactive' mode, testing items which had been submitted by some supplicant, often at the supplicant's personal expense. Few inventors were reimbursed for their work, especially if, as in the case of Captain S. F. Stokes, their design failed to be accepted. Some were lucky, but the example of Colonel H. H. Hemming shows that even a wonderful technical breakthrough might have to be paid for out of the inventor's own pocket. In his case he could not make progress on his flash-buzz system for locating enemy guns until he had a chance to visit the electrical shops of London while on leave. He then bought a mountain of switches and buzzers, perfected an elegant system which remained in use for thirty years, but could not persuade the authorities to pay for his prototype until after the armistice.[14] One is left with an impression that inventors and inventions were generally unwelcome in the army.

Yet if viewed in another light, the very same set of underfunded lash-ups may also be seen as an unmistakably modern phenomenon. They represent a mobilisation of hundreds of active brains around a single theoretical problem — the problem of trench warfare — in a way that had no true parallel in the nineteenth century, but which would be seen again several times over during the long subsequent history of warfare in the twentieth century, from the Manhattan Project of the early 1940s to the 'electronic battlefield' of the late 1960s. One cannot help being impressed not only by the all-embracing scale of the undertaking, but also by its breadth of scope and the active willingness to 'think the unthinkable'. Whole new classes of weapons were designed, built, developed and improved, ranging from the tank to chemical warfare, within a remarkably short timespan. To take just one small example of this phenomenon, we find that already in 1915 a distinct category of 'optical munitions' had for the first time been officially recognised and was routinely included in the weekly reports of munitions production, differentiated from the more familiar categories such as shells, rifles, machine guns and so on.[15] During the Boer War there had admittedly been a few heliographs, and officers had had to find their own binoculars; but now the whole field of prismatic compasses, sights, scopes, hyperscopes, searchlights, cameras and every other form of polished glass was suddenly opened up as almost a new arm of the service.[16]

The key break from the past perhaps lies in the fact that the year 1915 saw a decisive turn away from the musings of gifted individuals — such as had been a feature of warfare from Leonardo da Vinci or the Marshal de Saxe through to Dr Gatling or Colonel Colt — to a more centrally organised harnessing of brainpower. During the course of 1915 there dawned a public realisation, albeit some six months later than it had dawned upon front-line trench fighters, that this war demanded a new type of scientific mobilisation such as had never before been seen. The first gas attack came in April 1915, just after the aeroplane and just before the tank had started to be accepted as real phenomena by influential people.[17] In particular, the Ministry of Munitions was set up in May in response to the 'shell crisis'; and it was this perhaps more than anything which changed the general mood from one of slap-happy amateurism to one of serious centrally organised determination.[18] The imposition of licensing hours upon the pubs used by munitions workers certainly brought an end to much of the party spirit with which the war had originally been greeted.

The increasing centralisation of invention found encouragement from a number of different directions. In the first place there had already been some tentative steps before the war towards state-sponsored laboratories, notably the National Physical Laboratory and the Imperial College of Science and Technology. These, in common with the army's Ordnance Board, the Royal Aircraft Factory and the other research departments and test ranges at Hythe and Woolwich, provided a preexisting framework which could be more widely exploited during the war itself.[19] Nevertheless a great deal of scientific talent remained outside these organisations and it would be at least a year before the Royal Society, for example, began to be properly harnessed to the war effort.[20]

Secondly there were the politicians, like Lloyd George and Churchill, who seized upon military unpreparedness for a long technological war as a stick with which to beat a complacent bureaucracy all the way through the pages of the Northcliffe press — and thereby further their own soaring ambitions. Not only did Churchill take the lead in organising landships while he was First Lord of the Admiralty, but he supported Lloyd George's moves to establish the Ministry of Munitions in the spring of 1915.[21] The new ministry soon included a 'Trench Warfare Department' and a 'Munitions Invention Department' which, if taken together, were responsible for most of the research into air defence and chemical warfare as well as such items as grenades and mortars. However, they still often duplicated the work of the War Office's research organisations until the latter were eventually brought under the Ministry of Munitions' control in November. This step roughly coincided with the consolidation of tank developments under the same ministry, and went far towards placing the analysis of all the army's weaponry on a unified basis. There were admittedly still some gaps, such as tactical wireless, nor was either air or naval research ever to be as well coordinated as that for the ground forces; but it remains true to say that by the time the BEF was ready for its big push in 1916, it was well served by a concerted technological mobilisation.[22]

Finally, there is no doubt that many important initiatives for new technology came from senior generals. Despite their popular image as reactionaries or even Luddites, they were often responsible for pushing forward new weapons which they hoped would help them to victory. In the case of the (W. F.) Stokes mortar, we have already seen how such important military leaders as Hunter-Weston and Rimington played an active rôle as sponsors for the new weapon; but it is worth adding that none other than Sir John French himself had also issued something approaching an official requirement or specification for such a mortar as early as 20 October 1914, in which he was supported and seconded by such figures as Lieutenant General Sir Archibald Murray and Major General Sir William Robertson.[23] The same would be even more true on 25 September 1915, when French rested a good half of his plan for the battle of Loos upon the entirely untested 'fifth arm' of chemical warfare, no organisation for which had existed at all before May. French had also preempted the Ministry of Munitions by setting up his own GHQ Inventions Committee, which in due course evolved into the RE Experimental Section.[24]

Nor would French's successor as commander of the BEF, Sir Douglas Haig, show himself to be any less avid in his quest for the latest weapons and the highest technology, regardless of whether it was a matter of gas, guns, tanks or aircraft. Haig was always anxious to multiply infantry (and indeed cavalry) by technology, and he undoubtedly played a very important rôle in pushing forward the causes of all those various sectarian groups. If he ultimately failed to convert the BEF into a fully mechanised 'Panzer Armee', as some of his critics seem to think he should have done,[25] that was surely more because he commanded it for too short a time for such a scheme to come to fruition, rather than because of any conceptual resistance to the general idea. Haig won his war in November 1918 with an army that was already incomparably more technological than the army he had inherited three years earlier, and he can scarcely be blamed for accepting an armistice six months before J. F. C. Fuller's somewhat dubious and over-optimistic 'Plan 1919' could possibly have been exposed to the test of battle.

In Fifth Army's attack on Bullecourt in April 1917 the infantry plan was entirely distorted by Gough's sudden — albeit in practice badly belated and ill-considered — enthusiasm for tanks. The result was a compromise which pleased no one, least of all the Australian infantry who had to lie out in the snow of No Man's Land for two nights, during both of which the tanks failed to arrive on time.[26] It would take over a year to win back antipodean sympathies to the mechanised arm; but at least this story does show us that even the most 'horsy' of the BEF's senior commanders was considerably more open to technological innovation than many of his critics have tended to allow. It gives credence to the claim by Major C. H. Foulkes, the creator of the gas brigade, that

> Throughout the war I found officers of high rank almost too receptive of novel proposals, especially when they were based on anything mysterious or scientific.[27]

That, in a nutshell, surely puts the matter into its true perspective. Crusty old generals were extremely anxious to embrace new technology not only because their own old technology had patently failed, but also because much of the new technology was mysterious, inexplicable and fear-inducing — or rather hope-inducing, if only it could actually be made to work. Perhaps the new 'boffins' did drip egg on to their ties, and did not speak the military language of Camberley or the Veldt, but at least they did seem to promise a decisively successful end-product as far as the trench war was concerned. And at the end of the day a 'decisive end to the trench war' was ultimately the only thing that mattered to the likes of Gough and Haig — and indeed to everyone beneath them.

It nevertheless remains true that Haig did miss some of the tricks, and did allow some of the available technologies to wander up blind alleys or remain underexploited. In this, however, his record was far less disgraceful than that of Wellington, who had scoffed at the rocket; of Napoleon, who had banished the rifle from his armies; and indeed of Ulysses S. Grant himself, who apparently did little to equip his men with the large number of machine guns that he might have been able to call upon. What Haig did have in common with these eminent historical personages, however, was the perennially familiar problem of conflicting institutional demands upon his resources. He could not do everything he might ideally have wanted, simply because his clients were too demanding, and often pulling in different directions. This was not merely a matter of the 'Easterners' like Lloyd George wanting to take away his forces for use elsewhere (or nowhere). Nor was it even primarily a matter of finite industrial production capacity at a time of infinitely escalating demand. Instead, it was to a considerable extent a matter of cap-badge clannishness (or empire-building) on the part of the various specialist organisations themselves, each of which believed that its own key to the battlefield was the best one. Although Churchill and Lloyd George had done much to bring order and centralisation to the process of research and procurement, they had been powerless to impose a similar unity of outlook upon the ways the weapons were used after they arrived in the front line. What often tended to happen was that a new technical speciality would be set up, with its own distinctive weapons, organisation and staff, and probably also its own specially coloured brassards. This speciality would then gain an identity and momentum of its own, and start to interpret tactics almost exclusively in its own terms, to the detriment of all-arms unity.[28]

It was perhaps well understood by everyone — and duly discounted — that the infantry should naturally demand that all auxiliary weapons which operated in their sector should come under their direct control, even if that happened to cut across the optimum technical uses to which those weapons could be put. It was equally accepted that the artillery was likely to interpret the whole problem in essentially gunner terms, placing the highest priority upon either the welfare of their horses or, more charitably, upon the number of shells that could be fired. To some extent it was also understood that the RFC should have adopted a vaguely detached attitude to the whole proceedings, counting their successes more in the number of Albatrosses shot or Fokkers killed than in the number of useful reconnaissance photographs taken. Yet a considerably less well understood fact is that equally exclusive technical considerations also usually applied to most of the smaller or less prestigious organisations which were increasingly bringing technology to bear upon the Western Front. The Tank Corps was apparently self-consciously modelled upon the RFC,[29] and it certainly bowed to no one when it came to making claims for its own importance. As we shall see in the next chapter, its chronological senior the Machine Gun Corps also had a bitter fight to establish itself as an independent outfit, free from direct infantry control, in the course of which there seems to have been a certain sacrifice of goodwill and cooperation on both sides. Something similar could be said of trench mortars, gas, the RE signals and even the humble hand-grenade.

The sum total of all these conflicting technical loyalties was an institutional jungle from which it was very difficult for the senior commanders of the BEF to pull out a coherent all-arms doctrine. Precisely because they were encouraging such rapid technological change and expansion, they lacked the time and resources needed to mould each new specialisation, as it arose, into a unified team. With the Germans, by contrast, this particular problem seems to have been a little easier. They had enjoyed a head start in modern technology before the war began, but then gave up the unequal race at least eighteen months earlier than the British. They therefore had a less comprehensive array of high technology to accommodate, but a better starting-point from which to do it. Even then, however, their solution still involved a divisive formal distinction between the specialist hi-tech 'Pioneer' (combat engineer) storm squads and the lumpen mass of infantry formations. This would later be smoothed out only to a certain extent by the distinction between a few élite or 'stormtroop' infantry battalions or eventually divisions, and a greater number of line divisions which did not have that type of training.[30]

Bombs, smoke and gas

The appearance of a trenchlocked No Man's Land almost inevitably meant that the various types of grenade or bomb — and particularly the hand-grenade — would be seen as symbols of the new warfare scarcely less potent than the machine gun herself. Bombs had never enjoyed much obvious application to mobile warfare in the open; but as close-range indirect fire weapons which could be made to bounce down into every hidden corner of an enemy trench, and then explode with high explosive force, they now suddenly took on a mighty new significance.

Bombs made such an appealing and logical answer to the problems of trench fighting that they enjoyed an instant vogue. For example, the 'fire-eating' Captain A. O. Pollard, VC, MC, and bar, of the 1st Honourable

Artillery Company, had started the war fondling his bayonet in secret — 'feeling its edge and gloating over it' — but by June 1915 he had come to realise that the bayonet had been decisively superseded by the hand-grenade. 'The bomb', he wrote, 'is essentially a close-quarter weapon. I was hoping to get to close quarters.'[31] A month later one of his letters to his mother included the following astonishing offer, still on the subject of bombs and grenades: 'They are very jolly things to play with. Would you like one as a souvenir?'[32] Pollard went on to become a persuasive advocate of and instructor in all the arts of the bomber. He won his VC on the Arras front in 1917 for making a successful counterattack, with just three companions, against a major enemy break-in. The four of them simply kept on throwing as many grenades as they could possibly lay hands on — German and British alike — until they were eventually reinforced.[33]

Especially during the first half of the war, before the safety features of the time-fused Mills bomb had been finally perfected, all bombs had a fearsome reputation for unreliability. This was not merely a matter of lethal accidents, such as the fourteen casualties encountered by Robert Graves at the Harfleur bull ring just after a sergeant had only too vividly explained the dangers of a percussion grenade by tapping it on a table top.[34] It was also a matter of simple military ineffectiveness, such as the thousands of Ball grenades which completely failed at Loos on account of nothing more threatening than the wet weather.[35] There were numerous types of grenade which had to be lit manually with a length of Bickford 'safety' fuse, like a firework, and this tended to be as complex a procedure as it was dangerous. Colonel Croft even claimed that 'bombs merely kill your own side in 9 cases out of 10'.[36]

The natural result of all its many forms of unreliability was that bombing gained a reputation as a very specialised art, requiring a relatively long training and indoctrination for its practitioners. Hairy-nosed enthusiasts of the Pollard type were therefore sent to bombing courses which qualified them to train up their own élite platoon of bombers, who might well eventually be promoted to a pampered life within their particular battalion or brigade, being exposed to the rigours of combat only for the duration of a few carefully planned trench raids, which they would be expected to spearhead.[37] It was only with manual SS 143 in February 1917, at a time when the fully safe version of the Mills was already universal, that every platoon was expected to contain not only a section of hand-grenadiers, but also a section of rifle-grenadiers. Thus almost a half of all infantry was officially designated as some form of 'bomber', and in fact every infantry soldier was expected to be conversant with the bomber's black art.

One particularly black corner of this art was the development of incendiary grenades, whether 'Thermit' — a very high-burning mixture of aluminium and iron oxide — 'P' for phosphorus, or 'Fumite' for smoke.

Each of these could set alight all woodwork inside the enemy's trenches and dug-outs, and in tactical terms they were much easier to use than flamethrowers. In the July 1916 analysis of the lessons of the Somme fighting they were preferred to Mills bombs for mopping up[38] — although it soon became clear that they were too much of a good thing, and left the captured dug-outs too burnt to be inhabitable by their new owners.[39] A move back to the basic Mills was then observed. One final development at this period which is also worth noting is the use of Stokes mortar bombs converted into hand-grenades, to give greater explosive power than the Mills.[40]

Within the bomber's multifarious skills the rifle-grenade seems to have received less attention than it properly deserves, being seen more as an unlikely contraption than as an integral mainstay of the platoon's firepower. Edmund Blunden, for example, dismissed the Hales grenade as a 'brass brutality'[41] — but then his own particular Hales bomb had led to an accident, and in any case he himself was a poet rather more than he was a tactician. The fact remains that SS 143 called for rifle-grenades to be as widespread as hand-grenades, and they could certainly be just as deadly. They had a range of between sixty and 200 yards, thereby effectively outranging German stick- and egg-grenades, which were themselves notoriously lighter and longer-ranged than the British Mills. The late-1916 British rifle-grenade normally consisted either of a Mills bomb in a special cup fitted to the muzzle of the rifle, or of a smaller grenade mounted on a rod which was inserted into the rifle barrel, and then shot out with a blank cartridge.[42] 'P' bombs could also be used effectively in this rôle.[43] If fired in a section volley to compensate for poor accuracy by sheer volume, a mixture of rifle-grenades could make a very impressive explosive display.[44]

Within each platoon, the rifle-grenadiers were intended to form half of the covering fire team, together with the Lewis gun, while the assault team was made up of the hand-grenade section and the rifle section. If it worked together effectively, this combination was expected to flush out even the most stubborn enemy machine gun position, and in practice this even seems to have happened quite often.[45] Rifle-grenades could also be used as a powerful defensive weapon with a long reach.[46]

The grenade, and especially the rifle-grenade, could be seen as a type of miniature artillery under the direct control of the infantry; but something very similar could also be said of the heavier type of 'bombs' which were fired by the many different types of bomb-thrower, trench-mortar (TM) or minenwerfer. These weapons emerged at about the same time and for very much the same reasons as the individual hand-grenades, although unlike them they soon passed out of platoon, company or even battalion control and started to be held at brigade level or higher (see Table 3).

Table 3: Effective combat performance of British infantry weapons

Weapon	Level held & description	Normal combat range (yards)	Normal rounds per minute
.303 Vickers	Brigade medium machine gun	2,000	250
9.45" 'Flying Pig'	Corps heavy trench-mortar	1,000	1
2" 'Toffee Apple'	Divisional medium trench-mortar	500	1
.303 Lewis Gun	Platoon automatic rifle	500	100
3" Stokes	Brigade light trench-mortar	400	6
.303 Lee Enfield	Personal magazine rifle	300 (-)	5
Rifle Grenade	Platoon HE weapon	200	1
Mills Bomb	Platoon/personal HE weapon	20–30	2
.455 Webley pistol	Officer's status symbol & 'persuader' — also used by specialist gas/tank etc. crew	3	6

By the start of 1916 the British had standardised around three types of trench-mortar — the light 3-inch Stokes held at brigade level and firing an 11-pound bomb around 400 yards; the divisional medium Vickers 2-inch 'Toffee Apple' firing a 60-pound bomb 500 yards (later superseded by a longer-range version); and the Corps-level heavy 9.45-inch 'Flying Pig' which fired a 150-pound projectile up to 1,000 yards.[47] Trench-mortars could create very much the same levels of damage as artillery, but by virtue of their positioning directly in the front line they enjoyed far greater local tactical flexibility. From an infantry point of view they dramatically bypassed the whole uncertain bureaucracy implicit in any appeal to the distant, unknown and distinctly quirky rear area agencies responsible for artillery support. The fire of trench-mortars could thus be adjusted more rapidly and accurately than that of artillery, and although their bombs moved slowly enough to be seen in flight, at least the drop-loaded Stokes could theoretically reach the exceptionally high rate of thirty rounds per minute — and in practice it could easily sustain fifteen rounds in the air together, adding a powerful reinforcement to more leisurely artillery bombardments or SOS fires, if not always offering a complete substitute for them.

Such a potentially devastating intervention in local battles did trench-mortars represent, that troops defending quiet sectors of the line were often anxious not to allow them to be used at all. The scale of outrage they could cause to the enemy tended to provoke exceptionally fierce levels of retaliation, and the trench-mortar operators tended to be regarded with great distrust by the infantry.[48] A deep problem of institutional conflict certainly arose from the separation of mortars into a distinct 'speciality', and they often seemed as remote as artillery itself[49] — indeed, the heavier types of mortar were actually served by Royal Artillery crews. However, the value of mortars in the attack was so great that these difficulties were often overcome in practice, especially in the use of the Stokes for suppressing individual machine guns or pillboxes.[50] It was a technique which proved no less effective for the British in 1916–17 than it would for the Germans in March 1918. In the case of very rapid advances, however, it was soon recognised that even the 3-inch Stokes mortars were too cumbersome to keep up quickly with the infantry,[51] while the heavier types would be left behind completely. The experience of the Hundred Days, in particular, led to the complete suppression of heavy mortars, and the provision of special carriages and extra transport for medium mortars.[52]

Apart from the direct infantry-support-mortars, there was also a limited supply of 4-inch Stokes guns for the Special Brigade, for firing gas, HE, smoke or Thermit bombs up to 500 yards (later up to 1,500 yards). These had been rushed into production for the battle of Loos, but they continued in use throughout the war and actually increased dramatically in capability.[53] By 1917 the 4-inch Stokes had become a classic and incomparable weapon, insofar as it was a major means of applying sudden and heavy concentrations of phosgene. By that time it had also been joined by the still more massed batteries of Livens projectors, sometimes with 1,000 camouflaged tubes arranged for simultaneous electrical firing in rather the same manner as a Second World War rocket or nebelwerfer barrage.[54]

Perhaps the most chilling of all the books to emerge from the Western Front was the account of the Royal Engineers' Special (gas) Brigade's operations written by its commander, Major (later Major General) C. H. Foulkes. Not only does Foulkes himself appear from his photograph[55] to have been as humourless, close-cropped and single-minded as any professional killer ought to be, but his literary style displays an odiously matter-of-fact detachment which turns out to be only too appropriate to his chosen subject matter. Foulkes takes pride in the notion that his formation probably caused several dozen times as many German casualties as any other formation of comparable size, in return for a relatively low friendly casualty list,[56] and his main complaint is only that the BEF's high command was generally reluctant to appreciate the significance of these facts.

The instant improvisation of the brigade out of almost nothing, from late May 1915,[57] was undoubtedly an impressive military feat, as was Foulkes's rapid but accurate appreciation of its strengths and weaknesses, and his vision of how it should evolve. His decision to eschew bacteriological warfare but go for chlorine as a stop-gap at Loos in September 1915, followed soon afterwards by his pure luck in stumbling across an extant phosgene factory near Calais,[58] must surely rank high in the list of key turning-points in the martial history of the British Empire. Within six months Foulkes had set up an organisation which could pump out many tons of gas over the enemy every week — often with deep local deviousness to ensure surprise — causing dozens, and more probably hundreds, of casualties on each occasion. For example there were 110 major British cloud gas attacks in the second half of 1916 alone, whereas the Germans appear to have managed only about fifteen on the entire Western Front in the entire war, with the last as early as 8 August 1916.[59] By the end of 1917 Foulkes had even organised a system for releasing his cloud gas directly from railway trucks brought forward on tramways, thereby obviating the laborious task of manhandling heavy cylinders forward through the trenches.[60] He also had a policy of gassing selected enemy regiments repeatedly and persistently, wherever they might be posted along the front. Thus the 1st Bavarian Reserve Regiment was attacked fifteen times, the 161st and 9th Bavarian Regiments were each attacked fourteen times, the 1st Guard Reserve Regiment twelve times, and the 156th and 10th Bavarian Regiments ten times each.[61]

By comparison with this virtuosity most of the Germans' own gas attacks appear to have been relatively clumsy and ineffective. They did admittedly score a palpable hit with the unexpected introduction of the blistering 'mustard' agent in the summer of 1917, which their opponents could not match for about a year. Mustard made an excellent defensive weapon, and at the start of the Passchendaele battle it helped defer the inevitable moment when the British artillery would win overwhelming firepower supremacy. It was, however, seen as too damaging to be used on a German attack front during at least the last six days before zero, so it could therefore be used only on the shoulders of an intended break-in, or as a diversion on some other front. In the spring of 1918 this pattern of distribution was easily monitored and identified by the wily men of Foulkes's gas brigade, and gave them a very impressive intelligence indicator of precisely where the true attacks might be expected.[62] Deprived of mustard on their attack fronts, moreover, the remainder of the German gas effort often degenerated into an excessive reliance upon dispersed area fire by artillery, especially using the innocuous 'Blue Cross' shell filling which had no more than nuisance value.[63] Artillery shells were further reduced in effectiveness since they could never carry more gas than about ten per cent of their total weight, whereas a Livens or Stokes bomb could take about fifty per cent.[64]

As a result of a frivolous (albeit potentially very fatal) mess jape perpetrated upon the British Gas Brigade HQ's kitchen stove, it was accidentally discovered that the German Blue Cross agent (diphenyl chlorasine mixed with phosgene) was effective only if it had first been superheated.[65] This led Foulkes into the development of the 'M' Device, which was a hand-held chemical flare designed to superheat the Blue

Cross but which could be lit simultaneously by thousands of infantrymen all along an attack front, to build up a cloud which could be resisted by no known German respirator. He saw this agent as a potential war-winner, and during the Hundred Days he made a number of attempts to organise a concerted wide-front attack with it, to be followed up by a general infantry assault and breakthrough. However, on every occasion he found himself frustrated by the high command.[66] Rather like J. F. C. Fuller with his cruiser tanks, in fact, Foulkes was forced to accept that although he thought he had a decisive breakthrough weapon close at hand, he was unable to use it earlier than 1919.

Among his many other claims to fame, Foulkes enjoyed the distinction of having been the friend and patron of Captain W. H. Livens, MICE, MIME, a kindred spirit who had 'a strong personal feeling in the war connected, I believe, with the sinking of the *Lusitania'*, combined with the expressed ambition of reducing the cost of killing Germans to a paltry sixteen shillings apiece.[67] After Willie Bragg, Livens was perhaps the ultimate 'boffin' of the BEF's war effort, having initially won fame by perfecting Captain Vincent's otherwise unworkable heavy-duty flamethrower.[68] This was first used just before zero on 1 July 1916 and intermittently, albeit on a very small scale, for about a year thereafter. At a total assembled weight of over two tons, it proved to be far too immobile, and at ninety-eight yards too short-ranged for all but quite specialised and long-prepared operations. It could have come into its own only if it had been mounted on a tank chassis, an ideal which had been pursued from the start of 1916 but which would become a reality only in the Second World War.[69] Its need was certainly felt in the trenches in September 1916, when one brigadier wrote that 'If liquid fire could be added to the tank's armament it would increase their offensive power enormously'.[70]

Perhaps the most famous British flame attack came in the attempt to take High Wood on the 'glorious 12th' of August 1916, in conjunction with the dreaded Barratt hydraulic forcing jack, 'mole' or 'pipe pusher'.[71] The pipe pusher was a device used in civil engineering for boring horizontal holes through the ground, big enough to take a drainpipe, cable or, in mining, sticks of dynamite. In the BEF it was being put to such uses on the line of communication[72] when someone saw its potential for digging instant trenches across No Man's Land ahead of zero. If the pipes were filled with ammonal and then detonated, they could gouge furrows six feet deep for the expenditure of considerably less labour than the 'Russian saps' that had been widely recommended — but rarely dug — before 1 July. Three machines were therefore set to this task on 19 July with some success,[73] and for a brief happy moment it looked as though the problem of trench warfare had been definitively solved by technology. Enthusiasts quickly added additional dimensions, extending the pipes a little further forward to explode beneath the enemy's trenches or, later still, to transmit gas through them.[74] The immediate result, however, was the not inconsiderable débâcle of 12 August. On that occasion the friendly heavy artillery dropped short and buried the flamethrowers, while the pipe pushers were turned back by tree roots and scored an 'own goal' on friendly trenches with disastrous consequences to the occupants. To add insult to injury, Livens's associated attempt to use a 'flaming oil rag projector' appeared to have no effect whatever. Obviously[75] some other new and equally unlikely contraption would be required, and sure enough it duly plodded out just one month later in the shape of a 'Tank' which seemed to many at that time to possess extremely dubious potential utility — and which certainly had a habit of scoring almost as many 'own goals' as pipe pushers themselves.

All was not lost, however, since Livens had taken to heart the lessons of 12 August, and had gone on to perfect his projector design until by October it had developed into a brilliantly lethal means of distributing gas, smoke or incendiary material.[76] A Livens projector barrage could transfer massive concentrations of many tons of phosgene from around 600 yards behind the British front on to a selected part of the enemy line, without warning, within a quarter of a minute. It was a much-feared weapon, not least because its emplacements were entirely inconspicuous and caused no disruption or danger to the British infantry which it supported. Foulkes himself, however, maintained his loyalty to the original 'cloud' method of gas dispersion over the projectors, and then to the projectors over artillery.

The main problem with all these devastating attacks, which Foulkes himself seems to have recognised very clearly, was that the high command apparently soon abandoned its interest in coordinating them with major infantry assaults. This had been attempted at Loos, but it had led only to nail-biting and recriminations once it was realised that the weather dictated the usefulness of gas, and that the gas in turn dictated the impact to be expected from the infantry. For many diverse reasons of both logistics and of *amour-propre,* therefore, it was considered to be unacceptable that the timing of the infantry assault should be allowed to depend upon such an unreliable factor as the weather. Generals needed more certainty, and therefore they relegated the gas weapon to a distinctly auxiliary status for use mainly on quiet fronts or in artillery shells.

7

Automatic Weapons

> The rifle is pre-eminently the weapon for killing any enemy slipping away over broken ground, while the bombers clear underground shelters; but the Lewis gun will also have its opportunities.
>
> Official manual SS 197, p.10

The struggle to control automatic fire

The story of the Machine Gun Corps very neatly highlights the way in which new weapons could become the focus of naked empire-building and institutional infighting. This corps was the very personal brainchild of Captain George M. ('Boss') Lindsay, a Boer War veteran infanteer who, like his contemporary J. F. C. ('Boney') Fuller of the Tank Corps, had been heavily influenced by the new wave of infantry thinking which emerged from that conflict. In the years before 1914 Lindsay had been an instructor of musketry at Hythe, where he was able to ponder deeply on all aspects of firepower and tactics, and to establish himself as a clear-headed and forceful personality who tended to make a lasting impression upon his students. The central tenet of his thought — not unnaturally for an instructor of musketry — was that every manoeuvre is essentially a fire problem, and that 'In war fire is everything'.[1] At first he saw this as primarily a matter of rifle fire; but before very long he became converted to the idea that the machine gun could develop far more fire more economically and with less exposure of the firing soldiers to the enemy's retaliation. At Hythe his duties included the training of machine gunners just as much as of riflemen, so he was ideally placed to see the true potential of automatic weapons as the key source of small-calibre fire.

Upon the outbreak of war Lindsay was posted to Chatham to organise musketry training for the new armies, and it was here that his ideas really took wings, since it soon became obvious that there was a serious shortage both of machine guns and of trained machine gunners. The training of riflemen might be difficult enough; but that of machine gunners was incalculably more so — yet it was also far more tactically vital. On 1 October 1914 he therefore submitted an ambitious paper for a gigantic 'snowball' scheme whereby trained machine gunners would act as instructors to produce more instructors, and so on.[2] There would be a geometrical expansion of machine gun expertise within the Kitchener armies, allowing the BEF to rival the Germans in this field. Before the war Lindsay had been highly dismissive of German practice, believing that their infantry lacked either fire discipline or fieldcraft,[3] and using this comparison as a polemical device to reemphasise those qualities for his British students. However, the first clash of arms persuaded him that at least in machine gunnery the Germans were greatly superior, mainly because they already possessed a machine gun corps while the British did not. Actually their riflemen's fieldcraft and fire discipline turned out to be just as poor as he had predicted, while British machine gunners at Mons aquitted themselves more than well. Lindsay soon had plenty of evidence in his hands for all of this; but he would never again point to the many obvious weaknesses in German tactics. Instead, his interest now lay in contributing to what would soon become a century-long myth of German tactical infallibility. He began to stress the superiority of German practice with the machine gun, since that weapon had now become his own particular hobby-horse within the British Army. Alas, there was nothing particularly objective about this process, and at one point he even made the self-serving statement that the Germans had 'forty times' as many machine guns as the BEF in August 1914, even though he must surely have known this was only because the German army at that time was almost forty times as big.[4]

When his 'snowball' scheme was in due course rejected by the authorities, Lindsay became convinced that they were wilfully closing their eyes to simple common sense and that he, Captain Lindsay, possessed a far higher insight than they did. He began a crusade that would continue for many years into the future. In his basic instincts he may very well have been right; but the end result, as with J. F. C. Fuller and the tank, turned out to be a very long and in some respects unfortunate power-struggle, which soon distorted the truth just as much as it bruised too many personalities along the way.

By 25 January 1915 Boss Lindsay had returned to the charge with his *Scheme for the Machine Gun Training of New Armies*.[5] This went several steps further than his original 'snowball' scheme, and called for the establishment of a true Machine Gun Corps (MGC), in which élite personnel would be organised into brigade machine gun companies which enjoyed full tactical control over their guns, free from the whims of infantry colonels who did not understand the true potential of the new weapon. The shortage of machine gun *matériel* at this time made it doubly important, in Lindsay's eyes, that each gun should be used to maximum advantage. However, one cannot also fail to notice the drift of his own ambitions towards an independent organisation which he himself might eventually hope to control.

The idea of a separate MGC, as an independent arm falling somewhere between the infantry and the artillery, rapidly took hold of Lindsay's thoughts during the first months of 1915. By this time he had returned to Hythe, although he was still deeply involved in training the new armies under the aegis of Eastern Command. He did win some victories, such as his suggestion of 15 February that each division's musketry officer should set up a machine gun school,[6] or his still more effective press campaign, which attracted the attention of Lloyd George just at the moment when he was about to set up the Ministry of Munitions.[7] Nevertheless, Lindsay remained frustrated at the slowness with which the authorities took up his main suggestion for an independent machine gun corps. GHQ in France appeared to be no more ready for change than the War Office in London, and both of the army commanders — Haig and Smith-Dorrien — seemed to agree. They insisted that at least four Vickers Medium Machine Guns (MMGs) should be organically intrinsic to every infantry battalion, with one gun as an inseparable part of each rifle company. These authorities believed that any creation of independent machine gun companies should wait at least until such time as more guns could be made available beyond one per company. They were therefore voting heavily against the idea of an independent corps in which machine gunners could train themselves or use their weapons separately from the infantry's hierarchy of command.[8]

With hindsight we can agree that the lack of machine guns in early 1915 was indeed a crucial difficulty, entirely comparable in importance with the shell shortage or the bomb shortage. For Boss Lindsay, however, it represented a particularly provocative challenge. For him the lack of machine guns made an important argument for their centralisation under the single authority of an MGC, in order to make the most of what little equipment there was; whereas for most line infantry officers it was an argument for greater diffusion to the lower levels of command, on the assumption that the machine gun was naturally an infantry weapon which had to be spread as widely as possible. On both sides there was a strong desire to improve the machine gun defence of the front line; but unfortunately this came to be interpreted in radically different ways according to which of the two alternative perspectives one happened to hold.

By June 1915 Lindsay was working at the GHQ Machine Gun School at St Omer, which was commanded by his friend and disciple Major C. d'A. Baker-Carr. Between them, they organised a number of questionnaires for all the machine gun officers in the army, in an attempt to make a detailed analysis of battlefield practice.[9] The result was that Lindsay became more convinced than ever of the need for an MGC, since infantry units on their own could not be trusted to arrange their guns in a properly interlocking and defilading defensive belt to cover the front line, with each battalion's guns helping to cover the frontage of neighbouring battalions. Nor could the infantry battalions be persuaded to hold some of their guns back in depth, with proper reserves. Lindsay took up a new rallying cry that 'The tactical handling of machine guns must be based on the principle of cooperation' — meaning centralised control of the layout of the whole machine gun belt — although he mixed this valuable concept with deep pessimism that the supply of guns remained too low and that a whole year had been needlessly lost from the necessary organisational reforms.

A further result of Lindsay's studies was that he began to refine his analysis of the use of machine guns in the attack, for which a number of different techniques seemed to be indicated. The first was the German practice of infiltrating silently forward at night, under the cover of standing crops or shrubbery,[10] to take a defender's position in flank while infantry attacked him from the front at dawn. This idea was boosted by the thought that a single machine gun team could be made inconspicuous in a way that a company of rifles never could, and it was a concept that would later be officially adapted for Lewis Guns in Fourth Army's instructions for the Somme battle.[11] However, the most famous application of this idea remains the Germans' own infiltrations of March and April 1918, although on those occasions they were doubtless greatly helped by the mist and the lack of adequate BEF defences. It is at least worth noting that the technique of 'machine gun infiltration' was already well known from the very start of the war, and was by no means an innovation of the notorious German stormtroops towards the war's end, as has sometimes been claimed. Within the BEF it nevertheless remained a very difficult ideal to live up to, since a bald, heavily shelled and deeply wired mudfield did not tend to offer the sort of cover that infiltrating machine gunners ideally required, especially if they had to follow their own creeping barrage. Following a number of unfortunate obliterations of machine gun teams on foot accompanying the initial waves of assault, in fact, it soon became orthodox British doctrine to leave these high-value personnel and weapons well to the rear until such time as the enemy's trenches had been very fully secured.[12]

Secondly, Lindsay took notice of overhead and 'long-range searching' fire, or what he originally termed 'long-range annoyance' fire, which was produced by massing random patterns of bullet streams from large numbers of Vickers guns against a single area somewhere behind the enemy's front line. There were two elements in this technique. The first was the idea of massing machine guns at one point; for example, at the battle of Neuve Chapelle in March 1915 Captain Ludwick, an Indian Army officer, had collected twenty guns to repel a German counterattack,[13] and at Loos in September 1915 the 9th Division collected fourteen guns to reinforce its artillery sweeping the approaches.[14] By 24 August 1916 at High Wood the 100th Machine Gun Company was able to fire over a million rounds from ten guns in a single half-day operation.[15] Secondly, there was the idea of indirect, overhead predicted and plunging fire, similar to the action of a howitzer, which would allow long-range shots to be placed fairly accurately upon any target selected from the map. Lindsay had already recommended this in his Chelmsford speech of 10 August 1915. When taken together, these two principles would later evolve into the type of carefully orchestrated massed MG barrage fire, using artillery principles and instruments, which by August 1917 had become the touchstone of BEF machine gun expertise — and which was technically far in advance of the practice in any other army, most notably the German.[16] Indeed, it was already being called 'barrage' fire in a British manual of July 1916.[17]

It would certainly be highly satisfying if we could attribute the practical development of barrage fire to Lindsay himself, but unfortunately this has not proved to be possible. By mid-1915 he had obtained most of the clues; but it is clear from such of his papers as survive that by this time he had become far more interested in questions of organisation and strategy than of technology or tactics. Besides, during the crucial period between the start of the Somme battle and the start of Passchendaele, he was diverted from these studies into hectic front-line service as Brigade Major of 99th Brigade, in the 2nd Division, which was involved in very heavy fighting at Delville Wood, Beaumont Hamel, Miraumont-Irles and Oppy. Instead, the true father of the machine gun barrage turns out to have been the equally energetic and forceful Brigadier E. Brutinel, the machine gun officer to the Canadian Corps, who had begun as Lindsay's friend and disciple when he first arrived in Britain, but who by mid-1916 had pushed ahead and solved 'the multiplicity of associated problems' needed to make the barrage a reality.[18] Brutinel apparently fired his first barrage with twenty guns on 2 September 1915, his second in support of a major trench raid in early 1916, and his best at Courcelette in September 1916. After that a grandiose machine gun barrage became a routine feature of all Canadian assault tactics and it quickly caught on in many British formations,[19] although Brutinel complained that the British 'schools' started to take notice of it only after the famous Vimy Ridge operation in April 1917.[20]

While Brutinel was busy designing the barrage, by the end of November 1915 the supply of machine guns had at last started to exceed demand. This allowed Lindsay to hammer home his main message that it was the training of the gunners themselves which needed to be reformed and centralised, rather than the procurement of weapons, and that they should therefore be accorded priority over the training of riflemen. He found a new flash of inspiration in what he called the 'strategic' value of machine guns, which was based on a very raw calculation that if one gun manned by ten men was worth 100 rifles, then massive savings in manpower could be made through an increased stress on the automatic firearm, allowing entire armies to be detached from the Western Front and sent to distant theatres such as the Balkans.[21] The key fallacy in this highly 'Lloyd Georgian' or 'Easterner' analysis was precisely the same fallacy that would soon be repeated in the case of the tank, namely the assumption that the new technological weapon — whether machine gun or tank — could perform absolutely all of the functions normally performed by the PBI which it was supposed to supplant. Of course the machine gun could lay down swathes of .303 bullets more efficiently than riflemen, and the tank could move through belts of wire considerably better than the man on foot; but it remains true that only infantry equipped with boots, backpacks, rifles and bayonets — and perhaps even with bombs — could really clear up a battlefield after all this high technology had done its work. It is perhaps ironic that both Lindsay and Fuller, who were both light infantrymen by origin, should have both so wilfully devoted so much of their professional energies to denying the rôle of the humble infantryman in modern war.

At first sight Lindsay's 'strategic' paper, recommending the 'Balkanisation' of the BEF, cut a very poor figure beside Brutinel's genuine tactical innovations with barrage fire. Yet if we take a longer view, we can see that Lindsay's essay did in fact touch upon some vital areas. One was that it strongly echoed the supposed German practice of basing defence (and even attack) upon machine guns rather than riflemen. The assumptions behind this idea may doubtless be questioned on a number of counts,[22] but it is interesting that it was being advocated so early for the British by no less a figure than George Lindsay, the 'high priest'[23] of BEF machine gun opinion.

A more important implication of Lindsay's 'strategic' essay was that it helped to nudge him towards a higher view that the machine gun should become a key element in a mechanised and mobile strike force. By the very fact of making the argument that 'One machine gun is worth 100 rifles in defence',[24] he was almost unavoidably leading up to a supporting argument that machine guns should also form an essential element in any attack. Indirect barrage fire from static emplacements, using essentially artillery techniques, was certainly one way that this could be achieved; but far more attractive to Lindsay the infantryman was the idea of physically pushing the machine guns forward alongside the leading waves of the attack itself, where they could use the far more economical direct fire. The machine gun manuals of the time were full of recommendations about how this might be achieved by soldiers on foot; but experience weighed against this practice, and Lindsay now started to see that it might be better achieved by more motorised means.

'Motor machine guns' — Vickers guns mounted on motorcycle sidecars — had already been widely deployed in 1915 as an adjunct to cavalry or cyclist units designed for the exploitation of breakthroughs. Every army corps had at least a battery of them as a mobile reserve, although in the conditions of static warfare they were often used merely to reinforce the infantry line, provoking a number of good men to leave in disgust and transfer to the RFC.[25] It was only in fluid operations that the motor machine guns really came into their own; for example, at Hermies on 23 March 1918 no.4 battery gave great assistance to 17th Division by its orderly phased retreat by sections, firing right up to the last moment. No.6 battery at Beaumetz around the same time reported 'the shooting was grand' against massed German attacks at close range, while on the Lys one NCO of no.7 battery actually made a charge with a single motorcycle combination, firing. He captured a German machine gun and had it turned against other enemy troops.[26] Nor were motor machine guns despised by the infantry. One subaltern operating east of Ypres in October–November 1918 thought rifles and bombs were the best mixture for busting pillboxes; but he conceded that motor machine guns would have been still better.[27]

In these circumstances it was scarcely an accident that the tanks themselves were originally designated as the 'Heavy Branch of the Machine Gun Corps'. The 'female' tanks were indeed nothing more than tracked and armoured motor machine gun batteries — perhaps unnecessarily complicated ones from Lindsay's point of view, but definitely within the sphere of his general thinking. We must remember that he would himself become perhaps the most judicious British analyst of the tank's possibilities during the 1930s, achieving a more balanced all-arms view than Fuller himself, and certainly a far greater technical knowledge than Liddell Hart.[28]

Lindsay was nevertheless running too far ahead of current practicalities with his 'strategic' paper at the end of 1915, and although the MGC was actually set up at around that time, his ideas were often misunderstood and met with a frosty reception. Despite the subtle niceties of his formulae, he was accused of wanting to create large formations exclusively of machine gunners, as a complete substitute for infantry armed with rifle and bayonet. The 'Sub Chief of the General Staff', for example, lectured him that 'Fire alone will never win the battle', and that 'If you turned all the infantry in the army into machine guns you would never get a decision.'[29] As if this were not already a sufficiently red rag to a bull, this staff officer compounded his crime by claiming that the brave Russians were showing how bayonet assaults unsupported by machine guns could overwhelm even the most technologically based army. In his opinion the MGC should be kept to a position comparable only with that of the Military Police: it should be the servant of the infantry, not the master of its own tactics. To this George Lindsay replied with a point by point refutation, denying that he wanted 'the whole army' to become machine gunners, but insisting that its machine guns should be organised in battalions rather than companies. Nevertheless it is also true that as early as May 1916 he wrote to Major General Burnett Stuart at GHQ that an autonomous massed 'machine gun force' was necessary as a motorised instrument of breakout.[30]

Meanwhile on 12 September 1915 Lindsay had been posted to the nascent machine gun training centre at Grantham in Lincolnshire as a GSO2. Although it would take three frustrating months before it had been properly organised and won official blessing, the establishment of this centre in effect represented the birth of the MGC itself. In theory the corps was constituted by a royal warrant of 22 October; but in practice it was so obstructed by Kitchener and others that it cannot be said to have enjoyed any genuine existence until the start of 1916, and even then it fell far short of Lindsay's hopes.[31] An offshoot of the Grantham school was set up in France, at Camiers near Étaples, only in March 1916. Still more significantly, the BEF's machine guns were organised not into a battalion within each division, but only into a company within each brigade. This was under the command of the infantry brigadier, and hence lost to any higher control by the MGC. It would not be until the last year of the war that they would eventually be organised in battalions as Lindsay had wanted from the start, and even then they would never fully win their own chain of command and proper scale of staff officers at higher HQs.[32]

It was particularly unfortunate that this last reorganisation was followed so soon by the German March 1918 offensive, since the new machine gun battalions do not seem to have fully shaken down. They sometimes failed to ensure a fully interlocking coverage of the whole front, and on other occasions they lacked decisive leadership. Signals were faulty, the belt-filling machines were unable to keep up with demand, and the lack of factory-boxed belts was sorely felt. A number of infantry commanders were happy to seize upon such failures as a convenient excuse for their own failings, and the MGC came under something of a cloud.[33] Long-range barrage fire was especially singled out for condemnation, since it seemed to distract the machine gunners from their more immediately useful rôle in delivering direct fire at closer range.[34] The MGC enthusiasts were nevertheless quick to respond in kind, and complained that the infantry failed to understand the correct mutual layout for Vickers and Lewis guns. They alleged many machine guns had been overrun only because they had continued firing to the bitter end in a vain attempt to cover the over-hasty retreat of the infantry.[35] Nor was the idea of grandiose barrages abandoned, and many were fired during the remainder of the war, including one with 111 machine guns for the Australian attack on Hamel on 4 July.[36] Indeed, there is evidence that the infantry really were reassured by 'the crackle of bullets overhead', and that the Germans were correspondingly dismayed.[37] In the fluid conditions of the Hundred Days the machine guns certainly made a very convenient form of mobile near-artillery, which could go some way to substituting for heavier guns.

There was nevertheless, alas, some substance to the claim that the MGC contained some men of dubious military quality and morale. In common with many other self-proclaimed technical élites, it liked to portray itself as an ardently innovative and professional body; yet it was also forced to draw most of its other ranks from proud infantry regiments with long histories, which were keen to unload their worst, rather than their best, material on to other outfits. In this the MGC seems to have suffered from a diametrically opposite process from that obtaining for the RFC or the tanks, which often drew great benefit from enterprising men and officers who were bored with humdrum regimental life and eager to volunteer for something more dangerous and exciting. In practice the MGC was therefore never really quite such an élite as it liked to make out, although in common with both the Tank and the Flying Corps it always clung stubbornly to an exclusive sense of its own importance.

As for higher organisations, it was only in the Hundred Days that full brigades of motor machine guns were finally assembled. The first of these was Brutinel's so-called 'Independent Force' (see Table 4) which spearheaded the later phases of the Canadian attack at Amiens on 8th August 1918, albeit with somewhat patchy success. It continued to grow, however, and enjoyed considerable political advantages over its British counterpart insofar as Brutinel himself, as Corps Machine Gun Officer, enjoyed an independent position equivalent to the Corps BGRA. This could be contrasted only too obviously with the British propensity to lump their machine gun officers in with such auxiliaries as the Military Police. Nevertheless Lindsay, using his personal influence with General Horne, the commander of First Army, was able to establish 'Lindsay's Brigade' on 27 August.[38] Horne, a gunner, had shown himself particularly sympathetic to the machine gun cause ever since the Canadian extravaganza on Vimy Ridge in 1917, and was especially enthusiastic for the idea of substituting 'mechanical power' for infantry.[39] Owing to the rigours of trench warfare, unfortunately, neither of the two machine gun brigades was ever able to come into action in the sort of breakthrough battle that they had been designed to spearhead; but as a milestone in conceptual development their creation must surely stand every bit as high as the original founding of both the Machine Gun and the Tank Corps themselves.

Table 4: Organisation of motor machine gun brigades

(a) *Brutinel's Independent Force (Source: OH 1918, vol. 4, p.42):*

1st and 2nd Canadian Motor MG Brigades (each of 5 × 8-gun batteries)
Canadian Cyclist battalion
One section of medium trench-mortars mounted on lorries
(Doubtless it also had wireless and medical support, not mentioned in OH)

Total: 80 machine guns and about 300 cyclist infantry

(b) *Lindsay's Brigade (Source: Lindsay papers, Machine Gun files E1, E2A):*

Brigade HQ with signal section, including wireless
Two MG battalions, each of 4 × 8-gun batteries + 8 armoured cars + motor transport
One cyclist battalion
One battery of 4 × 6" trench mortars
Medical support with motor ambulances

Total: 64 machine guns, 8 armoured cars and about 300 cyclist infantry

---

The rise of the light machine gun[40]

Perhaps the starkest example in the BEF of both the triumph of technology and the limitations on its application may be found in the adoption of Automatic Rifles (ARs). On one side the British seem to have held the international lead — both chronologically and numerically — for actively deploying ARs in the infantry fighting line, in the shape of the Lewis gun.[41] Yet against all this they never fully succeeded in integrating their automatic weapons into a single tactical policy.

The Lewis gun itself was only one out of very many ARs being designed by arms manufacturers during the two decades before the war; but it was remarkable both for its relative technical efficiency and for the celerity with which it was adopted by the BEF. It was originally invented and produced in 1911 by an American working in Belgium, and its makers migrated as war refugees in August 1914 to the BSA factory in

Birmingham. Their gun was adopted first by the RFC but then, as early as November 1914, by British ground forces. Being an air-cooled gas blowback weapon, the Lewis weighed only 28–30 lbs plus 4.5 lbs per magazine — rather lighter than a water-cooled MMG, and apparently ideal for use within a small group of active infantrymen. It was being procured in large numbers by the spring of 1915, although it suffered production delays throughout that year and achieved a weekly output of more than 200 only in December.[42] Sufficient were nevertheless issued for each infantry company to include either two or four guns by the battle of the Somme in mid-1916,[43] which meant either one gun or half a gun per platoon. Brigadier Jack indicates that so great a value was attached to Lewis guns in September 1916 that he would not send a pair of them into No Man's Land at night unless they were tied to ropes, so that they could be hauled back if lost.[44]

Despite a proliferation of technical courses for them, Lewis gunners remained in relatively short supply throughout the war. Nevertheless, the continuing proliferation of the guns themselves allowed the infantry assault manual of February 1917, SS 143, to assume one gun per thirty-six-man platoon, serviced by a whole section of nine men carrying thirty drums of ammunition.[45] By the battle of Passchendaele this scale of issue had been definitely consolidated in practice,[46] and during 1918 the figure eventually often doubled to eight guns per company, or two per platoon. Under conditions of combat attrition this would normally amount to a very modern-looking layout of something like each platoon containing two four-man Lewis gun sections and two seven-man rifle and bayonet or Mills bomb sections.[47]

Unfortunately, the Lewis appears to have received a disappointingly suspicious welcome from the front-line soldiers, since it was seen as displacing the greatly appreciated Vickers MMG, yet failing to offer sufficient compensating advantages in weight, simplicity or reliability. Charles Carrington, for example, complained that the ammunition pans for the Lewis were heavy and very hard to fill, yet could be fired off in only five to six seconds.[48] The Lewis needed a wheeled carriage, 'resembling a coffin', to move around,[49] while postwar research has shown that its ingenious — but intricate and heavy — air cooling ducts were entirely unnecessary to the efficient functioning of the weapon[50] — and of course, like every other piece of military equipment, it all had to be cleaned and was prone to malfunction if it got dirty.[51] One subaltern said, 'This gun was a very shoddy affair after the Vickers',[52] and he pointed out that whereas the Lewis was merely a 47-shot weapon, the Vickers could fire no less than 250 rounds per belt. There is plenty of other evidence to suggest that when the British Army got the Lewis gun it felt it was not so much 'gaining an automatic rifle' as 'losing a superlative belt-fed medium machine gun'.

Automatic Weapons

This perception was mightily reinforced by the fact that the Lewis gun started to appear in large numbers at just the same time as machine gun enthusiasts were having the four Vickers guns per battalion removed to a specialist company under brigade control. This meant that the Vickers became successively more remote from the infantry line and infantry control, in very much the same way that artillery had become successively more centralised during the Napoleonic Wars, and again during the American Civil War, in the interests of developing its full tactical effect. Other analogies include the centralisation of tank forces during both the First and Second World Wars, or the successive removal of close support aircraft to more 'strategic' tasks. All these measures were carried through against the resistance of infantry who wanted these high-value weapons for their own immediate close support. In the Great War there was a similarly widespread suspicion that the new science of massed machine gun barrages was actually a massive waste of resources. Brigadier Croft of 9th Division was certain that the barrage lacked the accuracy and precision claimed for it,[53] while the historian of 2/RWF said in early March 1916 that the brigading of machine guns was an 'Insane Act'.[54] Edmonds, the BEF's official historian, wrote that:

> The Lewis gun was little more than a cumbrous, heavy and not too reliable automatic rifle — in fact, the firepower of infantry battalions and brigades had just been lessened by the reorganisation of the machine gun companies into divisional battalions.[55]

Throughout the war the infantry battalions were always scheming to get their hands on more Vickers guns for the front line, and by hook or by crook they often succeeded.[56]

Much of the trouble arose out of a confusion over just what a 'machine gun' was supposed to be. Before the war the tendency had been to classify all automatic weapons as 'artillery' pieces, to be used in battery for long-range barrage fire. The linkage with artillery was reinforced by the fact that almost all machine guns in 1914 were still carried on wheeled carriages, often horse-drawn and accompanied by traditional wheeled ammunition limbers. In the Boer War, moreover, Wolseley's concept of a heavy Pom Pom, firing 1-pound explosive shells of the minimum calibre allowed by the Geneva Convention, seemed to be a cheap — albeit not very successful — British imitation of the rapid-fire French '75' field gun. Admittedly the normal layout for a machine gun team was usually on a somewhat smaller scale than in the artillery — for example, the Belgians used a pair of large dogs to pull their Maxims, rather than horses — while the original wheeled Lewis gun trolley, first seen in the trenches in 1915, was designed for a single mule, but usually pulled by men on foot.[57] However, the general image was still that machine guns tended to be presented more as artillery weapons than infantry ones, and this impression was further deepened by the complex calculations and fire control apparatus required for the burgeoning science of barrage fire around 1917.

Against this perception, a minority of British officers who happened to be machine gun enthusiasts before the war could nevertheless point to a considerable body of experience in colonial campaigning, where machine guns had been used as purely infantry weapons. In a string of (spectacularly!) annihilating battles against the Zulus, the Dervishes, the Afghans, the Hausa[58] and even the Tibetans, British theorists of machine guns had come to understand that these were not artillery weapons at all, but should be deployed in very small numbers right up at the front, as an integral part of the infantry line. The debate raged over whether one, two or three Maxim guns made the optimum number for a section — and professional opinion eventually came to rest at two — but the whole thing eventually muddied the waters by leading the infantry to suppose that they always had a right to automatic guns in close support.

The recognition of machine guns as infantry weapons certainly represented progress of a sort; but after the first year of war the minds of machine gun officers had been seized by the still more radical suggestion that the Vickers should be taken out of infantry battalions just as much as it had already been taken out of artillery batteries — with the MGC being seen as a 'fourth arm' that could assist in the transactions of artillery, infantry and cavalry alike, but without having to identify itself too closely with any of them.

The mainstream of machine gun opinion also made an assumption that the only weapon worth considering *as* a machine gun was the incomparable Vickers, owing mainly to its capability for firing on fixed lines from a stable platform. The gun was thereby freed from the erratic vagaries of aiming that afflicted riflemen. Half a century of musketry fire-control training had failed to solve this problem for any infantry apart from the exceptionally well trained; but with a machine gun on fixed lines — 'a nerveless weapon', as J. F. C. Fuller apparently termed it — the fall of shot could be accurately and scientifically predicted. Smaller machine guns which were not fitted on tripod mounts, such as the Lewis,[59] possessed none of this higher science, and could be contemptuously dismissed as mere 'rifles' by the enthusiasts of the MGC. They were *automatic* rifles, to be sure, but rifles all the same, and hence considered as entirely infantry weapons. In the eyes of the machine gun purist, not even the 39 pound German 08/15 Maxim Light or General Purpose Machine Gun (LMG or GPMG) counted as a true machine gun at all. Although it was basically just the same weapon as the classic sled-mounted German 1908 Maxim, its lack of a fixed mounting disqualified it from serious consideration as a long range or indirect fire barrage weapon.

Despite repeated lectures by the MGC,[60] this almost theological distinction was generally lost on the infantry, who persisted in regarding any automatic weapon as a 'machine gun' and demanded simply that it should be as light and reliable as possible, and capable of direct fire, regardless of the technical nomenclature. At least at first, if not throughout the war, they made very little practical distinction between the Lewis and the Vickers in terms of how they should be used. Thus Vickers guns were often still pushed right up into the infantry firing line for short-range direct fire, and in one attack the Vickers section was even thrown forward to the enemy trenches and pinned down there all day, but scared to go back because of the fire of Lewis guns posted behind it.[61] Conversely, Lewis guns were often used very defensively, in the static firepower role generally attributed to the MMG, for example in the 2/RWF defence of Red Dragon crater in the spring of 1916.[62] Equally when it was raided near Potije, Ypres, in early 1917, Edmund Blunden's battalion reported that one of its Lewis gunners had fired heroically from the shoulder — although this was based on the sadly post mortem evidence of the gunner's cadaver.[63] Even in their more theoretically 'pure' rôle accompanying an attack, furthermore, Lewis guns tended to be used less offensively than defensively, either protecting an open flank[64] or consolidating the defence of an enemy trench that had just been captured.[65]

The Lewis gun was doubtless very useful in all these defensive rôles, especially once it had become more widely available to, and better understood by, the infantry. However, the real potential breakthrough in tactics that it represented lay more in providing covering fire within the front line of an attack, and suppressing enemy pillboxes while the assault troops moved forward. This idea may already be found in some of the tactical instructions for the Somme battle itself, which call for Lewis guns to go out into No Man's Land ahead of zero, to rake the enemy's parapets and snuff out his MGs during that fatally awkward pause between the lifting of the British barrage and the arrival of the British infantry. This was an eminently sensible suggestion, and one wonders how the massacres of 1 July could ever have happened if only it had been properly put into effect.[66]

The general level of understanding did nevertheless quickly improve, and Charles Carrington was able to claim that good divisions had already worked out effective Lewis gun tactics by the end of the Somme battle, even without the help of high command directives. He said that he understood the system better than the official tactical schools, and that official codifications even came rather late into the field.[67] On 30 December 1916 the US military press also reported, albeit with evident self-interest, that:

> The merits of the Lewis gun are long past any discussion in British army circles after so many months of testing. It has also added a

much-needed redeeming quality to the fame of American products in general.[68]

In the specific case of tackling pillboxes, the SS 143 manual of February 1917 was already urging that the platoon Lewis be used as a major source of covering fire, and there was also apparently a very impressive high command recommendation of around November–December 1917 for Lewis guns to be used as pillbox busters.[69] In the event they were undoubtedly often used in this rôle in 1917 and 1918, alongside trench-mortars or volleys of rifle-grenades,[70] and it is certainly misleading to suggest that they were ineffective because they lacked long range, or that the BEF never properly mastered their use.[71]

It is also possible to find the Lewis gun being used as a mobile centre of firepower during the attack, being fired from the hip as recommended in the 1918 manual.[72] On 20 November 1917 the 36th Division at Cambrai covered some trench-clearing from a mound with its Lewis guns, but also sent in 'heavyweights' who fired from the hip alongside infantry armed with Thermit bombs.[73] Equally the Duke of Wellington's Regiment in the 62nd Division is reported to have fired from the hip in street fighting at Anneaux on the same day.[74] Brigadier Jack further describes an assault at Ypres in July 1917 that was beefed up with extra Lewis guns hand-held by two men, although this system was described as lacking in accuracy.[75]

Despite these instances of the Lewis being used in a blatantly offensive rôle, and the infantry's eventual appreciation of what it could offer, there was often a reluctance to accept the full MGC dogma concerning the proper relationship between the AR and the MMG. This would perhaps have been easier if the Lewis had been still lighter than it was, and the Vickers heavier, making a clearer distinction between the two. The Germans, by contrast, suffered from no such problem, since both their medium and 'light' machine guns were even heavier than the Vickers, and very cumbersome. Some British infantry officers were specifically contemptuous of the MG 08/15's weight, saying 'The light automatic [the Lewis], we thought, was the weapon of the future and the heavy machine gun, to which the Germans were faithful, ceased to intimidate us'.[76] This led the Germans to make widespread use of rechambered captured ARs — first the Madsen, then the Lewis[77] — and finally in 1918 to adopt small numbers of 'sub' machine guns, in the shape of the Bergmann. The British never took this last step during the Great War, despite a number of trials with equivalent guns, mainly because the Lewis had finally been accepted by the infantry as the excellent weapon it was.

8

Artillery

> Our artillery was extraordinarily good and nothing could possibly live under its fire. The ammunition is very different now to what it was on the Somme.
>
> *The Diaries of Lord Moyne,* p.157, referring to the capture of Messines, 1917

The evolution of precision munitions

By far the most effective set of new technologies in the Great War lay in the development of artillery. A number of highly significant innovations had been successively introduced from the 1880s onwards, and they were joined by others as the war progressed. The guns could already inflict enormous damage on infantry as early as Mons and the Bataille des Frontières in August 1914; yet it is easy for us to forget that their full potential came to be realised only around the second half of 1917.

The primary element in the original late nineteenth-century 'artillery revolution' was the appearance of a comprehensive new generation of powders.[1] This extended the range and power of the guns when new low explosives were used as a propellant, at the same time as it greatly enhanced the battering effect of their projectiles when high explosive was used as a shell filling. Whereas the new powders had brought only relatively straightforward increments to the performance of small arms, with low explosive propellants, this double application to the heavier weaponry quickly produced a geometrical increase in the overall effectiveness of artillery.

The increase in artillery range was dramatically greater than the equivalent increase for small arms. Whereas the normal use of rifles in the

American Civil War had averaged around 127 yards, their use in the Great War had probably not greatly increased, even though their rate of fire certainly had.[2] In the American Civil War the normal use of artillery, by contrast, had been with direct fire at somewhat less than 1,000 yards; but the guns of the Great War habitually used indirect fire from camouflaged positions at ranges between 5,000 and 10,000 yards (see Table 5).

Table 5: British artillery weapons

Sources: Gander and Chamberlain, *British Artillery Weapons*: I. V. Hogg and L. F. Thurston, *British Artillery Weapons and Ammunition, 1914–18* (Ian Allen, London 1972)

Weapon	Type	Level held	Shell weight (lbs)	Max. range (yds)
13 pdr	Horse artillery	Cavalry Div	12.5	5,900
18 pdr	Field gun	Infantry Div	18.5	6,525
4.5"	Field howitzer	Infantry Div	35	7,300
60 pdr	Heavy field gun	Corps	60	12,300
6"	Siege howitzer	Army RGA	112.5–118.5	5,200 (travelling) 7,000 (siege mount)
6"	Siege gun	Army RGA	100	14,200
8"	Siege howitzer	Army RGA	200	10,500
9.2"	Siege howitzer	Army RGA	290	13,935
9.2"	Railway siege gun	Army RGA	380	21,000
12"	Siege howitzer	Army RGA	750	11,340
12"	Railway howitzer	Army RGA	750	15,000
15"	Railway howitzer	Army RGA	?	?
18"	Railway howitzer	Army RGA	2,500	22,300

The guns would thus be kept safely out of the direct infantry battle, which they could mightily affect without unduly exposing themselves to counterbattery retaliation. This factor was of great importance in extending the power of a defence, since the loss of a defender's front-line trenches did not now imply the destruction of his main source of firepower. The guns could sit back and torment the attackers at every stage of their operation, even though their own front line had been lost. In these circumstances counterbattery fire programmes quickly took on an especially high priority, although before the end of 1917 their techniques often tended to lag behind the protective measures that could be taken by a resourceful opponent.

Another effect of the new powders was that the concussions of the exploding HE shells inflicted a dire effect on the nervous system, and it was no accident that the Great War term for combat stress became standardised as 'shell shock':

> Bursting shells do seem to have a curious effect on people. Some are sort of mentally paralysed, and some even physically so. I've seen a man alive and unhurt, but absolutely incapable of lifting a hand or of speaking. It's curious.[3]

However, the full potential impact of the artillery revolution had been understood only very imperfectly before the war, even after the experiences of South Africa and Manchuria. The remarkable French failure to grasp the potential of guns heavier than 75 mm calibre (i.e. the famous soixante quinze, unveiled in 1897)[4] is only the most blatant example of a general myopia that affected every European army to a greater or lesser degree.

A part of the problem was that new powder was by no means the only artillery innovation of the immediately prewar era. The soixante quinze herself incorporated another very important one, namely a sophisticated hydraulic recoil system which permitted the gun to remain laid very accurately upon the same target throughout the firing sequence. Although the barrel would move sharply backwards under the pressure of each departing shell, the gun carriage — and hence the weapon's seating on the ground — did not budge. Combined with composite cartridges and an easily manipulated breechblock, this arrangement meant that a previously unimaginably high rate of fire could be achieved without loss of aim. Not only did this obviously imply that a single gun could henceforth do the work of a whole battery, but it rather less obviously meant that the target need no longer be kept under absolutely continuous observation. If you could hit it once, you could keep on hitting it: nor did you need to expose your gunners to the enemy's return fire. An observer in a forward trench — or even in a balloon or an aeroplane — could do the job as well as was necessary.[5] By 1917 techniques had even been refined whereby the fire could be very precisely predicted merely from the map or an air photograph, without any observer at all, and without preregistration. Some of the most impressive technical achievements of the war included the comprehensive survey and regular aerial photographing of the entire battle zone, linked to vastly improved flash and sound location of enemy guns. There was also a quite new emphasis on detailed analysis of weather conditions, the firing characteristics of each individual gun and each individual batch of shells. Taken together, these innovations gave gunners confidence that they could accurately identify the positions of their own guns and of any target they chose to engage, at the same time as they could predict the exact way their shells would fly.[6] The time-honoured urgency to achieve a direct line of sight from gun to target throughout the firing process was thus quietly but very completely removed.

The possibilities of heavy and accurate indirect fire struck deeper into the prewar assumptions of tacticians than anyone fully realised at the time, and we are strongly reminded of the electronic battlefield revolution of the 1970s and 1980s, when the slogan was 'If it's there we can see it, if we can see it we can hit it, and if we can hit it we can destroy it'.[7] In theory, at least, this formula was almost true already in 1917, when the standard 18-pounder field gun had achieved an accuracy in predicted shooting of 80 yards at a range of 4,000 yards.[8] The implication was that both infantry and gunners had to go to great lengths not only to camouflage and conceal themselves against enemy aerial and other observation, but also to exploit dead ground at every opportunity and to dig deep into the earth for physical protection. The armoured prewar gun shields, optimistically provided to protect dashing horse artillery charges from frontal rifle fire, turned out to be very inadequate as cover against heavy shells falling ballistically from above. It was found that true security could be gained only by putting a very great distance between yourself and the enemy's guns, or by having either a heavy slab of concrete overhead, or at least six feet of earth — and preferably fifteen feet.[9]

It is not generally appreciated that the greatly improved accuracy of guns, survey and target acquisition systems during the Great War was matched by a significant corresponding increase in the complexity of the shells themselves. Whereas a Napoleonic shell had been within the manufacturing capability of any village blacksmith with a pyrotechnical imagination, the 1914–18 version required advanced metallurgical and chemical skills, combined with extensive industrial plant. Munitions manufacturing had become a highly specialised occupation, in which even the slightest slip was potentially fatal. To continue the analogy with more recent precision-guided munitions, a 1917 artillery shell represented the equivalent, in terms of the technology of the day, of a Walleye Smart bomb, a Maverick or even a Cruise missile. Fine tolerances and sophisticated 'software' in the fusing were essential if the finished product was to do the job for which it was intended.

In Napoleonic times the shrapnel shell had been seen as a relatively rarefied and complex piece of equipment, whose mechanism was a state secret, whereas the simple fragmentation shell had been much easier to produce. By 1914, however, that order of precedence had been reversed. Shrapnel had then become easier to manufacture than HE, because it was both metallurgically and chemically more straightforward, including its fusing and initiation mechanisms.[10] Indeed, the first HE shells were tested for the 18-pounder field gun only in October 1914,[11] and they continued to be in short supply for over a year thereafter. Even at the end of 1915 it was found that a backlog of thousands of HE shell casings were being stacked unfilled in the factories, since filling them was such a complex undertaking. But at least the relative ease of manufacturing shrapnel provided a quick way out of the shell shortage,[12] even if many unreliable shells continued to be delivered to the guns up to around the end of 1916 (see Table 6).

Table 6: Total British shell production per month
(Approximations taken from *Abstract of Statistics*, pp.446 ff.)

Date	Shrapnel	HE, Smoke, Chemical
March '16:	952,708	818,932
Sept '16:	1,885,234	3,279,776
Jan '17:	2,868,645	4,129,945
June '17:	2,936,406	5,604,791
Dec '17:	1,113,266	3,023,840
June '18:	2,709,888	5,302,103
Oct '18:	2,890,030	6,367,528

Overall total = about 6m shells per month for much of the time.[13]
Peak totals = 8,121,026 in May '17; 8,541,197 in June '17; 9,012,314 in May '18; 8,001,991 in June '18; 9,257,558 in Oct '18.

Shrapnel was also found to be the best available means of cutting barbed wire, which was often a *sine qua non* for infantry attacks. The importance of this factor had already been fully understood during the winter of 1914–15, but the evolution of reliable wire-cutting techniques came only slowly. Tests were made with machine guns, but they were unsatisfactory.[14] More heroic solutions such as Bangalore torpedoes[15] or hand-held wire-cutters could sometimes work, but only for as long as the enemy was prevented from shooting the men using them. Before 1917, moreover, the fuses available on HE shells delayed detonation for too long to cut wire effectively, so the projectile buried itself in the earth and the force of its explosion was channelled directly upwards. That left only artillery shrapnel or trench-mortar bombs. Both of these could cut the wire well enough, although progress always still had to be constantly monitored by personal reconnaissances, which might prove dangerous and, perhaps more importantly, time-consuming — thereby sacrificing the element of surprise. Even then it might not be totally successful.

These difficulties in wire-cutting were highlighted in the Somme battle, where even the best-prepared attacks still often encountered uncut wire. By the end of 1917, however, succour had come from two directions, the tank and the number 106 fuse. In the case of the tank, the wire-cutting was instantaneous, and could therefore achieve surprise by dispensing with any preliminary bombardment. The infantry had only to stand far enough away to avoid the whiplash of the wire strands as they broke, before picking its way through the gap to assail the enemy.[16] The main problem was that tanks were few and far between, and would usually be available only in small numbers if they were available at all.

For most 'everyday' operations it was the 106 percussion fuse for HE shells which provided the best means of wire-cutting. This detonated the shell immediately it impacted with the ground, before it had dug itself into the earth. The blast therefore went outwards horizontally, rather than upwards vertically, proving to be better than shrapnel at breaking wire barriers. It still required a certain preliminary bombardment before the job could be completed, but its duration could now be reduced from days to hours. A still more useful effect of the 106 fuse was that it permitted HE barrages to be fired which did not also make craters that would impede the infantry's advances,[17] and did not produce a 'backsplash' which deterred the infantry from hugging the barrage.[18] The gradual arrival of 106 fuses during the course of 1917 was therefore universally hailed as a major advance over the bad old reliance upon shrapnel, and 'As a man-killer it was incomparable'.[19] Apart from anything else, shrapnel had been entirely ineffective against fortifications, whether to destroy them or merely to terrify their occupants, whereas any type of HE represented a considerable improvement. A number of forward-looking tacticians had already started to insist on less shrapnel and more HE by mid-1916; and by mid-1917 the increased availability of reliable HE shells made this trend general throughout the army.[20]

As for more specialised smoke or gas shells, they also appeared relatively belatedly owing to technical difficulties in production, and they were only hesitatingly adopted by battle commanders who often lacked trust in such scientific wizardry. Both types were available to the BEF in very small numbers towards the end of the Somme battle, but it took most of 1917 for them to become widespread.[21] As a delivery system they produced a very inferior volume and density of chemicals as compared with projector or cloud attack; but they did have advantages in pinpoint accuracy at all ranges, and were relatively immune to fluctuations in the wind. Thus smoke shell could be added to a creeping barrage to help hide the infantry advancing behind it, or it could be concentrated very precisely on a flank to blind an overwatching machine gun. Equally in the case of gas shell, the enemy's rearward artillery lines could be neutralised while his front-line positions were left free of the poison, and therefore accessible to attack by friendly infantry. By 1918 both smoke and gas shells were being very extensively exploited by tacticians, and it is likely that they would have occupied an even greater proportion of the shells fired in 1919, if the war had continued that long.[22]

The increasing complexity of ammunition may be traced by a look at the changing composition of creeping barrages as the war progressed. At first they were entirely shrapnel until the canny gunners at the head of 9th (Scottish) Division[23] saw the advantages of HE for their attack of 14 July 1916, and set a trend which was eventually widely followed. By the time of Arras they had added smoke to the HE,[24] and continued to experiment with this medium throughout the rest of the war. By the end of 1917 it was normal for a creeping barrage to contain roughly one third shrapnel, one third HE and one third — or usually rather less — smoke (see Table 7).

Table 7: Proportions of shell types in selected creeping barrages

(Sources: Ewing, *History of the 9th (Scottish) Division*; Farndale, *History of the Royal Regiment*; I. V. Hogg, *The Guns*; OH 1916, vol. 2; 1917, vols 1 and 2; 1918, vols 4 and 5)

Place & Date	Attack troops	Shrapnel %	HE %	Smoke %	Comments
Montauban 1 July '16	7th & 18th Divs	100	0	0	Almost the first creeping barrage, and one of the most celebrated
Longueval 14 July '16	9th Div	0	100	0	Time-delayed no. 101 fuses, to avoid tree-bursts. The 1st creeper with HE
Arras 9 April '17	3rd Army	50	50	0	The start of each phase marked by salvoes of 100% shrapnel
Pointe du Jour 9 April '17	9th Div	0	75	25	The first creeper with smoke
Menin Road 20 Sept '17	2nd Army	50	50	0	75% of the HE to have no.106 instant-percussion fuse
Cambrai 20 Nov '17	4th Army	33.3	33.3	33.3	The first predicted barrage (albeit 'lifting' not creeping). New phases marked by salvoes of 100% smoke

Meteren 19 July '18	9th Div	0	75	25	% of smoke left variable up to last moment, as per wind speed
Meteren 18 Aug '18	9th Div	0	0	100	Smoke reduces to 25% after 1st minute
Meaulte 22 Aug '18	12th Div	0	93	7	Successful
Chuignolles 23 Aug '18	Australian 1st Div	45	45	10	Successful
Gomiecourt 23 Aug '18	3rd Div	75	25	0	Successful
Boyelles 23 Aug '18	VI Corps	84	0	16	Generally successful — mixed result
Drocourt-Quéant 2 Sept '18	XVII Corps	45	40	15	Successful
Hindenburg Line 27 Sept '18	Canadian Corps	50	40	10	Very successful
Riqueval 11 Oct '18	6th Div	?	?	25	Enemy alert at 4.30 pm: Failure

The shift from destructive to neutralising fire

The creeping barrage was a vital innovation in Great War tactics because it represented a decisive shift from 'destructive' fire to 'neutralising' fire. Its aim was not to cut wire, destroy batteries or collapse enemy dug-outs, but simply to persuade the enemy's infantry to cower in the bottom of its trenches, and stay there, for the duration of the attack. The defenders had to be blinded, dazzled and deafened rather than killed. Nor, if the aim was to achieve surprise, did this need to be continued for any great length of time. Provided it could be maintained for the crucial few minutes during which the attackers were in the open and vulnerable, and provided it covered the whole of the crucial hinterland from which the defender might bring weapons to bear, an assault force could hope to move forward without encountering effective opposition on the ground. In 1915 it had been almost exclusively 'destructive' or 'killing' fire that was being envisaged by tacticians; but in the course of 1916 the essential rôle of 'neutralising' began to be better understood, and this would eventually add much greater value to attacks while obviating the need for lengthy and laborious 'destruction'.

Apparently the very first creeper was used by the 15th Division at Loos in 1915, as part of a general development of various types of 'lifting' barrages at around that time.[25] However, the technique was still felt to be very novel on 1 July 1916, leading to a number of local variants all along the front, few of which could really be counted as genuine creepers. Yet the successful applications around Montauban by XIII and XV Corps did seem to fill the bill pretty closely, and within a few days the idea had caught on famously. It soon spread across national boundaries even as far as the French Army,[26] and by September the creeping barrage had become an accepted commonplace of all tactics. Any 'debate' about it sometimes almost seemed to have degenerated into merely a matter of generals repeating exhortations to their doubting infantry that they should hug their barrage ever more closely. Given the generally poor responsiveness and signals intercommunication with the artillery at this time, however, there was probably no realistic alternative to the formal barrage. Modern complaints that concentrations should have been substituted for barrages may therefore be suspected of a touch of anachronism.[27]

Different individual commanders would continue to tinker with their own personal improvements to the creeping barrage right through to the end of the war, but a number of salient common features were quickly established as characteristic of the 'classic' creeper. The general idea was that within each phase of an advance a wall of bursting shells would be established just ahead of the attacking infantry and then lift regularly, at so many yards every few minutes, with the assault troops rushing forward as closely behind the shells as they could manage with safety. Indeed, the ideal would be for the infantry to go faster even than that, since it soon became accepted that it was better to lose a few casualties to a friendly barrage than to lag so far behind that the enemy was allowed time to reman his parapets and reactivate his machine guns. Once the attacking infantry reached its immediate objective, however, there would be a planned pause for consolidation and mopping up, with the wall of shells halting just beyond them to form a protective screen.[28] The advance could then be resumed for the next phase at a predesignated time.

The wall of shells would itself consist of a deep moving zone of several lines of explosions; perhaps as many as seven lines in all, with a total depth of up to 2,000 yards. The lines nearest to the attacking infantry would be laid on by 18-pounder field guns and 4.5-inch howitzers, those at greater depth by medium or even heavy guns and howitzers. At least in the initial phases, there might well be additional barrage fire from mortars and/or machine guns, to complicate the enemy's difficulties still further. Some of the deeper lines of fire might also be moved backwards towards the attacking infantry, to break up counterattacks, even while the shallower lines were still advancing. Such a 'back-barrage' would further add to the uncertainty in the enemy's mind, and help persuade him — or to 'drill' him, as Plumer would have it[29] — to keep his head well down. Clearly the effective coordination of all these diverse elements into a single fireplan made for a very ambitious administrative operation, and it is no accident that it has often been likened to the orchestration of a major musical score.[30]

The timing and speed of barrages were matters of constant debate, but a few examples from the innovative 9th Division may help to illuminate some of the chief issues (see Table 8). The record shows that an advance of 100 yards every two minutes was quite possible provided the going was not too heavy, the artillery was sufficiently numerous and the troops were properly prepared; but if those conditions did not apply, it could easily lead to failure. Where the mud was particularly thick, as in the Passchendaele battle of October 1917, not even a slow advance of 100 yards in eight minutes could guarantee success.

Table 8: Speed of creeping barrages in some 9th Division attacks

(Speeds expressed as average number of minutes per 100 yards advanced. Sources: Ewing, *History of the 9th [Scottish] Division*; OH 1918, vol. 3)

Place & Date	Zero hour	Speed in phase 1	2	3	Comments
Longueval 14 July '16	3.25 am	3	3		Very successful, due to complex forming-up in No Man's Land during the night
Warlencourt 11 Oct '16	2.05 pm	2	2		Poor knowledge of enemy positions/ too little preparation. Disaster despite a 'Chinese' barrage
Pointe du Jour 9 April '17	5.30 am	3	4		Thought too slow, but very successful. Includes an MG barrage (which became standard thereafter)
Greenland Hill 12 April '17	5.0 pm	2	2		Infantry lost barrage: barrage badly placed
Gavrelle Road 3 May '17	3.45 am	2	2		Badly prepared. Barrage lost
Greenland Hill 5 June '17	8.0 pm	2	2		Includes MG and Stokes fire, and a 'Chinese' attack; success achieved
Frezenburg 20 Sept '17	5.0 am	4	6	8	Multi-layer barrage, very successful even though mud slows down everything
Lekkerboterbeek 12 Oct '17	5.35 am	8	2		Very thin barrage. Mud kills the attack within 100 yards
Meteren 19 July '18	7.55 am	2	2		Very successful surprise attack

Perhaps the most nerve-racking question for an attacker would be his attempt to preserve the element of secrecy and surprise for zero hour.[31] This particular item was clearly crucial to any attack; but there remained many different ways in which it might be approached. Most obviously, perhaps, it might be done by choosing a moment for zero that was outside the classic 'stand to' moments of dusk and dawn. The darkest parts of the nocturnal wee small hours were especially recommended as candidates for 'alternative zeros'; but it was later found that mid-morning or mid-afternoon could also sometimes prove to be even more surprising to routine-bound trench-dwellers, who tended to catch most of their sleep at those times.

Secondly, surprise might be enhanced by avoiding an obvious build-up to a crescendo in the preliminary counterbattery or wire-cutting bombardments. Such a build-up had been a common feature of the early creepers on the Somme; but by the time of Arras in April 1917 it was thought to give too much warning to the enemy, and so it was generally discontinued. The logical next step, however, would have been to do without any preliminary 'destructive' fire at all, or even to start the infantry assault before the creeping barrage opened;[32] but such a step was highly controversial. It was quite rightly pointed out that infantry attacks could be hung up only too easily on uncut wire, and could only too easily prove suicidal if the defender's batteries had not been effectively countered. Many tacticians were therefore desperately keen to cling on to at least a little preliminary destructive fire, however veiled or furtive it might have to be. They found it very difficult to put their entire faith in a comprehensive programme of predicted, neutralising fire, especially when it came to counterbattery action. Tanks might admittedly make an adequate substitute for artillery in opening gaps in enemy wire, but the tankies were if anything even more anxious than the infantry to see enemy field guns suppressed before they ventured forward.[33]

A compromise was found in many of the battles of 1917, whereby heavy preliminary destructive fire was continued, but the timing of the assault was concealed behind a series of false starts to the creeper. Such 'Chinese' attacks had already been used at Festubert in May 1915 and were well known in 1916. They had proliferated on 1 July, mainly with smoke candles mixed with cloud gas to confuse the Germans as to the true time and place of zero, without exposing the assault troops prematurely.[34] Something similar continued on and off throughout the Somme battle, until the principle was extended to very extensive artillery bombardments, sometimes mixed with the display of dummy men to suggest that the assault was already happening.[35] Nevertheless, this was all essentially a halfway stage towards the concept of the entirely unprepared attack, which became accepted as soon as the technical practicality of accurate predicted fire had been convincingly demonstrated. This happened at Cambrai on 20 November 1917, when hundreds of tanks were available to help in wire-crushing. However, the true surprise rested less in the tanks than in the silent assembly of masses of artillery which opened fire only at zero. General Tudor of the 9th Division had personally insisted on this innovation, which may fairly be claimed as an even greater technical breakthrough than the tank, and an idea which the Germans never mastered, even in their spring 1918 offensives.[36]

Table 9: Speed of some creeping barrages in the Hundred Days

(Speeds expressed as average number of minutes per 100 yards advanced. Sources: OH 1916, vol. 2; 1917, vol. 1; 1918 vols 4 and 5)

Place & Date	Zero hour	Attack troops	Speed in phase* 1	2	3	Comments
Amiens 8 Aug '18	4.20 am	Canadian Corps	3	5	8	4 minutes per lift in final phase. Very successful attack.
Ditto	4.20 am	III Corps	4	4		Infantry loses barrage in mist
Meaulte 22 Aug '18	4.45 am	5th Royal Berkshire	2	2		Success (elsewhere that day 4 minutes per lift, but less success in mist)
Béhagnies 23 Aug '18	11.0 am	2nd Div	80 seconds			Intentionally fast, to accommodate the Whippet tanks; but too slow for them, and too fast for the Infantry!
Bapeaume 29 Aug '18	5.30 pm	5th Duke of Wellington's	4	4		Great success 'with the bayonet'
Beaulencourt 1 Sept '18	2.0 am	110th Bde	6	5		Very slow in the dark — but v. successful
Hindenburg Line 28 Sept '18	2.30 am	IV Corps	5	5		Surprise achieved in murky rainy night
St Quentin Canal 29 Sept '18	5.50 am	46th Div	2	2		Very successful canal bridging attack, in mist
Gonnelieu 29th Sept '18	3.30 am	21st Div	6	6		Failure: barrage too slow
Selle 23 Oct '18	1.20 am	XIII Corps	4	4		6 minutes per lift in woods. An unlikely success in unpromising conditions

*Note 'Phase 3' was often out of range of the creeping barrage, which was primarily the responsibility of field artillery. Phase 3 was also planned to be much slower than earlier phases.

(Note that the sample shows some of the more interesting creepers: not a representative sample of successes and failures. The wide variety of start times is, however, pretty representative)

A similar use of predicted fire was often used in the BEF offensive during the last three months of the war, most notably at Amiens on 8 August.[37] All arms had by this time mastered the various techniques associated with the creeping barrage, and could often organise one at pretty short notice. An analysis of some of the barrages fired in this period shows a high rate of success despite a generally slower pace than the 100 yards in two minutes that had been favoured by the 9th Division (see Table 9). As deadlocked static warfare gave way to 'semi-open' or even 'open' operations, however, the creeping barrage was often dispensed with, in favour of lifts or concentrations of fire on specific targets. Although certain divisions clung to them religiously, in general relatively few true creeping barrages were fired once the Hindenburg Line had been breached towards the end of September.

Apart from the appearance of improved munitions and barrage techniques, the general threat from artillery was multiplied still further by the simple fact that the gunners, the guns and the shells themselves inexorably multiplied in numbers as the war progressed. By the end of hostilities there were more than half a million gunners in British service worldwide, or around one in seven of all British troops, including some thirty or forty per cent of the BEF.[38] In some cases there was 'a squad of mathematicians around every gun, doing sums, reading graphs, setting sights, calculating fuse lengths and getting in each other's way'.[39] The average number of guns available to support each BEF division rose from an average of sixty-eight in August 1914 to about 100 in November 1918.[40] The latter figure would usually be higher still in practice, since divisions were by then often leaving their own guns in action for about twice as long as their infantry, firing in support of some other division's infantry. Still more important, the weight and size of the guns available was steadily rising. Only about 5 per cent of the BEF's guns of August 1914 were held at command levels higher than the division, and these were all 6-inch howitzers classed as 'medium' rather than 'heavy' weapons. By the end of the war, however, the number of tubes held within each infantry division had actually fallen from seventy-six to forty-eight; but there had been a terrific proliferation of army or corps artilleries, including all calibres right up to 12-inch, 14-inch and 15-inch heavy pieces, which more than compensated for this shortfall.[41] Overall, no less than thirty-five per cent of the BEF's guns were classed as either 'heavy' or 'medium' by November 1918,[42] although perhaps surprisingly the equivalent figure for the French, who had started the war almost entirely devoid of heavier guns or howitzers in the field army, had risen decisively to fifty per cent.

As for the expenditure of shells, it varied enormously according to the tactical situation, with extensive hoarding during 'quiet' periods[43] interspersed with frenetically massed bombardments during decisive moments. Sometimes the entire British Empire shell production for a month might be fired off on a single attack frontage during a single week (see Table 10).

Table 10: Expenditure of artillery ammunition, by weeks

(Western Front only. Source: *Abstract of Statistics*, pp.408 ff.)

Date	Shrapnel	HE (incl. smoke, chemical etc.)
1916		
Low weeks around Jan '16 — averaging around	55,000	50–60,000
Rising in spring to around	80,000	50–60,000
Unchanging till late May	c.100,000	c.120,000
Week to 25 June '16	245,361	160,845
to 2 July (start of Somme battle)	1,208,575	1,177,396
Thereafter averages through autumn	c.5–600,000	600,000
Late Nov	c.100,000	300,000
1917		
Jan onwards	c.140,000	390,000
March about	250,000	340,000
Week to 8 April (start of Arras)	767,630	1,376,013
to 15 April	1,055,295	1,691,631
to 22 April	340,874	653,089
to 29 April	920,093	1,490,939
to 6 May	824,791	1,329,542
Thereafter about	333,000	800,000
to 3 June (start of Messines)	542,313	1,025,112
to 10 June	1,386,852	1,870,963
to 17 June	618,059	956,836
Thereafter about	300,000	500,000
to 29 July (start of 3rd Ypres)	627,429	1,403,416
to 5 Aug	1,209,486	1,572,470

to 12 Aug	631,217	1,185,808
to 19 Aug	1,029,886	1,658,484
to 26 Aug	716,474	1,330,721
Thereafter about	360,000	630,000
to 2 Sept (resumption of 3rd Ypres)	1,245,444	2,033,832
Thereafter about	800,000	1,200,000
Start Nov–9 Dec about	500,000	840,000
Late Dec/early Jan about	130,000	350,000
1918		
Mid Jan–end Feb about	70,000	210,000
March rising gradually	108–315,000	291–733,000
3 weeks 18 March–7 April (German offensive)	n/a	n/a
to 14 Apr (Battle of the Lys)	518,519	1,120,286
to 21 Apr	819,688	1,263,782
to 28 Apr	940,926	1,209,246
Thereafter about	450,000	800,000
(Exception: week to 11 Aug – Battle of Amiens)	674,665	1,257,212
18 Aug–1 Sept about	1,100,000	1,810,000
Thereafter very variable	425–812,000	656–1,459,449
(Exception: week to 29 Sept – Hindenburg Line)	1,220,948	2,241,815

As the war progressed, tacticians began to understand that a definite number of guns and shells would be needed to neutralise or destroy any given length of the enemy's trenches, and that this number could be discovered by a direct mathematical formula. Trial and error showed that there should be less than ten yards of friendly frontage for every gun, and 400 pounds of shell — or more — for every yard of enemy trench within the area to be attacked (see Table 11). At Neuve Chapelle in March 1915 this recipe had been discovered accidentally, but it was not then fully recognised for what it was. Densities were therefore allowed to fall short of what was needed in most of the subsequent battles during 1915–16.[44] The attack at Loos was particularly underprovided, although the first British use of gas did go some way towards making up the shortfall. The first day of the Somme was still less successful along much of the front, and in many of the later Somme actions a greater density of British artillery tended to be counterbalanced by the shortness of the attack frontage, which allowed the Germans to make a big artillery concentration of their own. Where there were successes, they tended to be accidental — or rather the result of instinctive overprovisioning by experienced natural tacticians.

Table 11: Yards of front per gun on 'first days' of battle

(Sources: Prior and Wilson, Farndale, Bailey, OH 1916, vol. 2, p.293; 1917, vol. 2, pp.138, 240)

Battle	Yards of British front per gun	Comments
Neuve Chapelle, 10 March 1915	6	Success (at first)
Aubers Ridge, 9 May	8	Failure
Loos, 25 Sept	141	Lack of guns supplemented by gas/smoke. Mostly a failure
Somme 1 July 1916	15	Mostly a failure
Somme 14 July (Longueval)	6	Big success
Somme 15 Sept (Flers)	8	Mostly a failure
Somme 26 Sept (Thiepval)	7.5	Success
Somme 3 Nov (Regina-Beaumont Hamel)	12	Some success
Arras/Vimy 9 April 1917	8–9	Big success (barrage density calculated)
2nd Bullecourt 3 May	7	Failure
Messines 7 June	7.5	Big success (all densities calculated)
3rd Ypres 31 July	8.5	Limited success; difficulty in CB
3rd Ypres 20 Sept (Menin Road)	5	Big success
3rd Ypres 4 Oct (Broodseinde)	5	Big success
Passchendaele, attacks later in Oct	5	Big frustrations, mud
Cambrai 20 Nov	10	Big success
21 March 1918 (Fifth Army)	60	Catastrophe in mist (but in defence)
(Third Army)	44	Failure (also in defence)
Hamel 4 July	15	Big success
Meteren 19 July	7	Big success
Amiens 9 Aug	35	Big success
Albert 23 Aug	35	Success
Hindenburg Line 28 Sept	12–21	Success

Note: Prior and Wilson add figures for how many pounds of shell were used against each yard of German trench (i.e. not the same as frontage):

Somme 1 July 16	132 lbs	Mostly a failure
Somme 14 July (Longueval)	660	Big success
Somme 15 Sept (Flers)	280	Mostly a failure
Somme 25 Sept (Morval)	400	Big success

Obviously a crude calculation of how many tubes were needed per mile of British attack front could not in itself be the complete answer, since there were many failures below the 'ten yard' threshold and some striking successes above it. Not even the additional refinement of calculating weight of shell per yard of enemy trench could absolutely guarantee success, as some modern authorities might appear to suggest.[45] This was because many other crucial variables also had to be taken into account, such as accuracy, the effectiveness of reconnaissance, deception, counter-battery and wire-cutting measures, or perhaps just basic matters of command and control. A good case could even be made for saying that the success or failure of attacks depended not upon the level of artillery support at all, but upon the degree to which the signalling arrangements had been properly laid out in advance.[46] Yet again, the effectiveness of a given number of guns was rarely anything like as high on the second and subsequent days of an attack as it was on the first, since unforeseen circumstances and last-minute improvisations would almost certainly interfere with effective fire. It gradually but inexorably sank into the minds of senior commanders that it was better to close a battle down quickly than to drag it out for many days after the first success.[47]

Having made all these caveats, it nevertheless remained true that a careful calculation of gun density as against friendly and enemy frontages could normally give planners a welcome additional margin of certainty. In the course of 1917 it therefore gradually began to be practised on a routine and formal basis throughout the army. At Arras in April it was applied to the creeping barrage, and by Messines in June it had been extended to the preliminary bombardments to destroy both fortifications and enemy artillery.[48] Other aspects of the gunner's art also began to be fed into the general planning on a more systematic basis, most notably the need to 'read' all terrain from the point of view of artillery observation.[49] It is only a pity that this consideration failed to prevent the choice of the badly overlooked Passchendaele battlefield in the first place.

In many respects the welcome developments during 1917 can be traced to the increasing hierarchical authority of each corps' senior gunner, the BGRA. In 1914 he had been only an adviser and not a commander, and had exercised no influence whatever over the heavy Royal Garrison Artillery (RGA) which might be supporting the corps' operations but which was commanded by higher HQs. During 1915 some corps started to give a distinct 'command' responsibility to their BGRAs, with attempts to coordinate all artillery action within the corps and to attach liaison officers to each infantry brigade HQ. Eventually the BGRA was formally recognised as the artillery commander (GOCRA) in every corps, although he was never given the rank of major general which would have automatically guaranteed him true equality with the infantry.[50] At the start of the Somme battle, especially, it was observed that each corps GOCRA was persuaded by his infantry commander to adopt a different local interpretation of Fourth Army's centrally issued artillery plan,[51] and some similar cases were still being noticed for a year thereafter.[52]

By the end of the Somme battle, however, there were distinct signs that matters were starting to improve. Each corps' GOCRA at last managed to establish a practical supremacy over his confrère from the heavy artillery, and this was soon formally recognised in official regulations. New counterbattery intelligence staffs were also established to coordinate all intelligence and planning for long-range fire, at both corps and army levels, and these were quickly completed by the addition of central artillery signals report centres.[53] These centralising arrangements laid the foundations for considerable technical advances in all fields of gunnery during 1917 and even, paradoxically, provided the background for a rational increase in flexibility and decentralisation during the fast-moving battles of 1918. By establishing clear channels of command and information flow, they allowed the most effective use of resources. Thus in the Hundred Days individual brigades were often able to arrange their own fireplans at short notice, and even individual battalions started to have sections of eighteen-pounders attached to them for direct fire and anti-tank tasks.

Of course things did not always run smoothly, for example in late 1916 the artillery Captain H. H. Hemming was a 'reconnaissance officer' commanding VI Corps' counterbattery intelligence staff, following his ingenious and innovative work in designing the 'flash-buzz' apparatus for locating enemy batteries by their muzzle flashes. As such he was detached from Third Army field survey battalion, commanded by his friend Colonel H. St J. L. Winterbotham,[54] and placed directly under the command of the corps GOCRA, Brigadier Johnny Rotton, whom he described as charming, lazy and essentially unemployed. Because the GOCRA in VI Corps was still only an adviser rather than a commander, he habitually lost every argument with his corps commander, General Haldane, with whom in any case he did not get on well. For Hemming this was nevertheless a comfortable and workable arrangement, since he was left free to relay his own intelligence directly to the firing batteries which needed it. However, when GHQ's enthusiasm for counterbattery efficiency led to the insertion of a 'Corps Counter-battery Colonel' between Rotton and Hemming, the whole thing instantly fell apart. The new arrival turned out to be the ferocious Colonel Fawcett, a famous Amazon jungle explorer who had no time for Hemming, his science or his RE personnel from the survey department, and who 'was probably the nastiest man I have ever met in this world'.[55] Fawcett immediately announced he would put the daily counterbattery intelligence reports straight into the waste bin unopened, and invited Hemming himself to 'go away and stay away'. The only counterbattery shoots he would allow were either against targets which were directly visible from British lines, and hence not properly the concern of 'heavy artillery', or against those which he had located through consultation with his personal ouija board (*sic*).[56]

The emergence of the deep battle

Such stories, however, reflect more upon the clashes of strong personalities which are inevitable within all organisations than upon the underlying institutional framework which was gradually being rationalised for the whole of the BEF's artillery. For example, the appointment of a counterbattery supremo ought by rights to have helped, rather than hindered, the work of that particular branch — but in VI Corps it just happened to go the other way. The technical achievement of most counterbattery intelligence staffs was at least very impressive, and made a big contribution both to the habitual suppression of enemy artillery in most British attacks from Arras onwards, and to the development of predicted fire. Hemming's own main contribution lay in the field of electrically aided flash-spotting by a chain of specialised observation posts, which were all wired up to react in unison to any observed German gunfire. Later, however, he extended his repertoire to include the whole spectrum of map survey and aerial reconnaissance. As a direct result of an ascent in an observation balloon, he came to the conclusion that neither this format nor verbal reports by aircraft pilots should ever be given very serious credence, but that actual air photographs could usually be fully trusted. When the Germans started to use flashless powders for their artillery propellants during 1917, moreover, he quickly conceded that 'flash-buzz' had effectively come to the end of its useful life, and that the true future of battery location now lay mainly with a combination of aerial photography and sound-ranging.[57]

Sound-ranging had become the personal speciality of Hemming's friend 'Willie' Bragg, who was already a Nobel laureate in 1915, and would later become Professor Sir Lawrence Bragg.[58] Like Hemming, Bragg was out of his element as an artillery subaltern, since he did not feel at home with horses and was too scientifically minded to fit very easily into the mess. In October 1915 he was asked to take over one of the sound-ranging kits developed by the French scientist Lucien Bull, which used a baseline of six microphones to pick up the sound of an enemy gun firing. Each microphone's signal was then translated into a movement of a wire inside a galvanometer, which was in turn recorded on cine film to show a differentiation of time that was accurate to one hundredth of a second. By comparing the time difference between the six incoming signals, the location of the enemy gun could finally be determined by trigonometry to within twenty-five yards.

In late 1915 Bragg started work on Kemmel Hill with a section of eight men, under the direct command of the Maps section of GHQ. Unfortunately the French microphones were sensitive only to high frequency noises and not to the low 'boom' of a gun's report, and it took a whole year to devise a new system for recording the latter. This represented a very major embarrassment to the whole battery location effort, as Bragg freely admitted in his postwar paper. At the time, however, he tried to discount it as much as possible, and to conceal the fact that his promising new technology did not actually work at all until the microphones could be improved. Indeed, one wonders how many other comparable fudges were going on elsewhere in the BEF during this period, and to what extent this contributed to disillusionment when they were finally called upon to 'deliver the goods' on the Somme. Yet — perhaps ironically — the fudgers often proved to be in the right in the long run, and their machinery was usually made to work properly at some point during 1917, although by then they may well have lost the confidence of their lay public, both military and civilian.[59]

Success in sound-ranging was finally achieved in the autumn of 1916 with the development of 'Tucker microphones'. These incorporated heated platinum wires designed to react to the lower frequencies, thus completing the coverage of every part of the shell's flight, from firing to arrival on the target. The operator was enabled to read off full details of the gun's characteristics, including muzzle velocity and calibre, thus marking the moment when sound-ranging really came of age. Subsequent refinements filtered out many of the effects of wind and other extraneous noises, and allowed any number of firing guns to be monitored at once. The apparatus achieved an amazingly high sensitivity, at least whenever any easterly wind was blowing, and it was soon also being used to calibrate both the muzzle velocities of the BEF's own pieces, and the effects of weather upon the fall of shot — both of which processes were essential for accurate gunnery. Eventually some forty sound-ranging sections were established, each the size of a large infantry platoon, and it was their boast that they could set up their equipment almost as fast as a battery could come into action. The Germans, by contrast, had been left very far behind, and were still floundering with crude stethoscopes right up to the end of the war.

Apart from Bragg's applied electronics and cinematography, the long-range artillery battle depended heavily upon an accurate survey of the whole battle area, replacing the pre-Crimean war French 1:80,000 and Belgian 1:40,000 ordnance maps. Major E. M. Jack took this work in hand in late 1914, as commander of 'Maps GHQ', and soon enrolled the help of Winterbotham, who was then still a captain but who had extensive expertise in surveying.[60] A 1:5,000 map of the Neuve Chapelle area was prepared in time for the battle there, but in general the work was found to be so demanding that each army HQ soon had to be given its own topographical section. Relief also came from an increased use of air photos, upon which a new and comprehensive series of 1:10,000 maps began to be based from the summer of 1915 onwards. The air observation system of 'zone calls', linked to a local target diagram or 'battery board', further helped reassure each artillery commander as to the accurate location of his own and the enemy's positions. All of this was already normal practice before the Somme battle.

The process of providing timely local revisions to existing tactical maps always posed considerable problems, and the necessary advanced photographic printing equipment was always in short supply at army HQs. Winterbotham took a lead during the Arras battle by printing daily updates of the Third Army situation map — but for this he had to use a printing machine that he had purchased himself. Nevertheless, the principle of multiple updating eventually became standard throughout the BEF, even when the maps had to be cyclostyled, and it took its place alongside the printing of one-off barrage maps and target maps for individual operations. So important was this work believed to be in early 1917, indeed, that each army corps was finally given its own specialist topographical section.

Essential to both mapping and tactical briefing was a plentiful supply of current aerial photographs. Good gunnery also demanded airborne observers to report the fall of shot, quite apart from other aerial tasks such as general reconnaissance of enemy rear areas or the protection of friendly positions from enemy air intrusions. All of this had already been pioneered — like so much else — by the time of Neuve Chapelle; but the difficulties were increased in the summer of 1915 when the Germans seriously started to dispute command of the air. Notwithstanding the popular stereotype of 'the Red Baron's supremacy', however, the RFC did manage to win and hold a notable superiority throughout much of the war, which greatly helped the work of the Royal Artillery just as it often blinded the German guns.[61] In fact there were only three occasions on which the Germans were able to mount significant challenges to allied air command, and of these only one represented more than a very fleeting success.

The first German aerial challenge came towards the end of 1915 and in the early months of 1916, when the Fokker monoplane appeared with a machine gun synchronised to fire forward through the propeller. This gave it a world lead in fighter technology, although in practice it was soon effectively contained by obsolescent allied 'pusher' aircraft until the synchronisation gear could be matched by the RFC in March. Thereafter the skies over the Somme belonged entirely to the British, particularly since the Germans believed in a very defensive concept of air warfare. They tried to establish a 'barrage' over their lines through which no allied aircraft could pass, but it was always full of holes and served mainly to restrict their own machines from flying over allied lines. However, at the end of September the Germans deployed a new generation of aircraft — notably the Albatross and Halberstadt — and moved towards a more offensive doctrine using massed formations of up to sixty at a time. The onset of winter weather served to limit the impact of these changes until the following spring, but in the battle of Arras in April 1917 the RFC found it had to all intents and purposes lost command of the air to the 'flying circus'.

Von Richthofen's moment was glorious indeed; but it did not last long. By the battle of Messines in June the British were well on the way back to regaining their lead. In that operation Plumer's Second Army was able to deploy no less than 280 ground wireless stations to receive counterbattery information from aircraft, and it habitually coordinated low-level fighter-bomber attacks against enemy infantry. This was a new technique which won for the aircraft the rather modern-sounding nickname of 'tanks of the air',[62] and it seems to have played an increasing rôle in every subsequent battle until the end of the war. Fighter ground attack was quickly recognised as an important element in any big operation, and it took its place alongside battlefield interdiction — the attempt to neutralise enemy transport just behind the battle area — which had already been a burgeoning feature of British operational art at Loos in September 1915.[63]

During the autumnal Passchendaele battle the bad weather restricted flying every bit as much as it restricted combat on the ground, although it did not stop the entire area being regularly photographed twice a week. This was all the more essential since the landscape kept being changed and deformed under the batterings of artillery and rain, so that only constantly updated photographs could make sense of it. In the month of August, also, no less than seventy-nine tons of bombs were dropped, which shows a dramatic increase over the four tons at Aubers Ridge in 1915, or the fifty tons during the whole five months of the Somme in 1916. It is clear that the RFC had by now started to take its ground attack rôle very seriously indeed — and the events of 1918 would only underline that lesson. During the spring retreats the RAF found many juicy targets in the marching columns following up behind the German spearheads, while in the summer battles of Reims and then Amiens it made serious bids to destroy key bridges just behind the enemy lines. At Amiens it dropped no less than 284 tons of bombs in the two weeks just before the battle, although it never did manage to destroy the key Péronne bridges. It then lost a total of ninety-six aircraft on 8 August itself — approximately thirteen per cent of its total strength — so perhaps the event should be remembered as 'the black day of the RAF' almost as much as it was the black day of the German army.[64]

The pressure was none the less remorseless, and the Germans found themselves condemned to a very reactive rôle. They tried to make a third and final bid for air command during the last three months of the war, once it had become clear that their ground forces were collapsing; but ultimately this amounted to very little. The allies had achieved both quantitative and qualitative mastery of the skies, and the chief limitation on their action turned out to be the weather rather than the German air forces.

Already by 1917 the BEF's effective air-artillery combination of photography and firepower meant that no German in a trench or a shallow dug-out, anywhere on the battlefield, could consider himself safe from the apparently random shell or bomb which might strike at any time. 'Attrition' took on a new meaning, even in the relatively quiet periods between major battles; and its effects gradually extended ever further behind the front lines. The range of accurate gunfire had become noticeably greater by the end of the battle of the Somme than at its start, and greater still at Third Ypres. By the second half of 1917 the probing long-range artillery was definitively supplemented by the increased activity of bombing aircraft, including at night. Taken together, these two sources of firepower successively brought many rear areas under attack which had previously been entirely safe, thereby blurring the essential distinction between 'combat' and 'rest' areas in the minds of many of the troops.[65] In tactical terms this represented almost the creation of a whole new combat zone, as the 'deep battle' joined the 'close battle' for the front line.

In view of all this, it is perhaps paradoxical that the Great War is often remembered as a war in which millions of shells were 'wasted'. The week-long preliminary bombardment on the Somme is best known for its failure to disturb the Germans in their deep shelters, just as the two weeks' bombardment before Passchendaele very obviously bounced off their concrete pillboxes. Later in that battle the mud became so deep that shells buried themselves relatively innocuously before exploding, and then seemed to do little more than merely add to the craters and mudholes. Many participants who had certainly witnessed the traumatic destructiveness of artillery at first hand could leave sarcastic comments dotted around the literature to the effect that 'to the bitter end of the war it remained the greatest wonder that so much ammunition could be expended without hurting anyone but the taxpayer'[66] or that 'An immense amount of metal can, in fact, be flung about trenches without doing much harm.'[67] Even Aubrey Wade, himself a gunner subaltern, believed that 'All those scores of shells didn't seem to have killed a single German.'[68] This was indeed a foreboding of Vietnam, where the Americans would use three times the weight of HE that they had used in the Second World War, for but little strategic benefit. Shelling merely for 'attrition' clearly left much to be desired when set against the probing standoff accuracy that would be demonstrated in 1991 in the Gulf.

But perhaps the question should be posed in terms of the effectiveness of the shells in fulfilling the task for which they were intended, since a large majority of them were never expected to kill any German soldiers at all. We can probably easily agree that the lengthy bombardments for 'destruction' in 1915–16 were often misconceived, and that Passchendaele in 1917 was a dreadful place to choose for any military operations at all; but we must also surely accept that many of the shells fired in the more successful late-war battles were intended for either immediate neutralising effect in creeping barrages, or for wire-cutting, or for diversionary effect in interdiction or 'Chinese' barrages. None of these applications were seriously expected to be very much more deadly than the bayonet charge itself, even though they were all sophisticated scientific techniques which each made a big contribution to the eventual victory.

Counterbattery and other 'deep battle' shoots were admittedly different, since they were intended to hit specific point targets manned by enemy soldiers. However, it was precisely because they were different and more difficult that they took longer to perfect — not only because of delays in instrumentation and gunner philosophy, but also because of von Richthofen's rampage in the spring of 1917, followed soon afterwards by the bad conditions and German mustard gas at Passchendaele. Yet by at least the time of Cambrai in November 1917 the British had achieved the ability to land predicted fire on their target almost with the first shot and even at long range. Whatever wastefulness there may have been in the battles before that, a new standard of economy and accuracy had certainly been attained during the final year of the war.

9

Controlling the Mobile Battle

> A General without a telephone was to all practical purposes impotent — a lay figure dressed in uniform, deprived of eyes, arms and ears.
>
> Wyn Griffith, *Up To Mametz*, p.185

Cavalry and armour

Throughout the whole period from August 1914 to November 1918, the gaze of the high command was fixed firmly upon the need to achieve mobility on the battlefield. For infantry generals of the 'bite and hold' persuasion this was merely a matter of ensuring that any attack succeeded in capturing its local objectives, preferably at a relatively light cost. They were concerned that the preliminary bombardment should be accurate and sustained, that the creeping barrage should fully neutralise the enemy, and that the attacking infantry should show local initiative and effective tactics during all three phases of assault, mopping up and consolidation. By 1917, as we have seen, most of the arrangements necessary for this type of operation had been properly worked out and could be put in place relatively quickly. New weapons and tactics had been developed — including smoke shell, ground attack aircraft and tanks — and the procedures for their use were starting to be widely understood.

For many other generals, however, the idea of mobility opened a tantalising higher vision of flat-out manoeuvres deep in the enemy's rear areas — and even of breakouts into 'the green fields beyond'. A surprisingly high level of movement became a habitual theme in their exhortations and often also in their preparations. For example throughout 1916 and 1917 many of the training schools seemed to emphasise a style of 'open warfare' which might be seen in some bright hypothetical future, rather than the static trench fighting which was the daily bloodstained reality of the present. Many were the hardened veterans from the front who found themselves flabbergasted that such a very different, or even 'make-believe', type of operation was being taught behind the lines[1] and many, too, have been the subsequent commentators who have castigated the high command for such an apparently blatant rejection of the real world in favour of the ideal.

There were, however, at least two perfectly valid reasons for teaching open warfare, even though neither of them was perhaps sufficient to justify any neglect of trench fighting in the curriculum. Firstly, a mobile breakthrough was envisaged as an integral part of the planning in many of the battles, and any given infantry division might from time to time be earmarked for this rôle alongside cavalry and motorised formations. One could even argue that the repeated failures to break out were to some extent due to poor training in open warfare on the part of the troops intended to act in that rôle. Secondly, open warfare did in fact become a reality on four separate occasions — the retreat from Mons in 1914, the advance to the Hindenburg Line in March 1917, the spring retreats of 1918, and then in the advances of the Hundred Days. In all four cases the BEF was essentially unprepared for the sudden shock to its system, and it would be only on the last occasion that it rose to the challenge at all convincingly. If viewed from this perspective, therefore, the high command might be criticised for placing too little emphasis on the mobile battle, rather than too much. At the very least we can agree that the rude lessons of the Mons retreat were so chastening that for the following four years its key participants must have experienced a very natural desire to do better in any future campaign of that type.

Much ink has been spilt in attempts to demonstrate that the very idea of breakthrough operations was a complete chimera in the conditions of the Western Front, and therefore that no attempt should have been made to keep the Cavalry Corps together as the nucleus of a mobile force.[2] Certainly defective communications were always a major bugbear which tended to prevent higher commanders from identifying any genuine gap in time to vector the cavalry usefully towards it. Nor was General Kavanagh, the commander of the Cavalry Corps, the BEF's most adept leader; and it is quite true that many of the individual cavalry charges ended in either frustration or ruin. Yet the same was also true of many infantry operations, even though the infantry was labouring under none of the logistic and administrative restrictions that the cavalry usually had to put up with.[3]

The main psychological problem was perhaps that because the cavalry was kept out of the line until a breakthrough seemed to be imminent, it became rusty within itself and undervalued by the infantry who felt it was not pulling its weight. Then when it did appear at the front it would be surrounded by excessively high expectations which would almost inevitably be disappointed. As a result of this process it became increasingly shunned as the war progressed, and both its efficiency and its fighting strength were allowed to dwindle. In that sense it became the most serious of all the casualties attributable to the infantry's ingrained feeling that everything that happened in the front line had to be auxiliary and secondary to the infantry itself.

It is nevertheless striking that many shrewd tacticians within the infantry and tank corps were able to rise above the fashionable mockery of cavalry, and continued to see its underlying potential as the only arm capable of sustaining a truly mobile breakout. The normally caustic Brigadier Croft, for example, testified to three occasions in the war when his Jocks opened a promising gap which he believed the cavalry could have exploited, if only they had been close enough at hand.[4] Nor was he alone, since on the second of these occasions, at Longueval and High Wood on 14 July 1916, the belief appears to have been pretty widespread that 'Now the Armée de Chasse may get on the move....'[5] Brigadier Jack, equally, believed cavalry could have made a breakthrough at Arras, and sincerely admired its achievements at Ypres in September 1918.[6] Still more surprising, Major W. H. Watson — who had fled from the cyclists in order to join the tanks in December 1916 — believed that Bourlon Wood should have been captured by cavalry on 20 November 1917 because 'it was defended merely by a handful of machine guns'.[7] The horse was certainly not obsolete as far as this technically minded officer was concerned, and at Amiens nine months later each of his tanks would be led to the battlefield by a far from anachronistic mounted officer![8]

It is easy to forget that the BEF cavalry had started the war with a level of tactical expertise superior to that of any other nation's mounted arm, and that it kept well abreast of modern developments throughout hostilities. Before 1914 the lessons of South Africa had been translated into a judicious application of fire and movement combined with sound logistics and horse care, making a newly sophisticated version of the 'mounted infantry' principle which had been so pregnant with promise during the closing stages of the American Civil War. From this starting-point the British cavalry soon went on to pioneer automatic rifles, battlefield wireless, the development of motor machine guns and the use of aerial contact patrols to gather intelligence.[9] Then by the start of 1916 it willingly accepted a rôle as just one element within an all-arms mobile force — for example, at High Wood in July the breakout was intended to be effected by a cavalry brigade battlegroup which included field engineers with bridges, two armoured cars, a machine gun squadron and a field artillery battery.[10] This was scarcely a 'reactionary' approach to tactics, despite the many difficulties of broken terrain, poor signalling and enemy defences arranged in both depth and strength. Indeed, it was actually put into successful practice on a number of occasions during the Hundred Days. For example, one Fourth Army cavalry raid on 9 November 1918 advanced no less than ten miles to pillage some German trains and supply dumps.[11]

The 'new cavalry' was surely following a strikingly forward-looking tactical conception, and it is ironic that some of its greatest detractors were precisely the same officers who claimed to 'invent' the revolutionary new idea of a mobile breakout in the Tank Corps' notorious 'Plan 1919', which relied upon the largely putative construction of a whole new generation of chaser or cruiser tanks.[12] One might suggest that they were arriving at the same conceptual starting-point that the cavalry had already reached some three years earlier, and that their scorn for the horse was more a matter of sour grapes than a reasoned analysis of the true requirements of a fast-moving breakout.

It should certainly be emphasised that before the arrival of a 'cruiser tank' in 1919 it was quite unrealistic for anyone to see the tank as an instrument of breakout. During 1916–18 tanks were never anything more than an auxiliary to the infantry for the break-in battle, with only a little greater speed than a foot soldier and, at something like a maximum of eight hours, considerably less staying-power. Its main utility was in flattening barbed wire and in offering moral support, sometimes taking over the psychological rôle of the creeping barrage in drawing the infantry forward into the firing arcs of the enemy defences.[13] It also often seems to have held a disproportionately central place, both during and after the war, in the complex web of German mythologies that were so ingeniously spun to explain away the crushing allied victory.[14] If this mythology later accidentally encouraged the Germans to invest in a weapon that would finally evolve into a serious instrument of breakout in 1939–41, then that surely had remarkably little to do with the technical capabilities of any tank that ever took the field during the Great War itself.

In the tactical conditions of the Great War the tank could sometimes knock out enemy machine guns or other strongpoints; but in this its record was probably no better than the infantry's own methods of pillbox-busting using Lewis guns, rifle-grenades or mortars. As successively improved and faster marks of heavy tank appeared on the battlefield, moreover, the crews seem to have faced an ever-growing risk of carbon monoxide poisoning or over-heating and general 'sea sickness'.[15] The tank's firepower was also downgraded by its officers' deep faith in firing while in motion,[16] which they originally derived from the analogy with ships at sea. Some of the tank crewmen who manned the 6-pounder cannons had been trained to fire from moving platforms by the navy's instructors at Whale Island,[17] and they were notoriously reluctant to accept the profound truth of the proposition that 'the ranging and laying of a 6-pounder from a moving tank is very difficult.'[18] Ultimately the tank of 1916–18 was considerably less potent in practice than its propagandists would like to have us believe.

Excessive enthusiasm for the tank appears to have come from three main sources. Firstly there were the armchair strategists at home, notably Churchill and Lloyd George, who were convinced that the generals had failed and therefore some 'way round' the trench deadlock had to be found. This impulse helped to expedite the procurement and supply of tanks, although it would also later help to distort the historical record in the direction of an excessive adulation of that particular weapon.[19]

Secondly, the infantry during the winter of 1916–17 had heard many inflated rumours of the tank's prowess, and was therefore often only too ready to believe it could achieve far more than was really possible. Yet the tank's début at Flers on 15 September 1916 had actually created as many problems as it solved,[20] and with only thirty-two machines crossing the start line it was in any case a pitifully small operation. But the newspapers had made much of it, as an only too rare ray of hope in a generally desolate landscape, and GHQ quickly gave its support to an expansion of the tank arm. But this expansion could only happen slowly, and no more than sixty tanks were available in small packets for the battles around Arras in April 1917. Nevertheless, at First Bullecourt on 9–10 April General Gough was more than ready to seize upon tanks with undue haste, and to distort his entire attack plan to accommodate them, even though only twelve machines were available for his Fifth Army.[21] The result was a predictable débâcle, since the tanks failed to turn up at all on the first night and then, after the whole attack had been delayed by twenty-four hours, they quickly suffered a 100 per cent casualty rate in machines and fifty per cent in crews. The Australian infantry whom they had been intended to support became instantly and extremely disillusioned with the tank, and would remain so for over a year thereafter. Indeed, this seems to have been the general view of the BEF during the summer of 1917. At Messines the tanks were regarded as a superfluous luxury, of much more use in drawing enemy fire and boosting friendly morale than in adding materially to the attack.[22] Then came the slough of Passchendaele, where over 190 tanks were swallowed up in the mud and many others were knocked out, albeit with a shorter salvage time. So unsuitable was the terrain, in fact, that the tanks suffered the ultimate indignity of being confined to the roads, where they were forced to attack in single file. Thus when the leading vehicle was hit, those following were faced with a choice between swerving off the road into the mud, and there ditching, or stopping helplessly on the road as sitting targets.[23] As Lord Moyne, then a major, succinctly put it, 'Of course they may be able to improve them but at present most people who've seen them in action say they'd rather do without.'[24]

The infantry's growing disillusionment and scepticism during 1917 led to the third burst of propaganda for the tank, as its commanders launched a frantic last-ditch campaign to win back general acceptance and credibility. By the end of June they had successfully advanced from being merely an experimental curiosity into the heady status of a separate corps, second only to the RFC in technological expertise; but like the Machine Gun Corps before them, they had been instantly deflated by the realisation that their wonderful new organisation was by no means universally acclaimed throughout the BEF. Although the tank corps was expanding painfully from nine battalions to eighteen during the summer of 1917, it was not universally loved. However, at Cambrai on 20 November it managed for the first time to collect together a genuinely imposing mass of vehicles for a battle which has conventionally been seen as the tank's true 'coming of age', and which certainly gave it its most potent propaganda.

Out of the whole war it was at Cambrai that the tanks won their best success, and a striking advance of some 4,000 to 6,000 yards was duly registered on the first day, for a loss of only 4,000 British casualties. This was better than the first day at Arras or Messines, and it certainly made a most welcome change after four frustrating months at Passchendaele. One can readily understand the psychological boost it gave to the British public, who fell to ringing their church bells for the first time in the war.[25]

Nevertheless, the success at Cambrai should not be attributed entirely, or perhaps even primarily, to the tank. The battle had originally been planned mainly as an artillery operation, to exploit the new techniques of predicted fire.[26] The tanks obviated the need for guns to cut the German wire, so to that extent they assisted the artillery fireplan; but it could be argued that their contribution extended little further than that. In operational terms they were certainly never intended to make the breakout, or trained for it, but only to accompany the infantry during the initial phases of the break-in. If anyone was intended to break out it was the cavalry, although the battle had been conceived on such a relatively small scale that many in the high command saw it as no more than 'a gigantic raid',[27] from which a true breakout could scarcely be expected at all.

One may also question the effectiveness of the tank at Cambrai on the grounds that the German defences were pitifully weak at the point of attack, with less than two infantry divisions and a mere thirty-four guns, all of which were effectively mastered by the unregistered British artillery within a few minutes of zero. Yet just as in earlier battles, it was found that the light screen of defending Germans in the forward areas could be strongly reinforced from the rear before the attackers were able to break all the way through the deep defensive zones. In the event the attack became stuck towards the end of the first day, and not even the presence of the mass of tanks seemed to prevent this from happening. By that time most of the tanks were knocked out, bogged down or tired out, so their assistance proved to be disappointingly short term.[28] If they had represented a sledgehammer to crack a nut, in fact, it was ultimately the nut which won. The Tank Corps' own rather negative conclusion immediately after the operation was therefore to complain that there ought to have been twice as many tanks as were actually used.[29] To the other arms this protest must surely have seemed somewhat curmudgeonly, and it is noteworthy that GHQ failed to sanction any repetition of the massed tank experiment in 1918, apart perhaps from a somewhat diluted version at Amiens.

Cambrai did at least help to crystallise the need for a concerted programme of tank training with other arms, even though this also meant spelling out the tank's limitations very clearly. Unfortunately these limitations were numerous. The tank was noisy and blind in its approach march, and at Cambrai even extreme care in choosing routes failed to prevent it from tearing up most of the signallers' delicately laid telephone cables.[30] Then again it could not operate safely in built-up areas or woods, which excluded a good proportion of the places where the infantry might need it most.[31] Badly cratered and muddy ground was equally no-go, as had been discovered only too humiliatingly at Passchendaele. Most telling of all, perhaps, was the early realisation that the tank was actually extremely vulnerable to enemy fire, particularly from field guns but also increasingly from armour-piercing machine gun rounds, special anti-tank rifles, and even a whole new breed of contact mines. The tank needed protection from all these threats if it was to advance over open ground; or in other words its way had to be opened by infantry, aircraft or, especially, artillery.[32] Right up to the end of the war the tank was therefore portrayed in the official manuals as only a semi-experimental auxiliary to the other arms. Thus as late as August 1918 the manual SS 214 was apparently quite happy to concede that the tank was still in the process of being perfected in a technical sense, and that its rôle independent of infantry still lay very definitely in the future.[33]

Admittedly the tank crews themselves formed a dedicated technological élite that was in many ways almost as impressive as the RFC; and certainly their training curricula seemed to cover a bewildering variety of subjects which ranged from wireless signalling to compass work, from mechanical maintenance to plasticine modelling, and from gunnery in all calibres to camouflage in all weathers.[34] In common with the RFC, however, this comprehensive virtuosity of the tank crews seems to have been accompanied by a somewhat unhelpful detachment from the everyday life of the trenches. Unlike the infantry, the tank crews would typically spend many comfortable weeks behind the lines working up their machines and training in general. Then they would be called to the front for a particular operation in which they might perform at high pitch for perhaps eight hours, often suffering very high casualties and emerging groggy from the fumes; but afterwards they would enjoy the luxury of retiring quickly to the relative calm and comfort of the 'tankodrome' several miles behind the firing line. They thus made an especially highly strung and privileged group, whose combat appearances were as rare as they were intense.

To illustrate the relative rarity of these appearances, let us pause to consider the record of the Tank Corps during the Hundred Days — the moment when it was at its absolute peak, both institutionally and tactically. Altogether there were some fifteen battalions of tanks (later sixteen), one of armoured cars, two of 'gun-carrying tanks' and five of supply tanks. On 8 August they had some 630 fighting tanks on their strength, with enough crews to man them all. By 31 August the number of crews had fallen to 405, although the fighting tanks available had fallen to 248.[35] During the interval tanks had been 'engaged'[36] some 1,184 times, or approximately twice per machine and crew (see Table 12). During the remainder of the war each machine would be engaged only two or three times more, making an average of perhaps one appearance per month.[37] So scarce did they become by the end, in fact, that their numbers were sometimes made up by dummy tanks constructed out of wood and canvas, which were pulled into battle by mules.[38]

Another important result of the scarcity of tanks was that only a few would be allocated to each attack. At Amiens on 8 August an entire tank battalion had often been committed in support of each division in the three corps making the attack, giving a total of 414 fighting tanks, including seventy-two Whippets, on a ten-mile frontage at zero.[39] This represented a density of one tank to every forty-two yards, which was well over twice as many as the tactical regulations required.[40] Even so, this was a lesser concentration than at Cambrai, when 378 fighting tanks[41] had supported six divisions on a frontage of 12,000 yards, for an average of two tank battalions per division and one tank to every thirty-two yards.

After Amiens tank numbers gradually declined, although the fighting was successively widening to include almost the whole of the BEF. Thus for the assault on the Hindenburg Line towards the end of September even more tanks were thrown into combat than had been used at Amiens, but they were more diluted because they had almost twice as many divisional attacks to support. Thus in none of these battles could the tanks really be said to have been 'massed'. Instead, they were 'penny-packeted' as a local auxiliary to the infantry, and indeed they were allocated their rôle in each operation by the infantry corps HQs rather than by the Tank Corps.[42]

Table 12: Tanks committed to action during the Hundred Days

(Source: Extracts from OH 1918, vols 4 and 5, wherever precise tank numbers are mentioned. Note that the very many attacks made without any tank support are not included here, whereas the many tanks 'committed to battle' but failing to arrive at the start-line are still included. Hence these figures suggest a higher level of tank participation than was actually the case.)

Week Ending	No of tanks committed	= Percentage of total tanks in the period	No of Div attacks using tanks	Hence av. no of tanks per attack
15 Aug '18	347	(21.6%, approx)	23	15
22 Aug	74	(4.3%)	7	10.5
29 Aug	266	(16.6%)	28	9.5
5 Sept	101	(6.3%)	10	10
12 Sept	—	(0%)	—	—
19 Sept	20	(1.2%)	5	4
26 Sept	31	(1.9%)	4	7.5
3 Oct	387	(24.1%)	39	10
10 Oct	194	(12.1%)	17	11.5
17 Oct	43	(2.7%)	5	9
24 Oct	56	(3.5%)	10	6
31 Oct	10	(0.6%)	0.5	10
7 Nov	75	(4.7%)	12	6
To 11 Nov	—	(0%)	—	—
TOTALS:	1,604	100%	160.5 = average 10 tanks per attack, 115 tanks per week in 11.5 divisional attacks.	

Analysis of Tank Losses

Of the above totals, 1,015(+) tanks rallied at the end of their attacks (including 105 out of 230 Whippets or armoured cars), making a total loss of about 589 tanks (including 125 lights), or 36.7% (including 54.3% of the lights). However, all but 1% of the disabled vehicles would eventually be returned to service.

Hence, of the average of about 10 tanks (including 1.5 lights) committed to each divisional attack where any tanks at all were used, some 6 tanks would rally (including 1 light), leaving 4 disabled (including 0.5 light).

Out of some 10,000 Tank Corps personnel, approximately 3,000 were made casualty in this period, or an average of about two casualties every time a tank was committed to action.

Penny-packeting by infantry commanders was a major source of complaint for tank officers throughout the war. They lamented the original rush to bring tanks into the Somme battle in September 1916, thereby sacrificing surprise without using sufficient numbers for a decisive blow.[43] At Bullecourt Watson had objected to Gough's use of his command over a wide front rather than a narrow one, and at Passchendaele almost everyone in the Tank Corps had regarded it as wasteful to use tanks at all. Even in the almost ideal conditions of Cambrai, as we have seen, the complaint was still that there were too few tanks, and this remained the constant cry throughout 1918. A theory of betrayal has thus arisen within tankie circles, according to which they were deprived of the tools they needed to finish the job, and subjected to too much interference by the infantry.

If we accept that tank procurement and training was pushed ahead as fast as it reasonably could have been during 1916 and the first half of 1917, we may agree that less penny-packeting on the Somme and at Arras might have allowed the massed tank blow of Cambrai to be launched several months earlier than November. It would not then have enjoyed the benefit of predicted artillery fire, but the tanks themselves would still have been a technological surprise, and would surely have encountered none of the anti-tank fire which turned out to be so troublesome. Yet without a year's tactical experience behind them, they might well have blundered into all sorts of mistakes of detail which in the event they were able to avoid. No new wonder-weapon has ever achieved decisive results on its combat début, apart from the atomic bomb.[44]

Even if we accept that the Cambrai battle might have been fought earlier and with greater success, it still seems incredible that it could have been converted into a full-scale breakout. There would still surely have been a swing of operational initiative back to the Germans during the winter of 1917–18, even though the high tide of their March offensive might have been less threatening and sooner overcome. However, the real cutting edge of the Tank Corps' critique concerns the situation as it might have evolved immediately after that. In the event the initiative was regained by the allies first at Soissons in July and then at Amiens in August, in both cases using plenty of tanks. But without penny-packeting, and with a greater emphasis on tank procurement during the preceding twelve months, it is argued that a far greater mass of tanks might have been collected for a still more crushing counterblow rather earlier in the summer. Perhaps a thousand machines might have rolled forward together on a genuinely narrow front in May or June, sweeping aside the exhausted stormtroopers before they had a chance to organise their defences. In this way the war might have ended at least four months earlier than it actually did.

Such an outcome would have required the high command to have nurtured a very single-minded belief in the tank, such as existed at the time only in the breasts of the most devoted enthusiasts within the Tank Corps itself. It would have meant overruling the competing claims of the Gas Brigade, the artillery, and the infantry — all of which had surely contributed far more to the war by midsummer 1918 than had the Tank Corps. And the tanks themselves were still a very uncertain technology, notorious for their limited range, reliability and speed. They might well have faded away very rapidly after the first day, just as they did in reality at both Cambrai and Amiens. It is therefore surely not reasonable to have expected GHQ to have placed all its bets on this one particular 'mechanical contrivance'[45] in 1918, since the penalties for failure would surely have been very high indeed. It seemed better to take a longer but more certain path to victory, and that should not be a matter for serious reproach.

One can scarcely blame Haig for exercising rational caution if one is simultaneously accusing him of spendthrift recklessness in multiplying doomed assaults. With the tanks, at least, he decided not to be reckless. Admittedly the German commanders of 1940 won the battle of France by opting for the opposite course and staking their all on the tank; but we must remember that they still regarded this as a very daring gamble, over which they agonised painfully and long. Besides, by 1940 the technology of the tank itself had immeasurably improved and mechanisation in general had taken several important strides forward. In the conditions of 1916–18, by contrast, the tank was still a highly dubious candidate for taking any central rôle in a grand offensive. It was probably better used in penny-packets, which helped forward as many infantry as possible, than in some desperate death or glory adventure which ran every chance of failure.

Signals and command

Less glamorous but far more important than the Tank Corps, and in many ways no less technical, was the RE signal service. Faithful to its corps motto of *Ubique* ('Everywhere'), it played a vital role in every battle and surely had to shoulder some of the heaviest responsibilities with the slenderest resources of manpower or *matériel*. It also underwent a radical technological evolution almost every twelve months, moving massively into telephones in 1915, then out of them in 1916–17, and finally into a complex array of different technologies of which many had broken free from wires.[46]

In common with many other technical branches, the signallers found they had to fight hard to establish their independence from the artillery and infantry. At the start of the war each battery and battalion had its own intrinsic signallers, leaving only brigade and higher HQs — and the line of communication — to the RE signal companies. Lacking any central chain of command of their own, moreover, the signallers were unable to supervise these functions properly. They were forced to preside over something very close to anarchy once they had lost much of their stock of cable during the retreat from Mons. Only despatch riders and the French civilian telephone net held the BEF together during the mobile operations which reached their culmination on the Aisne, and it was only then that signals reorganisation could start in earnest.

By the end of 1914 the signallers nevertheless managed to win a number of important victories. They had started training schools for each division and brigade, and were well on their way to winning control over the signals within each infantry battalion, although not yet in the artillery. They were starting to apply a proper command hierarchy between their own companies, and they set up a central depot and school at Abbeville which would soon become the model for those of many other technical branches. They regularised both the despatch rider service and the military post, and they started to cast covetous eyes on the various independent signalling organisations within GHQ. Still more important, they created a system to ensure a continuous flow of new equipment and cable, although in practice it never did fully succeed in meeting demand. Magneto telephones of the continental pattern were in particularly short supply over the winter of 1914–15, and so 'kleptomania...became a confirmed habit'.[47]

1915 saw a gradual process of education for senior officers in the etiquette of using the scarce signals assets. Sometimes this was a matter of persuading them to keep their telegrams short, so as not to overload the system; but more often it was a matter of urging calm and politeness when talking on the telephone through an exchange. This was an uphill struggle, since generals felt they outranked the humble switchboard operators by such a gigantic margin that every excess of bad language or time-wasting was seen as legitimate, and some operators were even arrested on the spot for very minor peccadiloes.[48] Still more difficult was education in the subtle art of utilising all available cable to its maximum efficiency; for example one artillery HQ was found to have no less than seventy-six phone lines, of which it was currently using only six. The others had simply been left where they lay when they broke down, which represented a wastage of many miles of precious cable.[49] An opposite example was the pillaging of Willie Bragg's specialised lines to his sound-ranging microphones by the infantry and artillery troops local to his installation, which caused him just as much disruption.[50]

Demand for signals continued to outstrip supply throughout 1915, as the great trench war moved forward remorselessly into top gear. There was a general multiplication of guns, headquarters, and chatter of every variety, at just the same time as both weather and enemy action were starting to take a visible toll on the cables and other signalling artefacts. New types of poles and trestles had to be designed for efficient stringing behind the combat zone, while by the middle of 1916 deep buries began to be seen as the only sure way to avoid shell damage within it. If it was deeper than six feet, experience showed that a cable was to all practical purposes protected from shelling;[51] but any shallower bury — such as the two or three feet common in early 1915 — would be liable to multiple breaks as soon as a major counterbombardment was fired. By 1917 this had become a doubly serious problem because the Germans were specifically targeting cable routes and exchanges.[52] The problem with a deep bury, however, was that it was very expensive in labour and could not easily be rerouted. It pinned down HQs and exchanges to particular spots, thereby adding to the immobility of static warfare. The problem was relieved a little during the Somme fighting by the adoption of a standardised grid or checkerboard layout, whereby each division would have a central cable running from front to rear, with side-branches at the level of batteries, brigade HQs and division HQs, and armoured test-boxes at regular intervals. If the front line advanced, new side-branches would be added further forward while existing branches were taken over by rear echelons. If the division moved sideways, it would take over the cables originally laid out by its neighbouring formation. This arrangement was found to be robust and practical, and a streamlined version of it was still being used in the fast-moving operations of the Hundred Days in 1918 (see Fig. 7).

Fig. 7: Schematic layout of BEF cable grid, as from late 1916

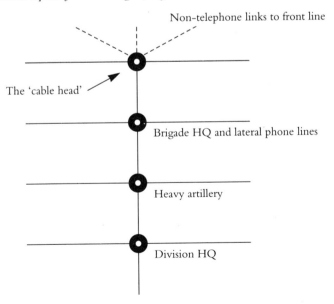

More worrying than the physical vulnerability of cable, perhaps, was the growing realisation during the summer of 1915 that the Germans could and did listen in to British telephone conversations within a range of a mile or more from their own front line. Breaches of security were only too often allowed to reach ridiculous proportions, despite many home-made systems of veiled speech or what Blunden called 'cryptographic transparency'.[53] Many specific instances of interception were gradually discovered, including one truly horrific example when a transcript of an entire set of divisional orders had been dictated verbatim down forward area telephones at Ovillers on the Somme.[54] It was also realised that the fast-growing complexity and density of the BEF's signals made a particularly attractive target for phone-tapping, whereas the German network was always much more sparse and inefficient, but by the same token harder to intercept.

The British reaction to the phone-tapping threat was slow but eventually comprehensive. The insulation of all cables was checked and double-checked against earthing; they were twisted for added security and then they were buried more carefully than ever. A major long-term offensive was even launched against insecure voice procedure — itself a revolutionary new concept! — but when all this eventually failed, the BEF reluctantly came to the conclusion that all telephones had to be suppressed within a mile of the German trenches. To many this seemed an excessive reaction, but in technical terms it was no more than common sense. By the end of 1916 there had admittedly been some improvements, helped by monitoring by an 'Inspector of Listening Sets', but in general the move away from telephones was gathering pace.

The alternatives included the Fullerphone, which still used cable but with lower direct current, and the Power Buzzer, which transmitted its signals through the earth. Both of these instruments were reliable and secure, and they came to be widely used,[55] although they surfaced only gradually out of a mass of experiments with other more or less unlikely systems. Pigeons and runners had already established themselves in 1915; but now in 1916 there were message-carrying dogs, bombs, grenades and rockets, aircraft contact patrols complete with klaxons, signalling panels and message bags, and a move away from the noisy Begbie Lamp to the far more elegant and focused Lucas Lamp. The last had been designed at the front itself by a gunner subaltern, in the best Heath Robinson spirit. Wireless also received a boost at this time, although its batteries continued to pose many practical problems and it still lacked mobility for ground operations.[56] Nevertheless, it was found to be particularly useful for keeping divisional HQs in contact with brigade HQs which had outrun their cable-heads, especially in the Ypres area where the high water-table often prevented deep buries. Wireless particularly came into its own during the mobile operations of 1918, by which time improved sets had become available.[57]

These developments were accompanied by an ever-increasing sophistication in the twin fields of signals intelligence and signals deception. Having been badly caught out by German surveillance in 1915–16, the BEF of 1917–18 showed itself determined to put the boot on the other foot. Significant efforts were made not only to listen in to enemy signals traffic, but also to create decoy targets as cover for friendly operations. The deception plan for the battle of Amiens has often been held up as a classic example of this movement, but there were many others.

In common with most of the rest of the BEF, on 1 July 1916 the signals service had lacked the full development and experience required for an operation of that magnitude. Yet also in common with the rest of the army, it was ready and willing to learn by its mistakes. By the end of the year it had transformed itself into a far bigger, more influential and more expert organisation, which was even starting to chip away at the artillery's monopoly of its own signalling. In the sudden advances of March 1917 the RE signals adapted surprisingly well to open warfare and simultaneously issued the 'definitive' manual upon which they would fight the remainder of the war.[58] For the set-piece battles of Arras and Messines the signals planning thus became meticulous and very effective, in common with the rest of the general tactical preparations; while at Cambrai there was an impressive signal silence to preserve surprise, followed by a connection of all sets at zero and a slimming down of the network to make it more suitable for rapid advances.[59] Thus once again the signalling arrangements remained well in tune with the shape of the battle as a whole.

Unfortunately the same could also be said of 21 March 1918 when, like the army in general, the signal service was suffering from serious shortages of manpower and equipment. It therefore put too much of its strength in the front line, and too little towards the rear. There were inadequate deep buries on the fronts that were attacked, and visual backups were negated by the heavy mist. As the retreat accelerated there was a veritable holocaust of the pigeon service, with no less than forty lofts falling into German hands; and large quantities of the other signalling apparatus suffered the same fate. Some corps telephone exchanges were left to operate as many as thirteen divisional HQs down a single central cable,[60] although this did at least have the beneficial effect of demonstrating how a minimal system could be made to work surprisingly well in mobile warfare. The lesson was well learned that HQs had to be sited according to the signals plan, and not vice versa. Perhaps still more significantly, the artillery was at last forced to combine its signals with those of the rest of the army. This was a major reform and long overdue.

All these experiences allowed the BEF to deploy a very sophisticated signals service during the Hundred Days, which was fully capable of leapfrogging forward to lay new networks very quickly, whenever the front line moved forward. The combination of leafy terrain and poor weather often made visual signalling difficult, but the signallers always managed to maintain at least adequate intercommunication between HQs by other means. The speed of the advance was further helped by the crushing superiority of British artillery over the enemy's batteries, thereby normally making it unnecessary to bury the cables, although a great deal of damage was still done by friendly traffic (not excluding the road transport of preinflated kite balloons suspended on cable-cutting tethering ropes!). At least the systematic salvaging of old wires allowed the demand for new supplies to be minimised, with further relief coming from a widespread use of wireless.

It would not be too much to suggest that the BEF's most successful battles, whether in static warfare during 1917 or in the mobile autumn of 1918, owed a very great deal to the dedication and expertise of the RE signal service. Signals provided the only means of bridging the formidably wide beaten zone which lay between the general in his château or spacious bunker, and the front line subaltern in his muddy shell hole or fire trench. In 1915 and often in 1916 the signalling resources had proved inadequate to permit coherent correspondence between these two extremities of the command chain; for example, at Trônes Wood in mid-July 1916 Colonel Maxwell ironically complained that '[Our Brigadier was] ensconced in a deep dug-out a mile away'.[61] Nevertheless, by the end of the Somme battle many of the problems of this type were well on the way to finding a solution. German phone-tapping was by then being defeated without any loss of responsiveness; the signalling organisation had by then achieved a proper centralisation of command within itself; and a standardised grid had been established which was robust and universally understood. Some brigadiers were even feeling confident enough to venture forward into No Man's Land by that time, and towards the end of the war it had actually become a formal principle that

> Division, brigade and battalion commanders must be right up where they can see the fluctuations of the battle and command. To command by telephone — basing action on the reports of junior commanders — is a crime. It is better to be cut off [from] communication with the rear than to be out of direct touch with the front.[62]

Even though this often narrowed a commander's overall view of the battle at the same time as it cut him off from his signalling support, the tactical benefits were thought to outweigh the risks. This was an exceptionally significant development for the supposedly 'over-rigid' BEF, and it even brought the additional spin-off benefit that it cut down the volume of signals apparatus that a commander would want to demand.

Despite all these concrete achievements, however, there was still always a certain inevitable delay in the transmission of messages between the front line and the higher command echelons, and in some cases this continued to pose frustratingly intractable problems right up to the end of the war. The most notorious example was the timing of cavalry interventions to exploit fleeting opportunities which had been created by the infantry. Alas, no sureness of touch was ever achieved in this area, since the vital gaps tended to open and shut within a timescale that was faster than the command loop could cope with. Indeed, this single factor is probably sufficient in itself to explain the generalised failure of any Great War attack to produce a decisive breakout. It should not, however, blind us to the fact that for most other purposes, when the action was moving only at the speed of the foot soldier or the tank, the signallers had essentially solved the problem by 1918. By the time of the Hundred Days the general in his château was able to receive sufficient notification of new developments to plan his future moves effectively, while the troops on the ground were kept sufficiently aware of his thinking. This had not been the case in 1915.

Part Four
The BEF's Tactical Achievement

10

Doctrine and Training

> Remember there is little time available and the men are not recruits...
> Lectures and demonstrations alone won't produce results.
> Maxse, *Hints on Training issued by XVIII Corps,* pp.15, 21

It has often been alleged by believers in Teutonic virtuosity that the German general staff habitually made a scientific analysis of all new tactical developments, and then systematically disseminated advanced tactical ideas throughout their army. The British, by contrast, are generally held to have bumbled into new tactics only in a local and accidental way, leaving it at least until 1918 before they made any attempt to centralise tactical analysis or training.[1] Indeed, to many commentators this supposed amateurism of the BEF is even a positive virtue, to be favourably contrasted with the unmitigated militarism of the Kaiser's sinister legions.

Such a picture, however, is surely very misleading. Not only did the BEF begin the war with a single set of centralised tactics, however artfully ambiguous they may have been, but it then made a prodigious official effort to send round detailed questionnaires and collect the lessons of each new battle, as a basis for revising the manuals. We have already glimpsed many aspects of this process in action, whether in the infantry or in the other arms; but it may now be useful to draw all the threads together within the framework of the BEF's training effort.

Captain Partridge and the dissemination of doctrine

On 13 August 1914 GHQ took a historic step which was destined to leave an indelible impression upon its entire war. This was no everyday matter of outmanoeuvring von Kluck, placating General Lanrezac or even persuading Kitchener to release the troops he was holding back for the defence of East Anglia. Instead it was the decision to implement a part of the mobilisation plan, dating back to January 1912, which concerned an apparently obscure branch of the Inspectorate of Communications. In a nutshell, the BEF was given an 'Army Printing and Stationery Depot', and Captain S. G. Partridge was appointed as its head.[2]

Unfortunately Captain Partridge's antecedents have not been discovered by the present author, but it is certain that from the moment of his appointment he became a vital influence in creating, centralising and disseminating the paperwork of the BEF at the full rigorous standard that was so obviously essential for twentieth-century warfare. He proved to be as energetic and judicious as he was loyal, holding the same post at GHQ throughout the war. He gradually rose to the rank of colonel, but he coyly refused every offer of a posting to exotic side-theatres such as Turin, Salonika or Cairo. It very soon became clear that Partridge was a brilliantly successful empire-builder — equivalent in his way to a Lindsay, a Foulkes or a Bragg — while in the matter of producing training manuals he must surely be ranked ahead of such expert tacticians as Jacob, Tudor, Fuller or even Maxse himself.

It was perhaps unfortunate that Partridge's first task was to unpack and distribute a mass of stationery stores arriving at the Channel ports at the very moment when those ports were in the process of being abandoned owing to the enemy's advance. The whole stationery establishment had to be moved hastily south to Nantes and then, when the invasion threat had ebbed, back again to Le Havre and Rouen, and eventually to GHQ in Abbeville. It was November 1914 before the muddle had been properly sorted out, and a number of important shipments were mislaid. For example, there is a particularly pathetic complaint that one truckload of 'typewriters, latrine paper and stationery'[3] was following Partridge around in his various wanderings which, whatever he may have thought of the 'bumf', seems to have created a fixation with typewriters which led him to spend much of the remainder of the war attempting to prevent the issue of these machines to the troops in the field. We should remember that in 1914 a 'typewriting machine' was an exceptionally attractive item, worth about a third of the price of a motorcycle[4] and standing as an almost equally potent symbol of modernity and high technology. In the hierarchy of office machinery it held a position almost equivalent to that of a mainframe computer today. Partridge clearly saw himself as the BEF's chief distributor of this technology, and hence the holder of a far more important office than his humble captaincy might suggest.

On 3 January 1915 Partridge secured a ruling from the Quartermaster General that correspondence in the field was to be discouraged, and therefore that typewriters were to be severely limited.[5] A week later he lamented that 300 had already been issued and that he had to organise a mobile team of mechanics to service them. By 11 March the number had grown to such proportions that he made an order banning the local hire of additional machines, especially since he was soon also having to find motorcycles to speed his mechanics on their rounds. As late as 28 September 1916 he was still fighting a rearguard battle and forbidding the mechanics to service privately owned typewriters; but by then it had been accepted that every infantry and cavalry brigade had a right to an official typewriter, as did many organisations on the line of communication. Only artillery brigades, for some strange reason, were barred from holding them until 11 August 1918.

It is ironic that despite his herculean effort of sustained parsimony, Partridge eventually suffered the undeserved shame of being denounced by Fleet Street. In February 1918 the *Daily Mail* alleged that no less than 35,000 typewriters were 'doing nothing but rust' in the trenches, and that there had been waste on a massive scale. To this he replied it was 'a gross libel', because he rigidly controlled all issues and no more more than around 5,000 machines had been released.[6] There were doubtless numerous unofficially held typewriters which had escaped his policing; but it is true enough that the BEF never had sufficient to please everyone who wanted one. The prevalence of handwritten unit war diaries surely offers very ample evidence of that.

If Partridge may be criticised for exercising a rather negative influence on the BEF's typewriting, his contribution turned out to be far more felicitous when it came to the organisation of printing and, in effect, publishing. At first he was concerned only with quite simple jobs such as standard field-service postcards for British and Indian troops,[7] or the manufacture of rubber stamps for issue to mail censors. All types of label, pro-forma, propaganda leaflet[8] and routine order were contracted out to obliging French printing firms scattered around the base areas. Very soon, however, his operations were being extended to secret documents which demanded secure distribution and, perhaps still more significantly, he began to take over the production of training manuals on a large scale. At the beginning of the war all manuals had been printed in London for the War Office Central Distribution Section (the CDS series),[9] and then sent to France for distribution, often via Partridge's department and sometimes being reprinted by him. Within about six months of the start of the war, however, he was originating and printing his own material, and from the beginning of 1916 this was accorded a separate Stationery Service (SS) coding.

At first this publishing effort was far from easy, since Partridge's manuals immediately encountered a nightmare problem of matching supply with demand, and of reconciling the diktats of mandarins in Whitehall with field commanders at the front, not to mention the fleeting whims of red-tabbed GHQ authorities or the inky practicalities of the printing staff themselves. An especially telling example came with the very first case on record, which was an aeroplane recognition card originally intended to be issued to every man in the BEF. A quarter of a million were duly printed on 6 December 1914; but the Adjutant General then promptly reduced the requirement to only one per officer, making 50,000 at most. Partridge took a furious complaint to his superior, the Inspector General of Communications, and was temporarily placated — only to find that on 8 January 1915 the War Office in London wanted 200,000 after all. A still more blatant incident was the printing of a quarter of a million copies of *Prevention of Frostbite or Chilled Feet* on 9 February 1915, of which the Adjutant General eventually found a home for exactly 103.[10] Obviously there was a web of competing authorities at work, with Partridge caught in the middle; and it is greatly to his credit that by the end of 1915 he had gradually come to impose his own system and coherence upon them. On many occasions he seems to have solved the problem by taking it upon himself to decide just what should be printed, in how many copies and with what degree of urgency.[11] This could not prevent the Director General of Transport from sitting on the proofs of SS 539 — *The Reorganisation of Transport* — for seven months up to July 1917, so that precisely one copy could eventually be printed before the whole thing had become obsolete; but it did at least help to minimise the effects of wasteful overproduction once a particular title had been agreed. During the paper shortage of 1918 Partridge even went one better by mounting a serious campaign of paper recycling.

Sometimes the printing jobs were very localised and ephemeral, such as a notice entitled *You Should Clean Your Teeth* for Number 9 Casualty Clearing Station on 25 April 1915, 5,000 *Instructions for Entertainment* on 24 May, or 37th Division's standing orders on 3 September. One is struck by Partridge's willingness to take on almost any commission, no matter how small or trivial, even down to mess menus or Christmas cards. By around this time, however, he was also starting his 'serious' work in producing extensive manuals for all aspects of trench warfare, ranging from translated French instructions to advice for grenadiers, machine gunners or RFC liaison officers. By 25 December 1915 he had already produced so many diverse titles that his printed lists of them had become formidable documents in their own right.[12]

Then came the preparations for the battle of the Somme, and a new burst of activity in the print shops. Not only was the BEF now laying down its own ideas and instructions for the offensive, but it was routinely issuing translations of captured German tactics.[13] From September the top secret monthly BEF Order of Battle was also produced by Partridge's organisation, few examples of which have survived, since each edition was to be destroyed immediately its successor was received.[14] At around this time, too, the equipment of his presses was being replaced by more modern machinery, and he was setting up an increasing number of out-stations and secondary depots of printed material. Still more important, perhaps, was his growing responsibility for the mass reproduction of aerial photographs and maps,[15] not to mention a regular monthly order, from 9 July 1917, for a million musketry targets designed to satisfy the high command's stress on intensive training in rifle fire.

Perhaps 1917 should be seen as the 'classic time' of the BEF's manual-production, as the lessons of the Somme were digested and disseminated. In previous chapters we have already mentioned a number of the most significant works in this vein;[16] but it may be worthwhile to emphasise here that in almost every case there was a gradual process of development and improvement as additional lessons were learned. Thus SS 135, *Training of Divisions for Offensive Action,* appeared in December 1916, but was largely based on Kiggell's SS 109 of May 1916 as revised by SS 119 of July 1916. Further revised editions were issued in both January and November 1918, making five different versions in all. Similarly SS 214 of August 1918, *Tanks and their Employment in Co-operation with Other Arms,* was a carefully considered replacement for SS 164 of May 1917, then of SS 204 of March 1918, and additionally of several less formal essays and instructions written mainly by J. F. C. Fuller. No one could possibly claim that the official formulation of this particular subject ever stood still for a single minute: there was absolutely no stagnation in the BEF's tactical development.

In some cases there was not merely an official manual to be constantly updated, but a periodical newsletter designed to keep recipients abreast of current trends and developments. For example, SS 139 consisted of a regular series of *Artillery Notes,* from March 1917 onwards, while SS 171 offered *Notes on Inventions and New Stores,* from April, which was joined in July by the SS 184 series, *Gas Warfare, Monthly Summary of Information.* SS 201 gave a regular tactical summary of machine gun operations from October 1917 onwards; and finally the newly established Inspectorate of Training issued a series of fourteen pamphlets, covering many aspects of tactics, which ran from September 1918 to March 1919.[17] This series reached a circulation of around 40,000 on such subjects as 'a day's training for a company' or 'attack formations for small units', but more normally ran at around half that number until the armistice cut it right down to under 5,000.

In view of general admiration for German doctrinal publications, it is paradoxical that the Inspectorate of Training had started its work in the summer of 1918 on the assumption that there were too many British training manuals, all recommending different methods. Its self-celebratory inaugural comic song expressed the point as follows:

> ...This is the tale of the Tower of Babel,
>   Where everyone sings on a different note
> And shouts his opinion as loud as he's able
>   And changes twice weekly the hue of his coat...
>
> ...And while the mad medley was gaily galumphing
> The Persons Above Them were steadily bumping.
>   Shades of St Rollox, of Castor and Pollux,
> You never espied such a powerful tide,
> Such a merciless flow of grandiloquent bollocks.
>   Pamphlets, oh dear, 300 a year,
> And nobody looked at them, no bally fear....[18]

Not only did the Inspectorate of Training itself greatly add to the number of such pamphlets, without ultimately exercising any particularly new powers for enforcing doctrinal uniformity, but it seems highly dubious that the existing pamphlets did in fact fall into the category of unread 'grandiloquent bollocks'. On the contrary, the papers of all the leading tacticians — not least Maxse himself — are full of detailed comments and proof readings of the key manuals from at least the start of 1916. The care with which they were updated is also eloquent of the consistency which GHQ was anxious to maintain between one manual and another — and hence the actual doctrinal uniformity of the BEF. Indeed, Brigadier A. Solly-Flood had been appointed to coordinate the tactical training of senior officers in France in October 1916, and this post had soon been expanded into a 'Directorate of Training' which was intended to embrace all training within the BEF.[19]

By the spring of 1918, however, Lieutenant General Maxse was doubly anxious to establish the credentials of his own newly created command, the Inspectorate of Training, since his former command, XVIII Corps, had been so ignominiously destroyed during the Germans' March offensive. He was indeed lucky that because he had shown himself to be an enthusiast for training and platoon organisation during 1917, he could be moved so tactfully sideways within the BEF when GHQ finally decided to kick him off the fighting line. Unlike Gough and others he was not sent home, but given a slender chance to redeem himself in what to many must have seemed like a classic 'non-job'. *The Song* certainly reflects his frustration at finding that his new staff numbered just twenty-two officers — surely a bitter slap in the face to the former commander of over forty battalions. Being every bit as bumptious and outspoken as any senior officer ought to be, however, he realised that he held a splendid Somme reputation and some influential patrons, and so he determined to put a brave face on developments.

Unfortunately Maxse's personal self-renewal in his new guise as Inspector General of Training appears to have led him too far into a denunciation of all the manuals and doctrines that had gone before. Examples of the non-observance of these manuals by the fighting troops, rather than instances of their inspired application, became a theme as insistent as Maxse's earlier complaints against disorganised or non-regulation platoons and sections. Solly-Flood, Partridge and their multiple works were scarcely mentioned; but local staff disagreements over tactical policy, and minor inconsistencies within existing manuals, were stressed. Barrington Ward, a gifted journalist, was drafted in to edit the training inspectorate's new manuals,[20] as if this work had not been done in the past, and the theme was developed that a long-overdue central core of all-arms 'doctrine'[21] had to be established within the BEF, with experts called upon from the RAF, the Tank Corps, the Engineers, the Machine Gun Corps, the Artillery and, above all, the Infantry. In this process Maxse doubtless did no harm to Britannia's overall war effort, which could only benefit from an enthusiastic sense of system or tactical doctrine — not excluding the establishment of the Training Inspectorate itself. Nevertheless, it is unfortunate that he also helped contribute to the myth that there had been no doctrine or coherent manuals prior to the summer of 1918. That was a convenient fable that he should never have propagated, since it was not only quite false, but in retrospect it also turns out to have excessively distorted subsequent perceptions.

There was perhaps an element of truth in the idea that each British commander was encouraged to interpret official doctrines in his own way and according to his own whims. Thus Haig would insist that 'a commander must be free to command'; and there was certainly no lack of quirky or individualistic ways of setting about this. However, it is clear that such initiatives tended to work according to fashions and conventions — even a 'culture' — that were developed at the centre. For example, Maxse's own 1916 claims to have personally invented new twists in the timing of creeping barrages[22] have to be set alongside many other similar competing claims, not least from the debate between Gough and Jacob. Maxse was by no means the unique tactical innovator that he might like us to believe; nor were his colleagues, whom he denounced for their quirkiness, working in the intellectual vacuum that he implied. In the specific case of creeping barrages on the Somme, for example, the idea had started with General Birch, the artillery adviser at GHQ, who had in effect invited a free discussion within each corps. Admittedly that did not amount to imposing a definite method upon each corps — and in the event that was highly unfortunate — but neither was it the same thing as a total lack of central direction. The system was designed to stimulate local originality through leadership from the centre. Sometimes it succeeded and sometimes it failed; but it is important to remember that it was never a complete free-for-all.

Apart from anything else, Haig was a firm believer in staff methods and regulation paperwork. His general staff took their job seriously, in the best traditions of the 1890s Staff College, and made sure that command was handled systematically. If this sometimes degenerated into excessive 'bumfing', then that was seen as better than too little direction or control. Thus the production of manuals — and the whole 'Partridge' story — already represented a prolonged and intensive effort of doctrinal analysis and dissemination several years before Maxse breezed into the Inspectorate of Training. Partridge himself may perhaps have been no more than the expediter and organiser of publications on behalf of the talented staff of GHQ — and of specialised interest groups such as the tanks, machines guns or gas — although one still suspects that he was often personally and directly responsible for important initiatives which greatly helped the general effort. Be that as it may, the list of SS publications certainly bears eloquent testimony to the broad scope of GHQ's didactic concerns, and to the fact that it invoked not only parochial British experience but also that of the French and Germans, in a way that could surely never be claimed in reverse. At least by the middle of 1916, the BEF was thus being offered a very great deal more than merely some badly digested version of prewar tactics, and it already deployed a large staff organisation that was devoted to thinking carefully about every aspect of what had happened, and teasing out useful lessons for future practice.

Training schools and other exhortations

An important but often overlooked feature of GHQ's thrust to improve standard tactics consisted in the extensive debriefs and questionnaires that were often completed immediately after combat. On some occasions this was done in an attempt to identify scapegoats and incompetents, as in the post-Loos persecution of the 21st and 24th New Army Divisions. On other occasions a very different motive came into play, as some particular organisation or HQ attempted to establish and document its own credentials in order to impress its rivals. The extensive analysis of machine gun tactics conducted by Lindsay and Baker Carr in the spring of 1915 may perhaps have slipped into that category at times, as may Foulkes's statistical investigations into British and German gas attacks.

Thirdly, some of the tactical debriefings were mounted so quickly after the event that they should properly be classified more as 'immediate situation reports' than as detached historical investigations. Despite a welter of subsequent abuse, generals and their staff officers really did often want to know the full detail of what was going on in the mudfields ahead of them, and they were avid to interrogate absolutely every individual who emerged from that zone, regardless of his nationality, state of health or extent of knowledge. To this end they also constructed complex paraphernalia for talking to aircraft and balloons, and on some occasions they even issued pro-formas to assault troops which listed every possible eventuality, dire as well as benevolent, so that a simple tick in the appropriate box would inform the higher HQ of whether the subject was hung up on wire, suffering a deadly counterattack, or even just possibly basking in the pleasures of total victory.[23] Finally, the paperwork required by distant HQs often became so voluminous and so trivial that it drove front-line soldiers to distraction.[24] A reasonable man does not want to be pestered by an urgent requirement to list his platoon's holding of toothbrushes at the moment when he is about to lead it in a headlong charge into the very muzzles of the German guns.

In spite of all these caveats, however, there can be no doubt that on many occasions the chief motive for collecting post-combat impressions was purely and simply an interest in genuine tactical science. Thus General Currie's debriefing of his Canadians over the débâcle of 8 October 1916 was as extensive as it was formative in that particular general's intellectual evolution.[25] At the battle of Messines, also, General Plumer sent out an extensive questionnaire designed to highlight lessons in four distinct areas; namely the effectiveness of tanks, cavalry and machine guns, and the techniques for organising areas that had been captured from the enemy.[26] The Tank Corps demanded a detailed log, called a 'tank battle history sheet', from every vehicle committed to battle.[27] This type of undertaking can only be seen as a very modern 'management response' to the problem of battle, since it helped to define the problem in a manageable way, at the same time as it reduced the bafflement felt by commanders and increased their ability to control events.

Once the generals had satisfied themselves that they understood what had really happened in combat, they could scarcely then be accused of keeping the lessons to themselves. On the contrary, they seem to have become notorious for their incessant stream of 'top-down' tactical exhortations and advice. Thus Dr Dunn of 2/RWF reported that 'A brigade conference was, as usual, a monologue by the GOC.'[28] Doubtless this type of lecture was often extremely irritating to the men who had collected the tactical data in their own persons in battle in the first place; but it may at least be worth noting that such monologues by no means amounted to a failure of central direction. A mass army, and particularly one which had been improvised as rapidly and chaotically as the BEF, was surely only too deeply in need of pithy guidelines from on high. Certainly GHQ's record can scarcely be faulted in terms of its post-combat publications, since it issued a long series of pamphlets entitled *The Experiences of Recent Fighting*, such as SS 111 (on gas), SS 119 (on the Somme 1916), SS 159–161 (on the spring of 1917), SS 172 (on Messines 1917) and SS 218 (on Hamel 1918).

Running parallel to GHQ's constant tactical analysis and production of tactical manuals, there quickly sprang up an archipelago of training schools designed to bring all ranks up to the high standards demanded by modern warfare. In principle this process would start in Britain with various types of basic training, although they were always suspected of being too sketchy, and too ideologically divergent from the orthodoxies of the army in the field.[29] This suspicion was still stronger in the case of colonial troops, who would probably have undergone esoteric local training schemes before being given a chance to forget them all in the course of a long sea voyage to France.[30]

Upon his arrival in France the soldier would be stamped by the BEF's own way of doing things at a 'bull ring' or base depot adjacent to one of the debarkation ports. The curriculum would consist of basics such as drill, route marching, musketry, gas precautions and a few elements of tactics. Then on the third stage of his induction he would join his unit, which might be fighting grimly in the front line, or more likely be engaged in labouring and 'resting' at some point more or less behind it, technically designated as in support, reserve or entirely out of the line. The soldier would then find himself sucked into a regular cycle or 'Jacob's ladder' by which his unit moved successively nearer to the front, then into action, and then successively further away again, stopping a few days on each rung of the ladder. In conventional theory there was a greater likelihood that training would take place in proportion as a unit moved away from the front; but by the middle of 1917 this doctrine was increasingly being questioned by General Maxse. He laid down a complete scheme whereby training should be conducted within each platoon at any time, even when it was fighting in the front line. All it required was ten minutes and a little imagination on the part of the platoon commander, provided only that the platoon was formally organised and could be united in one place under a single leader.[31] Against this view, however, may be set Brigadier Jack's sceptical opinion that

> These very detailed Training Programmes seem to regimental officers as little more than eyewash to alleviate the suspicions of Superior Authority that without them troops will perform no training, or the wrong kind of training.[32]

Doubtless a very great deal of informal training actually did take place in the trenches, on the basis that survival itself depended upon speedy adjustment to local conditions. To most BEF soldiers, however, 'training' was seen as a distinct activity which involved a formal course of instruction behind the line. If it was training for a specific assault or raid, the troops designated to take part would be pulled back a few miles and given intensive study of maps, air photos and models of the terrain. They would then take part in 1:1 scale reenactments of their duties, in a field specially taped out to the correct proportions. An attempt would be made to simulate such aspects as creeping barrages, the loss of key personnel at critical moments, or the enemy's expected counterattack during the period of consolidation.

In other cases the training might be simply a repeat of the type of curriculum practised at the bull rings, designed to tone up the general military efficiency of a complete unit, or at least a selection of its NCOs and junior officers. This might take place in a battalion, brigade, divisional, corps, army or even GHQ school. Practice differed very widely from year to year and from one formation to another, although there was an attempt to standardise in 1918 with fixed schools no lower than corps level. It is at least fair to say that there was already some sort of school structure applicable to most of the BEF within less than a year of the war's start. Since the turnover of personnel was so high and so unremitting, moreover, it became a concern that could do nothing but grow. This point was highlighted in the winter of 1917–18 by the uphill struggle involved in working up a complete division to full efficiency. This task was deemed to require a stretch of six weeks continuously out of the line; but in the event only one division could be found which fulfilled that criterion.[33]

Another very important stimulus to the growth of training schools was the incessant multiplication of technical specialities. Every new piece of equipment demanded its own group of knowledgeable operators, whether it was a matter of pigeon lofts, tank camouflage sheets, Nissen huts or aero engines; and every operator somehow had to be trained to his speciality. The infantry battalion certainly changed enormously during the course of the war, from an undifferentiated mass of riflemen into a leaner but far more structured group of signallers. grenadiers, Lewis gunners, gas experts, snipers and scouts. Once again, each of these skills needed its own training course, especially since by the end of hostilities an actual majority of all infantrymen was theoretically expected to be at least reasonably conversant with almost all of them.

At one higher remove there were subjects such as machine guns, trench-mortars and artillery cooperation; and beyond them again came such items as air-to-ground liaison, counterbattery location, combined tank-infantry tactics or — the ultimate mystery of all — 'staff writing'. Each one of them could boast its own school, its temporary course, its weekend seminar, or at least its half-day firepower demonstration.[34] Soldiers of any rank might be assigned to almost any of these subjects at almost any time, often on an apparently random basis; although naturally officers tended to be disproportionately heavily represented. Indeed, the more senior he became, the more likely an officer was to spend time in training rather than in the trenches.

The proficient soldier might even hope to graduate to the elevated status of an instructor in such a school or course, on either an accidental or a knowingly deliberate basis. The schools doubtless included as many malingerers among their staff as they did genuine enthusiasts for their particular subject, and motives varied very widely. For example, George Lindsay assertively carved his way into first the GHQ machine gun school at Camiers, and then into the initial foundation of the Grantham school, essentially on the basis of raw talent combined with deep inner convictions about how machine guns should be organised.[35] For Colonel Croft, by contrast, his influential spell in charge of the 9th Division's tactical school in early 1917, in which he elected to teach 'shooting and thinking',[36] represented little more than a diverting incident which lay between two far more serious combat actions in the front line. Then again, GHQ allowed Ronnie Campbell to base his entire Western Front career upon his bloodcurdling 'bayonet' lecture, thereby neatly excusing him from any duty of a more actually perilous nature. Equally Brigadier Cumming drifted into command of the Grantham machine gun school simply because his superiors found it a tactful way to 'rest' him from what had been seen as his failure at Bullecourt in the spring of 1917.[37]

Sometimes the instructors abused their position, and the literature contains many tales of sadistic NCOs at the bull rings whose regime was one of unrelenting physical punishment for their charges.[38] More subtly abusive were the many schools which taught inappropriate or outdated lessons that were of little value to the students. However, they did at least offer a respite from front line duty, and we often read accounts of how training courses seemed to possess an idyllically peaceful atmosphere after the harsh shocks of trench life. They also spiritually refreshed students by throwing them together with men from other units and arms — even different nationalities — whom they would not otherwise have had a chance to meet. Edmund Blunden even went so far as to suggest that 'Probably the underlying cause of the numberless "schools" in the BEF at this time was as much the desire to give officers and men a rest as to instruct them'.[39]

There is nevertheless a very different way of looking at all this, according to which the schools were neither abusive nor merely restful, but actually an important means for disseminating new ideas throughout the BEF. Perhaps this description should not be extended to the many basic training couses, or even to schools for the simpler specialities; but in the case of the more complex subjects it seems fair to suggest that they really did constitute a 'university of higher tactics'. They were surely every bit as capable of dissecting and disgorging the current state of the art of war as were, for example, the staffs of élite divisions and corps which might find themselves grouped together at the spearhead of some

Doctrine and Training

particular battle. Thus the archipelago of schools that was spawned by the BEF, and supported by a monumental production of manuals, surely amounted to a very significant contribution to the ultimate victory.

11

Conclusion

> The Germans can make it rain when they want...
> ...they drain all the water from their trenches into ours.
> Respectively Edmund Blunden, *Undertones Of War,* p.173
> and Wyn Griffith, *Up To Mametz,* p.60

Every major historical event is reevaluated and reinterpreted for every new generation. In the case of the Great War its image provoked a shudder of revulsion during the 1960s, since it seemed to represent only too close a parallel to the costly new war of attrition that was then unfolding in Indochina. By the 1980s, however, the American military establishment was pulling itself together and starting to draw more positive lessons from Vietnam in the shape of a new 'manoeuvre warfare doctrine'. This was linked to high precision firepower and the idea that combat should be possible without significant friendly casualties, both of which ideals would be spectacularly realised in the Gulf in 1991. The 'bad old' generalship of Vietnam could thus implicitly be contrasted with the 'good new' tactics of more recent times and, perhaps predictably, a parallel critique was also allowed to creep into writing about the Great War. During the past decade we have therefore seen a long series of books contrasting the 'bad old' château generals who designed the battles of 1914–18 with a breed of 'good new' tacticians emerging towards the end of the war — most notably the German stormtroops who so nearly broke through in March 1918.

Within the BEF this focus has been centred on a number of technologically innovative formations such as the ANZACs, the Canadians or the Tank Corps; and such figures as Monash, Brutinel and Fuller have been awarded honorary membership of the same club as the storming

Teutonic maestros.[1] However, the parallel with recent times apparently does not permit more than a very small part of the British Army to be similarly honoured within this canon, since it is felt that a contrast with earlier rigidities must be retained. The bumbling British high command of 1915–16 cannot be portrayed as having evolved into an effective director of set-piece attacks in 1917 or of mobile operations during the Hundred Days in 1918, since that would imply a technological awareness and acceptance which is generally deemed to have been alien to its very soul.[2] Indeed, it might even be taken to imply that the bumbling US high command of the 1960s had actually laid the foundation stones for Airland Battle and operation Desert Storm....

Yet when seen from a modern perspective, this entire orientation now appears to be highly suspect, not just because the weapons and tactics of Vietnam were obviously very directly responsible for those of 1991, but also because British tactics on the Somme were already in essence the same as those used by both sides in their decisive manoeuvres of 1918. The exclusive charmed circle which is conventionally drawn around the Great War's 'tactical innovators' — whether Germans, ANZACs or tankies — turns out to be quite illusory, since tactical innovation was a game that almost everyone was playing, even including the woolly old cavalry generals themselves.

Within the BEF the colonials and Tank Corps certainly held no monopoly of good ideas, since GHQ was itself busily encouraging tactical and technological innovation on a much wider basis throughout the war. As early as Neuve Chapelle in March 1915 the essentials of artillery preparation had been demonstrated, while at Loos in September there was a major attempt to base an entire battle on an untested new weapon, gas, which would quickly lead to impressive technical advances by men such as Foulkes and Livens. Soon afterwards the Machine Gun Corps would be formed in ways suggested by the insistent Lindsay, and the full benefits of air superiority would be systematically worked out during the summer of 1916. By the beginning of 1917 the whole shape of modern infantry tactics had been settled, and the artillery was well on the way to technical reforms which would place it ahead of its German confrères for the remainder of hostilities.

Although we have not made a special study of the Germans, we can at least apply some perspective to their famous stormtroopers if we remember that they emerged from precisely the same type of trench-raiding in 1915 that the BEF was conducting with a rather higher frequency and intensity. There was nothing magical about this technique: it was simply an all-arms attack in miniature, and as such it made excellent training for larger operations later. The same ground rules applied to both sides equally and, although there were some minor differences in weaponry, both sides seem to have taken very much the same lessons well to heart.

Thus the Germans began by creating specialist *Sturmabteilungen*, then *Sturmbataillone*, and finally in the winter of 1917–18 they expanded this type of training to embrace a large proportion of their whole army. However, in this evolution the British were generally ahead. They too had organised élite raiding and bombing teams in 1915, complete with recommendations for light machine gun infiltrations, intense mortar barrages and 'mission orders'. These techniques were developed and refined in the course of the Somme fighting, allowing the specialist teams to merge back into the mass of the infantry by the winter of 1916–17 once their skills had become more widely accessible. Some formations were doubtless better at using the new tactics than others — more 'élite' or less disrupted by recent losses — but in principle the same methods were freely available to the whole BEF. It is also interesting that the British went at least one step beyond classic stormtrooper tactics in the matter of hand-grenades. This weapon remained the German staple throughout the war whereas the British, although they had already fully adopted and analysed it before the battle of the Somme, thereafter decisively rejected it. They found it less efficient for the assault than the rifle and bayonet working in conjunction with creeping barrages and a highly developed long-range counterbattery programme.

The BEF's platoon training manual SS 143 of February 1917 is in essence a stormtrooper's handbook, and it includes instructions for a miniature all-arms battle in which every section plays a distinct part. This revolution in tactics was applied to the whole BEF rather than just to a portion of it and, contrary to much recent transatlantic disbelief and denigration,[3] was routinely applied in practice throughout 1917. Admittedly the learning process had been long and painful on the Somme, but it was also sustained and eventually fruitful. Besides, the Germans themselves encountered many horrific difficulties whenever they put their own theories to the test of battle. Their Verdun offensive started well but soon bogged down; their Somme counterattacks degenerated into an overrigid and excessively expensive system, and on the first day at Arras they were caught with their reserves too far away to affect the issue at all. At Messines they were knocked for six. Certainly they took careful heed of these lessons and tried to introduce reforms; but even in purely defensive battles — a genre in which they did undoubtedly achieve greater expertise than the British, mainly by dint of far more practice and deeper digging — they were still marching to a very uncertain trumpet even in the ideal defensive conditions of Passchendaele. Particularly when good weather around the end of September 1917 allowed the BEF to demonstrate the full power of its methods, the Germans found that not one of their theoretical systems actually worked. They chopped and changed desperately from close-in reserves to distant and deliberate counterattacks; but could find no adequate answer[4] until the rain intervened to save them. In this they may perhaps be compared with the North Vietnamese, whose war effort had been largely disarmed in tactical terms during 1968–9 — until they were saved by the intervention of American internal politics.[5]

The perceptual problem for the modern reader comes, as much as anything, from the fact that the British failure and exhaustion at Third Ypres was followed by ultimate failure at Cambrai on 30 November, and then by a series of spectacular German victories in the spring of 1918. The facts seem to be eloquent that the Germans during this period held a tactical mastery that was vastly superior to that of their opponents. It must be remembered, however, that both at Cambrai and in the spring of 1918 the Germans were to some extent knocking down straw men. On both occasions the British were caught outnumbered, overexposed and essentially unprepared for defence. They enjoyed very few of the defensive advantages which the Germans had enjoyed on the Somme or at Ypres, but were suffering from serious shortfalls in preparation time, manpower and guns. It is scarcely surprising if the attack did well against such opposition, nor was it necessarily proof of high expertise on the part of the attackers. On the contrary, there is much evidence to show that in the spring of 1918 the Germans often used very poor tactics, shambling around in 'cricket match crowds' and accordingly suffering anachronistically high casualties.[6] This specifically tactical fact is surely no less important than Ludendorff's failures of strategy, or the effects of the allied blockade, in explaining both the eventual defeat in the early summer and the total collapse in November.

The point has too often been missed that 'infiltration' attacks are possible only where the enemy defences are crumbling or incomplete, for example in the conditions of March 1918 or, for the allies, in the Hundred Days. Where the enemy defences are strong and continuous, no infiltration is possible at first, since no soft spots can be identified in advance. Only an initially linear frontal attack can be expected to succeed and even André Laffargue, 'the father of infiltration tactics', had already realised as much in 1915. By the same token, successive linear waves were needed to push through whatever gaps were then discovered or created. The Germans were not therefore offering a true alternative to the 'bad old' linear assumptions. They possessed no secret panacea with their talk of infiltration, merely the good fortune to encounter some incomplete British defensive systems in misty weather. That was surely a matter as much of luck as of good management, particularly since they seem to have been less aware than the British of the value of smoke as an artificial substitute for mist. The myth of the stormtrooper would thus appear to have been a myth for a considerable portion of the German army in 1918, albeit perhaps not for all of it.

When we turn from the Germans to the BEF, we have seen that many of its non-colonial divisions could be every bit as cohesive and combat-effective as any on the Western Front. By the second half of the war — and that is a crucial proviso — there was a high and ever-growing level of tactical skill throughout the New Armies and territorials, as much as among the regulars. This was greatly helped by a sustained effort of training and doctrinal analysis which was quite as impressive, in its way, as anything produced by the German General Staff. Questionnaires were often issued after attacks, carefully studied and then incorporated into revised training guidelines. Instead of the 1915 stopgap translations from French models, by 1916 there were scores of original BEF publications being written to cover every branch of warfare, all regularly updated and widely distributed. Equally 'Maps to sections' may have been a specifically Canadian cry,[7] but the mass distribution of aerial photographs had been initiated first by the British, and indeed the work of standardising platoons and sections had itself been the special concern of a Coldstream Guardsman, Ivor Maxse. Far from being merely a symptom of rigidity and overcentralised control, as some have seen it, this impetus was actually a precondition for effective decentralisation and the stimulation of local initiatives. It would produce its benefits in myriad small unit actions in which infantry was able to fight its way forward using its own weapons to overcome such enemy strongpoints as had survived the artillery barrage. Admittedly it did not always work, and many were the units which found themselves flung into the cauldron without adequate preparation or leadership; but the point is that a workable system and doctrine did exist, and could be exploited by any formation which possessed the necessary cohesion.

Quite apart from the various American, Canadian and Australian motives for denigrating the BEF's many tactical achievements, the mainstream of conventional criticism has originated from at least two other biased sources. The first, Basil Hart, is easily dismissed. He seems to have plagiarised the reformed infantry tactics of 1917 by making short popularisations of them for the Cambridge volunteers in 1918, then basing his career upon an expansion of them in the official manuals of the 1920s. Little of this tactical work appears to have been original, although he would later make constant attempts to portray it as a revolutionary indictment of the Great War commanders. He would also go on to describe wider applications of the 'indirect approach' which he claimed to see in the German infiltration tactics, although these had probably come to him originally through GHQ's own translations of German documents, and from Maxse's late 1918 advocacy of 'soft spots' as the technique of the British stormtrooper. Maxse himself was not averse to sniping at other British tacticians who had come before him or — like G. C. Wynne and even G. M. Lindsay — to overpraising the Germans. It is perhaps only too easy in war to imagine that one's opponent is a giant and that one's predecessors have been pygmies.

Conclusion 197

A more broadly based criticism of BEF tactics was developed by believers in the tank, which from the very start was often invoked as a modern and mechanical reproach to the 'bad old' style of generalship. It was initially adopted by politicians who saw only the many weaknesses of GHQ rather than its considerable strengths. They were therefore wrong-footed for a moment when GHQ itself seized upon the tank with unseemly enthusiasm in 1916 and early 1917. However, the result was a series of unfortunate incidents in which the tank was committed prematurely and penny-packeted for little tactical benefit. This led to a growing disillusionment among the infantry, who increasingly saw it as an optional extra which should not be allowed to distort the fundamental shape of operations, in very much the same way that they had reacted to the ambiguous results of the gas attack at Loos. There was then a reciprocal resentment of the BEF's high command on the part of the pro-tank lobbyists, although they were soon able to draw solace from the successes achieved by massed armour on the first day of Cambrai. From this example they have subsequently argued that the opportunity for a much bigger and better tank breakthrough was missed during both 1917 and 1918 — not to mention 'Plan 1919' — and they have consistently refused to accept the justifiable reasoning behind GHQ's frequent reversion to penny-packeting throughout 1918. Instead, they have been happy to invoke the events of 1939–41 as a vindication of the rather anachronistic view that tanks could have won the Great War almost unaided, if only they could have been given the proper official support.

The case of the air force is perhaps instructive in this context, since it was established on generally the same basis as the Tank Corps, but earlier and with far more lavish underpinning in terms of both machinery and personnel. Unlike the Tank Corps it even broke completely free from the army in 1918, and was thus enabled to expand, rather than contract, during the interwar period. It certainly enjoyed many advantages which the Tank Corps could only envy from afar during 1916–18, and which would later become a source of recriminations. If ever GHQ can be said to have nurtured a true technological *enfant gâté*, in fact, it must surely have been the RFC–RAF. Yet for all its many undoubted achievements on the Western Front, not even this pampered air force was able to become a genuine 'breakthrough' or decisive weapon for ground warfare until around the year 1991. Until that moment it was always ultimately doomed to remain an auxiliary, even for the Germans in 1940 or the Israelis in 1967; and during the two world wars it was often condemned to be little more than a handmaiden to the artillery and intelligence services. This interpretation has perhaps been disputed just as hotly as the somewhat similar case of the tank; but to the present author it seems to be no less valid for all that. Even if still more resources and political clout could have been put at the disposal of the air forces during the Great War, it is extremely doubtful that they could have offered very much more assistance to the ground troops than they actually did.

If we now return to the tank, we may agree that there were indeed many moments when it was badly used, misunderstood or underresourced; but we still have great difficulty in finding ways in which it could have broken free from its essentially auxiliary status. Before the mass deployment of genuine cruiser tanks supported by all-arms mechanised combat teams, there could not possibly have been a true armoured breakout. At Cambrai and Amiens the pre-cruiser tank surely demonstrated the absolute maximum achievement of which it was capable — and in both cases this fell short of any such breakout. At best it could crush wire, bust strongpoints and boost friendly morale; but with a fighting range of only eight hours, it was effectively denied access to 'the green fields beyond'. For want of anything better, that elusive zone still remained the responsibility of all-arms teams based upon the horsed cavalry, which was itself labouring under as many technological limitations, and as much obstruction and denigration by the infantry, as was the Tank Corps.

Since the innovatively 'mechanical' arms of aircraft and armour were both doomed to be auxiliaries, and for a number of reasons the cavalry failed to find its true place as the decisive weapon of breakout, we are left with a battlefield that was in effect dominated by the infantry: 'The infantry man is the Queen Bee of all that box of tricks — air force, gunners, sappers, tanks and so on — which composes an army'.[8] Perhaps this fact was itself the real worm in the apple, insofar as the infantry would not tolerate any rival, and thus liked to define every other arm not merely as 'auxiliary' but also as 'subordinate'.

The principle of all-arms unity was the part of tactics that was most difficult to achieve, and it was scarcely helped by the infantry's justifiable pride in its own importance and exceptionally heavy sacrifices. There was an élitism about the infantry which was further compounded by the cap badge loyalties of the British regimental system, not to mention the informal pecking-order of élite or 'spearhead' formations which grew up during the war. Thus the 9th Scottish Division — perhaps the finest and most 'tactical' of all — could congratulate itself on having won its many victories without the help of a single tank.[9] In other formations there was sometimes a positive strategy of avoiding all use of the Royal Artillery, who were nicknamed 'dropshorts' for more than light-hearted reasons; while elsewhere there was a similar wariness about being supported by trench-mortars, or by gas, or even by the Machine Gun Corps.

The integration of each new weapon into the infantry's battle was almost always a long and difficult institutional undertaking, fraught with arguments and administrative nonsenses of every sort. In the case of signals, for example, it took over a year to establish a proper chain of

Conclusion

command under the specialists, and almost three years to impress the need for voice discipline upon the loquacious infantry. With handgrenades the specialists were hived off in 1915 to form their own separate 'monastic order', and they were not formally reintegrated with the rank and file of riflemen until February 1917. Trench-mortars never did succeed in forming a separate corps, and sniping with telescopic-sighted rifles remained underdeveloped. Machine guns themselves were a particular source of trouble, since the infantry was reluctant to allow them to be organised as a separate speciality at all, and suspected that the innovative Lewis automatic rifle was merely a cheap substitute for a 'real' machine gun. Once again it would be early 1917 before the full potential of the Lewis was properly understood and embraced at platoon level, and only towards the end of that year that the corollary would be realised, in the shape of a full independent battalion of Vickers guns within each division, capable of providing barrage fire as easily as direct fire. This in turn offered a handy scapegoat for the failures of 21 March 1918, since the newly formed and hence inexperienced machine gun battalion was also the only battalion within each division that could be disowned by the infantry. The bad feeling created by this controversy is eloquently attested by the indecent haste with which the Machine Gun Corps, built up with such difficulty over four years of war, was abolished after the armistice.

In tactical terms the battlefield was often dominated by artillery; yet even here there was a great gulf of misunderstanding which had to be bridged before the infantry would accept a proper relationship with that arm. The hesitant experiments of 1915 and early 1916 have been fully covered in recent literature,[10] and in a sense they did indeed represent the most formative time for the infantry's and artillery's understanding of each other. However, the really decisive moment in infantry–artillery cooperation appears to have come at some point during the Somme battle itself, when both sides finally began to see just how powerful they could be in combination, particularly with a 'neutralising' creeping barrage followed up by a line of self-sufficient fighting riflemen. During the Somme there turned out to be many technical reasons which tended to prevent this understanding from bearing the good fruit of which it was capable — but in 1917 it would certainly flourish with a vengeance.

The central tactical irony of the Great War is perhaps that the infantry, which traditionally held itself so proud and incomparable, depended for so many of its successes in 1917–18 on the artillery, and then not even on a particularly responsive artillery. Forward observation posts were often notable for their absence or desultory service, and tactical signalling across No Man's Land notoriously took a long time to establish. In these circumstances it was difficult to call for alterations in a formal preprogrammed fireplan, which therefore became an element of inflexibility at the centre of any attack. Nevertheless the British creeping barrages were usually both heavier and technically superior to their German equivalents, and were often sufficient not only to blind and numb the defender, carrying the attack through to its objective, but also to break up incipient counterattacks on the far side. It was certainly better to have their support than not to have it, and only against relatively light opposition in the open warfare of the Hundred Days were creepers sometimes dispensed with.

Just as important to the infantry's chances of survival was an efficient counterbattery programme, a field in which the Royal Artillery came to excel. Working closely with aircraft, survey and sound-ranging teams, the heavy guns gradually achieved pinpoint accuracy and came to master the enemy's artillery. Predicted fire allowed the element of surprise to be retained, which went far to achieving success at both Cambrai and Amiens. Overall, a complex 'deep battle' was orchestrated to cover every threat, in a way that seems to have left the famous Colonel Bruchmüller rather far behind.

Despite the various institutional frictions, therefore, the principle was gradually established that tactical success could normally be guaranteed to any commander who worked hard and methodically at coordinating all arms into his plan. It perhaps took far longer than it ought to have done for this realisation to dawn, and many wastefully premature attacks were launched in defiance of the principle, even as late as the Passchendaele battle. Nevertheless, by the end of 1917 it had been repeatedly proved that carefully prepared assaults could be very successful even against strong defences, and that it was better to call them off quickly after the initial success rather than to persist in an improvised way thereafter with tired troops.

Unfortunately this tactical mastery achieved at such heavy cost by the BEF has tended to be obscured by the exhaustion of the troops in the mud of Ypres, and by the disappointments of the following spring. We should remember, however, that these were operational, strategic and political failures rather more than they were tactical ones, and therefore that the BEF was at heart a considerably more efficient fighting machine than it was ever allowed to demonstrate during the winter of 1917–18. In the Hundred Days, by contrast, it would finally and decisively appear in its full stature and sweep all before it.

Appendix 1

Some limitations in the university approach to military history

During my training as a historian in the late 1960s I was exceptionally fortunate to encounter a number of people who seemed to possess a very rare and precious historical gift. This was that they combined a certain inspirational talent in their teaching with the ability to construct an entire past society, in all its multifarious complexity, and hold it intact within their minds. Theodore Zeldin is perhaps the best-known member of this group, for his particularly systematic exposition of all the diverse aspects of *France 1848–1945*; but I personally saw more of Peter Lewis and his fifteenth-century France, of Trevor Aston and his Anglo-Saxon England, and of my own contemporary John Brewer, with his eighteenth-century Britain. Others who seemed to be cast in somewhat the same mould included Brian Harrison with his English nineteenth century and Richard Cobb with his Revolutionary France. I always felt there was something rather special about these people because they aimed for a complete overview, excluding absolutely nothing. They seemed to be going for the 'big prize' of the polymath, combining deep mastery of technical historical apparatus with almost a philosopher's appeal to 'the general reader'.

These are the sort of people who particularly stand out in my memory as the ones who set out to construct mentally a whole society, complete with its political culture, its artistic culture, its religion, sex and journalism, its commerce, gastronomy and metallurgy, its millinery no less than its agriculture or its penal reforms. Many of the other historians whom I encountered preferred to concentrate on one or perhaps several of these aspects, but not on all of them together in quite the same way. They doubtless still knew a very great deal about all the other facets of their chosen era, but they never entirely seemed to embrace them all with the same enthusiasm. They could therefore be classified as 'specialists' rather than 'polymaths', although they were of course no less legitimate as historians for that. If the polymaths were in the first league, the specialists were by no means despised or rejected and they undoubtedly had a great deal to offer, not least to the synthesisers or generalists themselves.

When we look at the specific field of military history, however, we find that a curiously discordant academic doctrine has grown up during the past thirty years or so. According to this doctrine it is in fact illegitimate to be a specialist in military history, and only the military polymath should command any respect at all. If one concentrates entirely upon the technicalities of what happens on the battlefield one is very liable to be dismissed, in the lubricious words of Professor Sir Michael Howard in a 1991 BBC radio interview, as 'childish, immoral and dangerous'. If, on the other hand, one prefers to look at what happens within the Ministry of War, the War Cabinet or the war orphans' homes, one stands a perfectly fair chance of being hailed as a great historian. In a nutshell, the problem for genuine military historians seems to be that the university has apparently decided that whereas history in the bad old days used to be dedicated exclusively to 'kings and battles', it must now be dedicated exclusively to 'politics and society'. This formula unfortunately implies that pure military history, including tactical history, no longer has any place within general history unless it can be artificially linked in some way to either politics or society, or preferably to both.

In order to win respectability within a university environment today, the military historian must apparently make it clear that his main subject of interest is not military history at all, but 'War and Society', 'Strategic Studies' or some definition of 'War Studies' (or its twin brother 'Peace Studies') which includes in its remit such things as international politics, international law, international morality, military economics, military sociology and military ethics. Strictly speaking one does not necessarily have to be a military historian at all, in order to succeed in any of these disciplines. Military historians may sometimes be tolerated and may work their passage if they are accepted, but only if they demonstrate that their interest in pure military history is secondary to some 'higher' social or political purpose.

Of couse it would be magnificent if all military historians could in fact be historical polymaths of the quality and type that I was privileged to encounter at school and at Oxford during the 1960s. But alas! I have to report that in my own experience such people are as rare within the 'War and Society' or 'Strategic Studies' community as they are within the field of general history as a whole. There are admittedly a few Geoffrey Bests and Christopher Duffys who enrich the wider field, but by and large the mere adoption of a 'socio-political' title is by no means the automatic passport to the laurels of the polymath that its users would perhaps like us to believe. Interdisciplinary studies can work well only if the practitioner is genuinely expert in all of the diverse disciplines he is attempting to integrate, and most aspirants are best advised to master one speciality in depth before branching out into others.

What tends to happen in the military field is that a student of pure military history finds he must compromise his interest in the subject in order to qualify for selection to the university. In the process he loses sight of the main target — the events of the battlefield — and wanders off into some other department which is already well subscribed by other university historians. The events of the battlefield are thus neglected by the best-qualified students, and relegated to an ambiguous hinterland inhabited mainly by *anciens combattants,* popularisers and, in all fairness to Sir Michael, even a few 'childish, immoral and dangerous' warmongers.

What the university seems to miss in this process is that by turning its face away from one very important part of human experience — the battlefield — it is wilfully leaving an important gap in its understanding of the total truth. It rests too hastily content with airy generalisations and conventional wisdoms about the battlefield which usually turn out to be wrong, and which therefore corrupt all the remainder of the story. This is not a trivial point concerning merely the colour of the feathers or the stitching of the epaulettes, but a crucial question of methodology which can easily lead to deep structural errors of interpretation.

To take just one example, A. S. Milward's treatment of the 'Blitzkrieg' economy of the Third Reich, *The German Economy at War* (London, 1965), was hailed within the university at the time as the ideal 'war and society' text, since it ostensibly contained both warfare and economics. It is indeed inspiring to read. However, this book now appears to be totally exploded: not only because it misread the German economy itself, but perhaps more crucially because it scarcely even mentioned the inner workings of the war. Its key premise that there was indeed a thing called a 'Blitzkrieg' itself turns out to be entirely mythological, since no such phenomenon played any part in German strategy or tactics at the time. Battlefield events were reported as 'Blitzkrieg' in the Western press; but apparently this was not a characterization that would have been recognised by either the soldiers or the politicians within Germany. Hence from my present point of view this would seem to amount to a highly relevant cautionary tale about the dangers of trying to become a polymath when one is really only a specialist — and about the crying need for the university to extend once again a welcome to the pure military historian.

Appendix 2

A Great War perspective on the American Civil War

One of the most interesting features of the Great War, to me, is the acute similarity between the evolution of the British Army in that war, and the evolution of the two opposing American armies in their Civil War of 1861–5. In both cases there was a massive unplanned and very rapid expansion of a small peacetime regular army into a full-blooded — but militarily innocent — 'nation in arms'. In both cases there were two tragic years of muddle and uncertainty before the resulting hotchpotch armies were able to shake out effectively, or come to terms properly with the tactical necessities of the battlefield. Still more frequent were the failures — in both 1861–3 and 1914–16 — to provide adequate munitions or to comprehend such fundamentals as gunnery, tactical mobility or even basic staffwork. Many were the officers who, like George McClellan in 1862 or 'Boney' Fuller in 1917, found it easy to steal the clothes and postures of the great Napoleon — but entirely impossible to orchestrate even one of his decisive victories. Each of them forgot that the Corsican Emperor's triumphs were really the result of the long period of French military reorganisation and battle experience which had preceded Marengo, and which had emerged only painfully from the initial muddle and confusion of the revolutionary *levée en masse*. Much of the fighting in both 1861–5 and 1914–18 should therefore be compared more with Valmy, Neerwinden or Wattignies, in 1792–4, than with Austerlitz or Jena more than a decade later. Indeed, we will not be stretching the comparison too far if we suggest that the spectacular German victory in France, 1940, was the true 'Austerlitz' to emerge from the 'Neerwinden' of Passchendaele; or that the battle of Berlin in 1945 would become the true 'Waterloo'.

Appendix 2

The Americans in their year of exhaustion of 1864 certainly shared many of the disillusioning British experiences of the winter of 1917–18, insofar as both were showing serious manpower shortages and signs of collapse after three years of heavy fighting. In neither case, therefore, was it possible to exploit the many newly discovered methodologies to the full. As would also be so ironically true of the American war in Vietnam by 1968, it turned out that a genuinely serviceable military solution to the problem was actually being created at the precise moment when the troops were finally losing their sharp edge, and decisive control of operations was being transferred away from the generals into the hands of politically minded men who had been empowered to express national impatience at the generals' slow progress. In all these cases there was, in other words, a marked change of attitude towards a long war at just about the moment when it reached 'half time'. In all three wars (1861–5, 1914–18 and 1965–73) there was a widespread cynicism towards the later stages of hostilities, if not an overt demoralisation, which served to dampen down the impact of the fabulous tactical innovations that had been introduced at such high cost through several years of tribulation and pain.

Leaving Vietnam aside for another forum, there were nevertheless several very important differences between the American Civil War experience and that of the BEF on the Western Front. In both cases a claim was lodged that 'the birth of modern warfare' had finally arrived: but in my earlier studies of Civil War tactics I tried to show just how little true modernity had really been available in the 1860s, and how even the trench warfare phases of that conflict had referred back to the sieges of Vauban rather than forward to the great *Materielschlacht* of the Western Front. It therefore seems particularly appropriate that in the present book I should have examined the converse arguments, by which the Great War can legitimately be represented as a major break with the past, and the foundation stone upon which most of the twentieth century's subsequent 'art of war' has been built. It is my contention that apart from the great size of the initial mobilisation, made possible by the railway, there was relatively little of importance in the initial battles of August 1914 that had changed tactically since 1870; and such changes as were actually seen represented only a similar rate of background evolution to that between Napoleonic times and the American Civil War. The warriors of the Franco-Prussian War would surely have found it just as easy to adapt to Mons and the Bataille des Frontières as the soldiers of Waterloo would have found it to fight at Gettysburg. However, the most important difference between the American Civil War and the Great War is that in the Great War this continuity of evolution was very rapidly shattered.

As early as October 1914 a very big change in the art of combat had already visibly taken place, and this almost immediately led to a clutch of technical advances, new doctrines and new inventions, which would come to a peak during the calendar year of 1917. Unlike the machine gun or ground mine in the Civil War, these were not mere novelties or playthings, but structurally serious tactical elements that were very widely disseminated and used. Taken together, they would finally alter the battlefield indelibly and forever, in a way that had not truly happened during the 'War Between the States'.

If the Americans of 1861–5 did often manage to adapt themselves to new conditions relatively quickly, and at least roll with such punches as modernity tried to throw at them, the soldiers of the Great War may fairly claim to have gone much further and faster in reforming their tactical outlook to meet a still more serious brand of modernity. The pace and scale of tactical and technological change was incomparably greater in the early twentieth century than in the mid-nineteenth; yet the participants still seemed to manage to roll with it, more or less, regardless. This picture has nevertheless not been sufficiently widely investigated by historians, who have generally preferred to look at the problem rather than at the many effective solutions that were found during the second half of the conflict.

While conducting this study I was struck by one final difference between the American Civil War and the British Great War, which is that the regular officer corps retained a greater grip on the latter than had been true of the former. Thus the cultural legacy of the pre-Civil War regular army may still linger within some sections of the US Army today; but it is nevertheless very vague and unfocused in its impact on the details of day-by-day military life in the late twentieth century. In Britain, however, the legacy of the pre-1914 regular army is still very much alive and well, even in its daily details, during the modern era of the Gulf War and the reopening of Eastern Europe. The Sam Browne belt and red tabs over khaki service dress; the British warm and the pace stick; the enforced muscular Christianity and distrust of foreigners; the stress on 'top down' presentations unsupported by analysis; the exclusivity of the regular army 'club' and its constituent regiments; the insistence on unbearded moustaches combined with an absolutely numbing hippophilia and deference to rank — all these things already seemed outmoded to members of the democratic new Kitchener armies on the Somme in 1916, yet they all still, astonishingly, remain firmly in place today. The explanation is that, in default of any grandiose social revolution or national collapse, the class-based British regular army was able to weather the Great War successfully, including a strong strand of cohesion and leadership at the junior officer level which transmitted itself successfully to hostilities-only entrants. At higher levels of command the regulars were also able to claim most of the credit for victory — at least in their own eyes — through spokesmen

Appendix 2

like Edmonds. They have therefore generally been able to consolidate their own prewar habits and suppress those of 'temporary officers', even though they recently seem to have adopted a few novel institutional features such as homophobia, the married pad and television.

Appendix 3

Armies, corps and divisions of the BEF

Armies

When the cavalryman Sir John French first led the BEF to war in August 1914 it ranked as equivalent to one of the five French 'armies', albeit a notoriously wayward and independent one that had originally turned up in the firing line rather later than promised and with considerably fewer troops. From a very early date it had to be *encadrée* by reliable French forces on either flank; nor was it ever entrusted with the same length of front as would have been given to a French army of comparable size. Statistically speaking, throughout the war each front-line soldier of the BEF usually had to hold about half as long a frontage as did each of his French comrades in arms. From the Marne to Second Ypres, moreover, and from the Somme to Amiens, he also shared his major battles with the French on many more occasions than the opinion-making British 'establishment' likes to think it is remotely seemly to remember.

Whatever qualitative comparisons may be made, however, the quantitative fact is that the BEF grew to the size of two 'armies' only on Christmas Day 1914, when the cavalryman Haig was given command of the First Army in recognition of his gallant defence, in command of I Corps, at the First battle of Ypres. The infantryman Smith-Dorrien took the Second Army because of, or despite, his brilliant tactical actions as II Corps commander at Mons, Le Cateau and Armentières, although in the political aftermath of Second Ypres he would soon be replaced by the still more brilliant infantryman Plumer, formerly of V Corps. Upon Haig's promotion to command the whole BEF in December 1915, First Army was given first to Monro, of Gallipoli notoriety, and then to the gunner Horne, after he had commanded 2nd Division in 1915 and then XV Corps on the Somme.

The BEF build-up continued fitfully through 1915, including the creation of a Third Army under Allenby — the Mons cavalry hero who had, like Plumer before him, subsequently commanded V Corps. At this time a convenient arrangement became established, and almost standardised, whereby the three armies would fight for most of the remainder of the war lined up alongside each other in the order, reading from left to right, of Second Army, First Army, then Third Army. In the December 1915 aftermath of the battle of Loos, however, the replacement of Sir John French by Sir Douglas Haig in command of the BEF meant that a new broom would start to sweep around many of the existing arrangements.

Haig enjoyed a flow of reinforcements which had suddenly increased significantly, both from the Kitchener and territorial forces being formed at home, and from the veterans returning from Gallipoli. A Fourth Army could thus be created to extend the right of the Third, and this was given to the footguardsman Rawlinson, who had previously been GOC IV Corps in the battle of Loos. His army was furthermore selected to deliver 'the big push' that would win the war, especially since it was to be supercharged by a second echelon army — what today might be called a 'mobile group' — that was intended to pass through from the rear to pursue the beaten enemy and exploit the victory. This vital supercharging force was created in May 1916 and designated as the 'Reserve Army'. It was entrusted to Haig's 'blue-eyed boy' Sir Hubert Gough, the controversial Irish cavalryman who in 1914–15 had commanded first a brigade and then a division in the Cavalry Corps, then 7th Division and finally Haig's own cherished I Corps, leading it at Loos, and thereafter preserving it from any further serious harm forever afterwards.

As would later become normal doctrine for the commitment of mobile groups, Gough's Reserve Army was thrust into the battle of the Somme while the fighting was still only a few hours old. Instead of riding forward triumphantly to Berlin, however, it found only deadlock on the sterile slaughter grounds of Thiepval, Beaumont Hamel and Serre. And there it rested, while Haig seemed to change his ideas about Gough's rôle almost overnight. The Reserve Army quickly ceased to be used as an *Armée de Chasse* against the Germans, but became a political means of restricting Rawlinson's rôle. By 3 July Gough, as Haig's special man, was given control of the less exposed or active half of the attacking forces, while a reduced Fourth Army was given the most difficult assault tasks. Fourth still remained senior to Reserve Army; but they were both now regarded as something of a cut above First, Second and Third Armies, which were to some extent discounted as merely 'garrison' or 'working up' organisations. In October the Reserve Army was redesignated as 'Fifth Army', thereby further regularising its status, even if it also represented a tacit admission that the race to Berlin had been postponed at least for the winter (see Fig. 8).

Fig. 8: Armies in autumn 1916

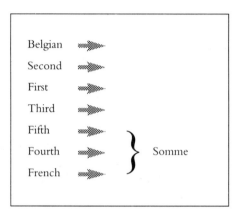

In spring 1917 the Somme victory was somewhat belatedly consummated by the German retreat to the Hindenburg Line, thereby for a time beguilingly reopening the prospect of 'open warfare'. This unfortunately threw the BEF high command into something of a crisis, from which it ultimately failed to produce an effective plan of exploitation within the available timeframe.

Meanwhile there was a feeling that the onus of breakthrough should now be passed to the 'northern armies' which had hitherto been only relatively lightly engaged; so in April Horne's First Army and Allenby's Third Army each took quite impressive swings at the enemy, at Vimy Ridge and Arras respectively. Even if Gough's Fifth Army on the right flank entered the fray only belatedly and without adequate preparation, this effort did generally represent quite a brilliant and innovative display of low-level tactics. However, it was a disappointingly linear and underexploited battle at its higher levels of command. It was improvised too hastily, and ironically it was precisely the absence of an adequate *Armée de Chasse* which finally prevented a really significant victory. To that extent Haig's overall conception should be seen as something of a regression when compared with his earlier optimism for the Somme. In the spring of 1917, however, he did at least take the hint and resolve to do much bigger and better in the summer.

'The big push' for 1917 was set to be located in Haig's own favourite

Appendix 3 211

stamping-ground in the Ypres Salient. Blighted gas-, bomb- and shell-trap though it might have appeared to most observers, it was a 'chosen battlefield' that GHQ had been systematically denied for at least eighteen months. In the heat of the summer its ground looked particularly firm, dry and inviting — and some of the mine galleries at Messines had undoubtedly been standing ready for use for over a year. Besides, there was considerable political capital to be made out of this battlefield's proximity to the enemy's notorious submarine bases, thereby once again bestowing on the British Army its sempiternal raison d'être — so shamefully mismanaged at Gallipoli — as an intrepid landward extension of the Royal Navy.

Fig. 9: Armies in autumn 1917

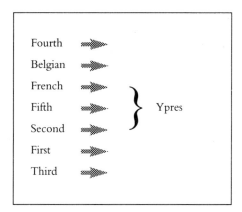

Many of the soldiers who eventually fought in the Passchendaele battle might perhaps be forgiven for believing that they really were in the Navy, or even in submarines, owing to the astonishing height of the water-table in the autumn, and the unrelenting persistence of the rains. As the battle was set up during the early summer, however, these potential difficulties had not yet arisen. Rawlinson's shattered Fourth Army was smoothly withdrawn from the Somme to a relatively quiet sector on the coast, where it was eventually wound up altogether. Meanwhile Plumer's Second Army — by now too long untested — was being carefully primed for its meticulous attack to straighten the line at Messines. This it achieved in exemplary style in June. The centrepiece in the new 'show' at Ypres was nevertheless reserved for the ebullient and thrusting Gough, whose Fifth Army had now been drawn out of its positions facing the Hindenburg Line, to be placed fairly and squarely in the Salient. Gough was very deliberately set up as the main man to win this battle and push through to a pursuit phase, just as he had been on the Somme. He did not, however, have a separate second echelon army to help him, but was expected to make both the break-in and the breakout himself (see Fig. 9).

Just as he had in the battle of the Somme, Haig very soon came to perceive that the Third battle of Ypres, which started belatedly on 31 July, was not running according to plan. With the same fleetness of foot as he had shown in 1916, therefore, he hastened to set aside his main front-line commander and insert a close neighbour instead. On 24 August Fifth Army's frontage was considerably reduced and Second Army's frontage proportionately extended. The 'breakthrough' battle was redefined as a 'siege', and Plumer became the new Gough — being tasked to organise most of the remaining attacks in just the same way that Gough himself had become 'the new Rawlinson' a year before. Meanwhile the real Rawlinson took a well-earned holiday at court, while the real Gough fought on for a while at Ypres but eventually took his Fifth Army off to the extreme southern flank of the BEF, where it was supposed to be well out of trouble. Plumer was left to preside over the final run-down at Passchendaele, while Byng — a cavalryman, formerly GOC Canadian Corps at Vimy Ridge, and now successor to Allenby in command of Third Army — was able to round off the year by an attack over hitherto unbroken ground at Cambrai. His spectacular assault with predicted artillery fire supported by tanks might just possibly have worked, but in the event it led instead to a bitter winter of nasty fighting and political recriminations.

For 1918 the four depleted BEF armies were once again lined up as they had been at the end of the Somme battle, but this time without a Fourth Army at the southern end of the line (see Fig. 10).

Fig. 10: Armies in spring 1918

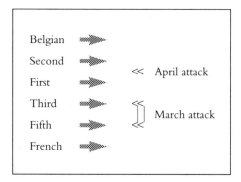

Appendix 3

Since the main German assault was expected in the North, furthermore, the Third and Fifth Armies were given relatively long sectors of front and relatively low levels of support. This made their defeat all the more devastating, at the hands of the 21 March 'Michael' or 'Kaiserschlacht' offensive which has been called 'a battlefield performance and achievement...with no peer in the war'.[1] Fifth Army ceased to exist as an organisation, despite many fully justified protests by its members that they had 'walked' to the rear, rather than 'run'. Not even Haig (or behind him, even the official historian Edmonds) could save Gough from a storm of personal criticism which was surely only too long overdue. He was sent home and his command disbanded, leaving only three BEF armies in the field.

Hard on the heels of this abrupt diminution of the BEF, however, came three very important developments. In the first place 'Michael' put the fear of God so fully into the hearts of the allied commanders that after three and a half years of (often pathetically self-regarding) cobelligerence they all now finally agreed to serve together under the orders of a single commander in chief in the shape of Marshal Foch, the notorious long-time crony of the still more notorious Henry H. Wilson. Secondly, Plumer and Horne did thankfully contrive to defeat Ludendorff's renewed 'Georgette' assault on the Lys in April, with rather fewer bones broken than had at one time seemed likely. This seemed to prove beyond reasonable doubt that the whole tactico-strategic crisis of the German spring offensive could in fact eventually be weathered and even brushed aside, even if it was troublesome to pick up the pieces and fight back over recently lost ground. The BEF was seen to be a far from spent force, after all, which was a most important reassurance to almost everyone who counted for much.

Finally, the BEF was gradually able to reorganise itself into five armies again, as Haig sought to inject fresh (or at least 'rested') talent into the two major sectors in which he had just been forced to concede wide swathes of territory. Rawlinson himself returned, to take over the remnants of Gough's old Fifth Army, boost them with new elements, and rename them 'Fourth Army'. This formation was allocated to what had by now become its 'customary' position on the far right flank of the line, on the Somme. It was there made ready to launch what would soon become a memorable counterstroke, from Amiens through to the Hindenburg Line itself, which would in many ways equal or even surpass the 'peerless' German offensives of March, and certainly regain far more ground than had thereby been lost.

Alongside this new Fourth Army an entirely new Fifth Army was also raised; and it too was designated to retake lost ground at the quick step. It was inserted between Second and First Armies, to the South of Ypres, in order to counterattack across the ground that had been lost in the April battle of the Lys, and eventually to extend through the whole Armentières position which had caused so many problems for so many years. This new Fifth Army was given to Birdwood — yet another cavalryman, albeit one who had enjoyed the extraordinary (if not 'peculiar') privilege of commanding the ANZAC (later I Australian) Corps ever since Gallipoli (see Fig. 11).

Fig. 11: Armies in autumn 1918

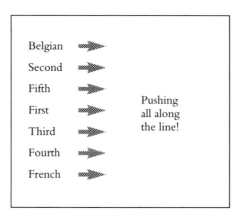

Corps

Within the BEF there was always a great movement of army corps from one army to another — and an even greater movement of divisions from one corps to another. Only within each division, and especially within each brigade, could there be found anything approaching genuine 'continuity' in the relationship between one formation or unit and another. The army corps did nevertheless represent a genuinely important level of leadership, and its commander — even if he might not be on his way up to command an army — tended to become an important, if not necessarily an overpowering, figure in the consciousness of most officers and senior NCOs who stayed with the corps for any length of time. As a member of a very small and select circle, moreover, he would also probably enjoy a pretty stable position in relation to his peers.

Appendix 3

The corps commands of the original BEF were at first very much promotion posts, with both Haig and Smith-Dorrien rapidly rising to command armies, and in Haig's case, of course, eventually to command the whole BEF. Monro, who came from 2nd Division to succeed Haig in I Corps, also went on to command armies, as did his own successor in I Corps, Gough. Rawlinson from the original IV Corps (founded in October 1914), Byng from the Cavalry Corps (also founded in October 1914) and both Plumer and Allenby from V Corps (originally founded at the start of 1915) also rose to army command. However, the footguardsman Pulteney in III Corps (founded September 1914) did not do so well, since he had to stay in the selfsame post until 1918. Something similar was true of the insipid Kavanagh, who quite soon succeeded Allenby in command of the Cavalry Corps, as also the rather more inspiring Haking, who took XI Corps when it was formed in mid-1915. Willcocks and Anderson, the successive commanders of the ill-fated Indian Corps, formed in September 1914 but effectively destroyed and disbanded as a result of the battles of 1915, were packed off after only a year in France, as was Rimington of the Indian Cavalry Corps. Nor did Fergusson, formerly of the regular 5th and then 9th ('K') Divisions, and successor to Smith-Dorrien in II Corps, last long. He was soon stellenbosched and replaced by Jacob — a rare 'Indian' officer to be so favoured, who grew into one of the leading tacticians of Second Army.

There were many rapid promotions during the vertiginously experimental first eighteen months of the war, and some personality clashes which led to top commanders — notably French, Smith-Dorrien and Monro — being stellenbosched. Most of the corps commanders by the end of this era nevertheless remained in place for a long period, and surprisingly few were sent back into the notorious twin sin-bins of 'Home Command' or 'India', indicating that Haig was perhaps rather more loyal to his subordinates than has sometimes been hinted. In view of this continuity, therefore, the first flood of BEF corps commanders must be seen as constituting a particularly privileged but rapidly upwardly-mobile élite that would not be reproduced among later generations of officers. Only Birdwood of the ANZACs and Cavan of XIV Corps subsequently rose to army command, and then only late in the war. The egregious Henry Wilson, who commanded IV Corps for a time in 1915, was exceptional since he pursued his erratic way to a field marshal's baton largely outside the normal BEF channels, through court intrigues and elevated liaison work with the French.

Even though few of the 'overlooked second generation' of corps commanders managed to rise any higher than that level, they nevertheless perpetuated a relatively unchanging chain of command throughout the war. As the BEF expanded rapidly through the winter of 1915–16, they were promoted to corps commands as follows:

I Corps:	Holland (formerly GOC 1st Division)
II Corps:	Jacob (formerly GOC Meerut Division)
IV Corps:	Wilson (formerly GOC 4th Division) then Woolecombe (formerly GOC 11th — 'K' — Division)
V Corps:	HD Fanshawe (formerly GOC 1st Indian Cav Div, then Cavalry Corps) then EA Fanshawe (formerly BGRA of 1st Div, then of 1st Corps, then GOC 31st & 11th 'K' Divs)
VI Corps:	Haldane (formerly GOC 3rd Division)
VII Corps:	Snow (formerly GOC 27th — Regular — Division)
VIII Corps:	Hunter-Weston (formerly GOC 29th — Regular — Division)
IX Corps:	Hamilton-Gordon ('Haig's idea of a joke')[2]
X Corps:	Morland (formerly GOC 5th Division)
XI Corps:	Haking (formerly GOC 1st Division)
XIII Corps:	Congreve (formerly GOC 6th Division)
XIV Corps:	Cavan (formerly GOC Guards Division)
XV Corps:	Horne (formerly GOC 2nd Division)
1st Australian Corps:	Birdwood (formerly GOC ANZAC Division)
2nd Australian Corps:	Godley (formerly GOC New Zealand Division
Canadian Corps:	Alderson (formerly GOC 1st Canadian Division) then Byng (formerly GOC 3rd Cavalry Division)

Some of these corps never thereafter changed their commander, although in the normal run of things many others did, just as some completely new corps commands were created. Perhaps surprisingly, rather few of these posts were filled by men who were new to corps command, and in some cases even a previously *dégommé* (stellenbosched) general was put in. An approximate, albeit not exhaustive list of the corps commands would include the following:

III Corps 1918:	Butler (GOC 3rd Bde in 1914, then staff jobs in First Army HQ) and Godley (formerly GOC 2nd Australian Corps)
IV Corps 1918:	Harper (formerly GOC 51st Highland Division, TF)
V Corps 1918:	Shute (formerly GOC 63rd Royal Naval Division)
VII Corps 1918:	Congreve (formerly GOC XIII Corps)

Appendix 3 217

IX Corps late 1918:	Braithwaite (formerly GOC 62nd Division, TF)
X Corps late 1918:	Stephens (formerly GOC 5th Division)
XIII Corps late 1918:	Morland (formerly GOC X Corps)
XV Corps 1916:	Du Cane (BGGS to III Corps in 1914, then staff jobs in RA and Ministry of Munitions), then 1917 de Lisle (formerly GOC 1st Cavalry Division)
XVII Corps 1917–18:	Fergusson (formerly GOC II Corps)
XVIII Corps 1917–18:	Maxse (formerly GOC 18th — 'K' — Division)
XIX Corps 1917–18:	Watts (formerly GOC 38th — 'K' — Division)
XXII Corps, late 1918:	Godley (formerly GOC III Corps)
Canadian Corps 1918:	Currie (formerly GOC 1st Canadian Division)
Australian Corps, 1918:	Monash (formerly GOC 3rd Australian Division)

Command of a cavalry division, the Guards Division or one of the first seven regular infantry divisions — or of 27th or 29th Regular Divisions — was clearly the most reliable path to corps command, and the earlier in the war, the better. 5th Division produced no less than three corps commanders, but it was definitely trumped in turn by the 1914 cavalry, which produced five. Overall this 'regular' group of divisions produced some nineteen corps commanders. Beside this, corps commanders came from only four New Army and three territorial divisions — usually ones with an 'élite' reputation — which was as many as produced by the Staff. There was, however, a strong representation of the colonial divisions, since some nine corps commanders rose through them. There were doubtless some strong political motives for this weighting; but it may also in a sense reflect the 'élite' status of those formations.

Divisions

The BEF which first landed in August 1914 consisted of Allenby's 1st Cavalry Division and five infantry divisions. These were soon joined by 6th Infantry Division, and then in October by 2nd and 3rd Cavalry Divisions; 7th and 8th Regular Infantry Divisions, consisting of troops returning from overseas; and a large Indian Army contingent of two infantry divisions — Meerut and Lahore — plus two cavalry divisions.
Total 10 infantry divisions by Christmas 1914

In 1915 not only did the regular 27th, 28th and Guards Divisions arrive in the battle line, but also the first of the non-regular divisions — both from Kitchener's New Armies (especially 9th Scottish [including the South African Bde], 12th Eastern, 15th Scottish, 20th Light, 21st, 23rd, 24th) and from the territorials (especially 46th North Midlands, 47th London, 48th South Midland, 50th Northumbrian, 51st Highland). The disgracefully unsung 'Colonials' of the Indian Corps had to be withdrawn after their heavy fighting; but 1st Canadian Division had already made its mark in action by this time, and was later successively built up into a corps that would finally include no less than four Canadian divisions. Meanwhile many other British divisions were starting to disembark through the Channel ports.

Total at least 26 infantry divisions by Christmas 1915

Over the winter of 1915–16 many divisions came to France not only from the fierce fighting at Gallipoli — notably the 11th Northern 'K' Division, the redoubtable regular 29th Division and the ANZACs — but also from the New and territorial armies which had been formed at home. In addition there were the Canadian 4th, 5th and Cavalry Divisions. The final line-up of infantry divisions in France by the end of 1916 — which with only relatively minor changes would remain in place until the Armistice — was approximately as follows:

11 Regular divs:	1–8, 28–29, Guards
26 'K' (New Army) divs:	9, 11–12, 14–21, 23–25, 30–41
14 territorial divs:	46–51, 55–59, 61–2…and the unique 63 (RN) Div.
8 'Colonial' divs:	1–4 Canadian, 1–2 & 4 Australian, New Zealand

Total about 59 infantry divisions by Christmas 1916

In 1917 42nd and 66th East Lancs and 67th TF Divisions, 1st and 2nd Portuguese and 3rd Australian Divisions were added; but 5th Regular, 23rd 'K', 41st 'K' and 48th South Midland Divisions moved to Italy, while 28th Regular and 39th 'K' Divisions went elsewhere.

Hence still a total of about 59 infantry divisions by Christmas 1917

In 1918 the 5th & 41st Divisions returned from Italy, while 52nd Lowland TF Division and 74th Dismounted Yeomanry Division returned from Palestine. But many divisions were shattered by the German spring offensives and either completely removed from the line (e.g. the two Portuguese divisions) or reduced to cadre status for at least a time (e.g. 66th Division effectively vanished after the March onslaught on Fifth Army, but reappeared as a spearhead of Fourth Army, complete with a

Appendix 3

reconstituted South African brigade, for the last two months of the war). The temporary attachment of 27th, 30th and 33rd US Divisions scarcely made up the slack during the Hundred Days; but by that time the enemy was even more shattered than the BEF, so the forces in France proved sufficient to finish the job.

Total about 56 exhausted infantry divisions by 11 November 1918

Notes

Chapter 1: Introduction

1 The phrase is taken from Field Marshal Lord Carver's account of another 'points victory' for the British, 'Operation Crusader', 1941. See his *Dilemmas of the Desert War* (Batsford, London, 1986), pp. 50–3. Between August and November 1918 the front did admittedly move forward fairly dramatically. The élite 9th Scottish Division, for example, made an advance of 26 miles out of Ypres between 28 September and 26 October, making a little less than one mile per day. Almost all of it, however, was concentrated into nine of those days – i.e. on 20 days out of the 29 there was an essentially static front: see J. Ewing, *The History of the 9th (Scottish) Division* (Murray, London, 1921), pp. 337–80.

Despite prodigious efforts, commentators such as Cyril Falls and John Terraine have failed to implant 'a feeling of victory' in the public perception of the Great War, which is still widely deemed to have been won – if it was 'won' at all – as much by naval blockade and economic exhaustion as by the action of ground or air forces.

2 The desire for a bloody revenge was felt even by otherwise exceptionally humane soldiers. See e.g. G. D. Sheffield and G. I. S. Inglis, eds, *From Vimy Ridge to the Rhine: the Great War Letters of Christopher Stone, DSO, MC* (Crowood, Ramsbury, Wilts, 1989), p. 140.

3 See e.g. 'Mark VII', *A Subaltern on the Somme* (Dent, London, 1927), pp. 95, 105; and Charles Carrington, *Soldier from the Wars Returning* (Hutchinson, London, 1965), p. 120.

4 Charles Carrington was struck by the great size of the battle, calling it 'six battles of Waterloo in a row, and four of them massive defeats', ibid., p. 114: but actually the major part of the Somme battlefield could have been fitted into just one field of Austerlitz, so it was not as big as all that.

5 Ewing, *The History of the 9th (Scottish) Division*, p. 64.

6 The expression 'the war to end all wars' has always been understood in its literal sense, i.e. the makers of the Versailles treaty really did hope to outlaw war – or at least certain higher forms of it – forever. Echoing a theme from nineteenth-century Utilitarianism, moreover, many observers believed the astronomical financial cost of the war made it economically illogical and therefore theoretically unrepeatable. Ironically, however, conventional usage would suggest that no such happy outcome was actually intended by the original

phrase 'the war to end all wars', since e.g. 'a party to end all parties' would normally indicate a 'very big' (or perhaps 'Great') party: a 'definitive' party upon which all future parties – and presumably many more are hoped for – should be modelled. In this sense, therefore, talk of a 'war to end all wars' could imply that it is hoped to follow the same blueprint in a later (Second) World War – as in fact happened.

7 The title of P. A. Thompson's book *Lions Led by Donkeys* (Laurie, London, 1927) gave a slogan to the whole movement; but see John Terraine's *The Smoke and the Fire* (Sidgwick and Jackson, London, 1980), pp. 170–1, for a discussion of the 1871 origin of this tag, when it was applied to the French army. Its attribution to Hoffman (Ludendorff's Chief of Staff), as given for example in Alan Clark's *The Donkeys: A History of the BEF in 1915* (Hutchinson, London, 1961), would appear to be unsubstantiated.

Denis Winter's *Haig's Command* (Viking, London, 1991) is a still more extreme recent assault on the competence and integrity of the high command.

8 P. Firkins, *The Australians in Nine Wars* (Hale, London, 1972), p. 130.

9 Lord Moyne, *Staff Officer* (Leo Cooper, London, 1987), p. 149. Ironically, it was Jewish activists who would finally kill Lord Moyne, although doubtless for reasons other than his personal opinion of Monash.

10 See Tim Travers, *The Killing Ground* (Allen and Unwin, London, 1987) for an interesting discussion of Haig's attitudes and style of command.

11 Winter, *Haig's Command* is eloquent testimony. See D. R. Woodward, *Lloyd George and the Generals* (Associated University Presses, East Brunswick, NJ, 1983), pp. 98ff., for the political turmoil following Kitchener's death.

12 Compare the front-line infantry's pathetic – almost spiritual – belief in the importance of success or failure in places like Rumania or Warsaw, e.g.

in Ewing, *The History of the 9th (Scottish) Division,* p. 151; J. C. Dunn, *The War the Infantry Knew* (new ed., Jane's, London, 1987), p. 64; G. H. Greenwell, *An Infant in Arms* (new ed., Allen Lane, London, 1972), p. 40. On some occasions it amounted to a sort of 'Cargo Cult' equivalent to that of the Mons Angels themselves. One is tempted to add that if it is indeed true that 'There are no atheists in a foxhole', then they can often look pretty silly once they get back to Blighty....

13 Churchill's technological futurism is found, e.g., in his paper *Variants of the Offensive,* 3 December 1915, in PRO WO 158–831, and it is certainly true that his own tour of duty on the Western Front was at an especially quiet time: Ewing, *The History of the 9th (Scottish) Division,* p. 60. For Lloyd George's bizarre 'eastern' strategy, see Woodward, *Lloyd George and the Generals,* passim.

14 The reader is referred to the modern continuation of this trend in the otherwise magnificent Lloyd George Memorial Museum, at his birthplace near Criccieth.

15 B. I. Gudmundsson, *Stormtroop Tactics* (Praeger, New York, 1989), p. 175. Compare M. Samuels, *Doctrine and Dogma: German and British Infantry Tactics in the First World War* (Greenwood, Westport, Conn., 1992), pp. 152ff., for a still more sustained attack on British officer quality, albeit one based mainly on Simkins's work on 1914–15. For a better-informed corrective, we look forward to G. D. Sheffield's forthcoming London PhD thesis showing the great strengths of the BEF's junior officers.

16 B. H. 'Liddell' Hart, *Memoirs* (Cassell, London, 1965), vol. 1, pp. 37–48. J. A. English nevertheless swallows Liddell Hart's claims at face value in his *On Infantry* (Praeger, New York, 1981), pp. 28, 32.

17 From South Africa we have the work of Ian Uys; from USA we have Bruce Gudmundsson; from Canada Dominick Graham, Tim Travers,

John English and Bill Rawling; from Australia we have Robin Prior and Trevor Wilson...and so on.

One is tempted to add that in terms of on-site battlefield monumentation of the Great War, the biggest monuments and signposts often seem to be associated with the nations which provided the smallest military contingents. Only the Portuguese are an exception.

18   Tim Travers, *How the War Was Won* (Routledge, London, 1992), pp. 86–8, for a timely revelation of German naïveté in tactics during the spring offensive. I am also grateful to Dr Geoffrey Noon for recounting his father's testimony of the same phenomenon at Arras in March 1918.
19   See, e.g., the many German manuals translated by GHQ in 1918 in the 24-part series *Notes on Recent Fighting*, in PRO WO158 70. Note that the importance of these captured documents was emphasised when the larger ones were to be given glossy red covers as a deliberate policy decision: war diary of Army Printing and Stationery Services, 31 August 1917, in PRO WO95 81.
20   Samuels, *Doctrine and Dogma,* represents a particularly extreme modern restatement of Wynne's bias.
21   The debate between believers in 'internal' or psychological factors and 'external' or physical ones is excellently summarised in Travers, *The Killing Ground*, pp. xvii-xix. In common with the present author, Travers is struck that less analysis has been done in the tactical history of the war than in its conduct at higher levels of command. Shelford Bidwell and Dominick Graham's *Fire-power* (Allen and Unwin, London, 1982) is a notably successful exception to this rule.
22   Carrington, *Soldier from the Wars Returning*, p. 106.
23   The non-mobility of the Second World War is well presented in John Ellis, *The Sharp End of War* (David and Charles, Newton Abbot, Devon, 1980).
24   Even Charles Carrington, who argued against the 'poets' and seemed to derive an intense personal boost from his heroic action at Ovillers on 15 July 1916, was patently collapsing under the strain by the time of his second *action d'éclat* at Ypres on 4 October 1917; *A Subaltern's War* (written by Carrington under the pseudonym 'Charles Edmonds', Davies, London, 1929), pp. 168–77.
25   In fairness, Travers's subsequent *How the War Was Won* does make some valid points about 1918. Equally, although Robin Prior and Trevor Wilson's *Command on the Western Front* (Blackwell, Oxford, 1992) concentrates mainly on 1914–16, the authors are currently preparing a new book on Passchendaele, just as is Peter Simkins on the later history of the Kitchener armies.

Other purportedly general works that are heavily biased towards 1914–15 include Arthur Banks's *A Military Atlas of the First World War* (new ed., Leo Cooper, London, 1989), and indeed Thompson's *Lions Led by Donkeys* itself.
26   Lloyd George already believed in the theory of French superiority as early as September 1916, albeit doubtless from unworthy personal motives – see Woodward, *Lloyd George and the Generals*, pp. 105, 146.
27   Even the regular guardsman General Ivor Maxse could make precisely this point in his *Notes on Army Re-organisation*, circulated 5 November 1918; see Maxse papers, Box 69–53–11, File 53.
28   For this and other problems of the campaign, see J. Terraine, *Mons* (Batsford, London, 1960); cf. General Lanrezac's memoirs, *Le Plan de Campagne Français et le Premier Mois de la Guerre* (Payot, Paris, 1920).
29   All this is well described in Peter Simkins, *Kitchener's Armies* (Manchester University Press, 1988).
30   Battle casualties were surprisingly high at first among staff officers, e.g. Major General Sir Thompson Capper at Loos; see Keith R. Simpson, 'Capper and the Offensive Spirit', in *Journal of the Royal United Services*

Institution, 1973, pp. 51–6. However, Lieutenant General Grierson, one of the BEF's two corps commanders in August 1914, was unable to survive even the undisturbed train journey to the front, during which he was said to have been killed by overindulgence in a Fortnum and Mason hamper.

31 R. E. Priestley, *The Signal Service in the European War of 1914 to 1918 (France)* (official Royal Engineer history, Mackay, Chatham, 1921), pp. 9–15 and passim. At mobilisation an infantry battalion could possess its own telephones only if the Colonel had seen fit to buy them out of his own pocket.

32 The first 'blooding' of the New Army certainly did not come on 1 July 1916, since quite apart from the statistical fact that this was actually the 'eighth day of the Somme' for the gunners, rather than the 'first', many of the Kitchener battalions had already spent many months in the line – see, e.g., Tony Ashworth, *Trench Warfare, 1914–18: The Live and Let Live System* (Macmillan, London, 1980), p. 196. The widespread feeling that the Somme was peculiarly the New Army's first really 'big show', and the vital test of its inner character, should also be moderated by the thought that it was also the first *really* big show for the rest of the army as well.

33 Ashworth, *Trench Warfare,* pp. 86ff., for 'thrusters'. Brigadier F. P. Crozier, author of *The Men I Killed* (Joseph, London, 1937), is cited as an example of the genre and it is true that – although certainly voluble – he can scarcely be called either relaxed or often even comprehensible. Compare Lord Moyne, *Staff Officer,* passim, on General Bethell, who had a somewhat similar character, quite unlike the calm but methodical Plumer, who was called a 'plodder rather than a thruster' in at least one newspaper report; Geoffrey Powell, *Plumer, the Soldier's General* (Leo Cooper, London, 1990), p. 193.

34 Ashworth, *Trench Warfare,* pp. 85, 178ff.

35 There is also, of course, a compensating lack of surviving eyewitnesses from the early war battles, due to combat attrition during the subsequent fighting. This is particularly true of the French Bataille des Frontières in August 1914, for which there is an extreme scarcity of detailed non-official tactical reports.

36 Robert Graves, *Goodbye to All That* (new ed., Penguin, London, 1982), p. 101, referring to the spring of 1915.

37 Ibid., pp. 78, 95, 175.

38 Ibid., p. 78. He estimates something like 15,000 casualties in the battalion, but the most recent regimental history (M. Glover, *That Astonishing Infantry* [Leo Cooper, London, 1989]) suggests it suffered around 1,200 dead, which even on a pessimistic multiplier of three wounded to every one killed gives less than 5,000 in total. All the same, this is more than six times the battalion's strength. In the ten-month NW Europe campaign of 1944–5 the average British battalion lost two or three times its strength, which is actually a higher rate of loss.

39 Ashworth, *Trench Warfare,* p. 15.

40 Martin Middlebrook,*The Kaiser's Battle* (Allen Lane, London, 1978), Appendix 9, pp. 405–7, shows that the average subaltern spent over six months with his battalion at the front, not the mythical 'three weeks'; and only one in five was killed. The popular fallacy nevertheless persists that every casualty was killed, hence that there were '60,000 dead' on 1 July 1916, etc. See my 'Small Wars and How They Grow in the Telling', in *Small Wars and Insurgencies,* vol. 2, No. 2, August 1991, pp. 216–29, for a discussion of how casualty counts can tend to get inflated in the popular imagination.

41 Prior and Wilson, *Command on the Western Front,* pp. 203–9, and analysis of the same point in their 'Summing Up the Somme', in *History Today,* November 1991, pp. 37–43.

42 Statistics in what follows are taken from the British Official History,

Military Operations, France and Belgium (OH; for full details, see Bibliography, p.258–76 below).
43 The 'French pantomime' aspect of the war has of course been best caught by Richard Attenborough's film, *Oh What a Lovely War.*

Wyn Griffith (no relation), *Up to Mametz* (new ed., Severn House, London, 1981), for the disillusionment with existing staff methods felt by one who was catapulted rather unexpectedly into the thick of them early in the Somme battle. Brigadier Sir James Edmonds was still dissatisfied with them in the Battle of Albert, OH 1918, vol. 4, p. 293.

Chapter 2: The tactical dilemma

1 H. R. Cumming, *A Brigadier in France, 1917–18* (Cape, London, 1922), p. 264, referring to the 1918 Hundred Days in Third Army.
2 Ibid., p. 192 (defence of the Aisne, May 1918).
3 Lindsay Papers, Machine Gun file A1-i.
4 E.g. OH 1918, vol. 4, p. 82, on the Amiens battle; and pp. 183, 292 on the Battle of Albert; also see discussion of Edmonds in Tim Travers, *The Killing Ground* (Allen and Unwin, London, 1987).
5 Haig, however, seemed to believe that his GHQ was responsible for 'tactics' (meaning what we would today call 'operations') as a field distinct from 'weapons' or 'techniques', which he felt could be left to specialists: Shelford Bidwell and Dominick Graham, *Fire-power* (Allen and Unwin, London, 1982), p. 72.
6 E.g. Lord Moyne, *Staff Officer* (Leo Cooper, London, 1987), p. 13, on the importance of Brigade Majors.
7 The availability of true 'châteaux' did of course vary enormously from one area to another. Joffre himself was at times perfectly ready to conduct the entire war from a humble village schoolroom, whereas BEF divisional HQs did occasionally find themselves ensconced in palatial ancestral homes complete with vintage wine cellars. More normally, however, the division would be run from at least semifortified farm buildings, and the brigade almost always from dug-outs. Telephone links were normally satisfactory upwards from the division, but notoriously intermittent downwards from the brigade.
8 The classic analysis is Colonel C. E. Callwell's *Small Wars* (first published 1896; new ed., Greenhill, London, 1990). Howard Whitehouse, *Battle in Africa 1879–1914* (Fieldbooks, Camberley, 1987) stresses that very high proportions of hi-tech weaponry were habitually deployed by European punitive columns.
9 Title from the stormtrooper Ernst Jünger's influential book *The Storm of Steel* (Chatto and Windus, London, 1929).
10 For artillery, see chapter 8, below.
11 E.g. Tony Ashworth, *Trench Warfare 1914–18: The Live and Let Live System* (Macmillan, London, 1980), p. 65; W. D. Croft, *Three Years with the Ninth (Scottish) Division* (Murray, London, 1919), p. 3.
12 The definitive official infantry complaint against all these irritants was surely Ivor Maxse's pamphlet *Hints on Training, Issued by XVIII Corps* (final version, HMSO, London, August 1918).
13 B. I. Gudmundsson, *Stormtroop Tactics* (Praeger, New York, 1989), pp. 46ff.; M. Samuels, *Doctrine and Dogma: German and British Infantry Tactics in the First World War* (Greenwood, Westport, Conn., 1992), pp. 11ff.
14 SS 119 pamphlet, *Preliminary Notes on the Tactical Lessons of the Recent Operation* (July 1916), p. 7: cf. the later 1916 perception that orders take a minimum of five hours to be passed from divisional HQ to front-line company: II Corps HQ, 'Miscellaneous Notes from Divisions, from recent fighting, 17 August 1916', p. 3, in *More Lessons of the Somme*, PRO WO158 344.
15 A similar timescale seems to have persisted into the desert war of the 1940s: Field Marshal Lord Carver,

Dilemmas of the Desert War (Batsford, London, 1986), p. 143.
16 The present author has himself sometimes been on the receiving end of this type of critique; but he is still not entirely convinced that it is morally different from a hospital orderly protesting that he knows better than the consultant!
17 Siegfried Sassoon, *Memoirs of an Infantry Officer* (new ed., Faber, London, 1965), p. 157.
18 Ibid., p. 160.
19 A. Thomas, *A Life Apart* (Gollancz, London, 1968), p. 89.
20 Ibid., p. 90. Compare OH 1917, vol. 1, pp. 219ff., which paints a picture of a much less easy advance by 6/RWK, with strongpoints having to be taken by volleys of rifle-grenades.
21 J. C. Dunn, *The War the Infantry Knew* (new ed., Jane's, London, 1987), p. 303.
22 Ibid., pp. 341, 293 for the last two examples; Robert Graves, *Goodbye to All That* (new ed., Penguin, London, 1982), p. 196, for Pinney.
23 J. Ewing, *The History of the 9th (Scottish) Division* (Murray, London, 1921), p. 187. It helped that they had previously been commanded by Major General Furse, who originally ordered production of the first smoke shell, and went on to become Master General of Ordnance.
24 Charlotte Maxwell, ed., *Frank Maxwell VC* (Murray, London, 1921), pp. 187–215.
25 Croft, *Three Years with the Ninth (Scottish) Division,* pp. 100–3, 106.
26 Ibid., pp. 22–3; cf. F. P. Crozier, *The Men I Killed* (Joseph, London, 1937), pp. 60, 93. In the Second World War the American combat historian S. L. A. Marshall stressed the need to identify the men in each platoon who would spontaneously shoot at the enemy in a firefight. Crozier's more refined version of this doctrine was to identify the officers who would be prepared to shoot their own subordinates.
27 See Travers, *The Killing Ground,* passim; Tony Ashworth, *Trench Warfare 1914–18,* p. 97, for the philosophy of scapegoating.
28 E. B. Hamley, *The Operations of War Explained and Illustrated* (first publ. 1866; 6th ed., 'brought up to the latest requirements', by L. E. Kiggell, Blackwood, Edinburgh and London, 1907), p. 390. Cf. my *Rally Once Again* (Crowood, Ramsbury, Wilts, 1987) for the American Civil War.
29 This never did apply to the Soviet Union, which has led to huge strategic difficulties ever since 1917. Her frontiers were simply far too vast for any linear defence to be effective, so she always had to rely instead on concepts of mobile counterattack. By (very stark and significant) contrast, the other global superpower – USA – always enjoyed the benefits of absolutely 'free security' at home. Throughout her history her almost equally long borders have been threatened by little more than the occasional Mexican revolutionary or the concentrated might of the Royal Canadian Mounted Police. In these circumstances she has never really needed a 'defence' force on the ground at all.
30 In August 1914 the two sides each threw something like 1,300,000 men into the battle for Belgium and France, making almost 3,250 per mile of front. By the autumn of 1917 Ludendorff had about 150 divisions – well over 2,000,000 men – making at least 5,500 men per mile. At Torres Vedras in 1810 Wellington had little more than 2,000 men manning each mile of front, which was enough to turn back Massena's army in disgust. It is fair comment, however, to point out that the lines were never properly attacked. If they had been, they might have needed considerable additional manpower reserves to keep the fighting troops topped up to full strength.
31 Note that 'a single trench line' actually meant at least two trenches close together – a front-line trench and its immediate supports. There would also be many associated side-trenches for listening posts, firing points, dumps etc., as well as the communi-

32 cation trenches (CTs) running from front to rear.
32 Robin Prior and Trevor Wilson, *Command on the Western Front* (Blackwell, Oxford, 1992), pp. 44ff.
33 Note that this 'ten miles' applies to a battlefield in which the Germans were constantly being forced to cede extra slices of ground, by *force majeure*. When it came to planning their 'customised' Hindenburg Line immediately afterwards, they seemed to believe that a satisfactory modern defence zone could be contained within just four miles. A very similar depth would also be seen in their conceptualisation of the defence at Third Ypres.
34 Naturally no one would wish a major hi-tech battle upon the San Francisco conurbation today; but if such a thing should by some mischance come to pass, it would certainly be fought with far more microchips than were used even in the 1991 Gulf War!
    Compare the extensive open field tests of gas warfare conducted in the centre of Runcorn, on Merseyside, in June 1915. This was the centre of the British chemical industry at the time – and one cannot help thinking that such irresponsible behaviour must rank as vaguely equivalent to, e.g., the testing of atomic bombs in central Detroit during the Second World War...and one is scarcely surprised to read that 'The bargees sailing past shouted abuse at us'; C. H. Foulkes, *Gas! The Story of the Special Brigade RE* (Blackwood, Edinburgh, 1934), p. 43.
35 A major lesson of the spring 1918 retreats was the need for prebelted machine gun (MG) ammunition, to save time coming into action: First Army report, 13 April 1918, in Lindsay papers, File D3, and technical test results in File D4. For further details of MG tactics, see G. S. Hutchinson, *Machine Guns: their history and tactical employment* (Macmillan, London, 1938).
36 See Table 10 (p.148–9) below, for the consumption of artillery ammunition.
37 The relationship between the interfering Haig and the pliantly vague Rawlinson is expertly delineated in Prior and Wilson, *Command on the Western Front*.
38 See especially Second Army minutes for 28 September to 7 October 1917, in PRO WO158 208.
39 The miseries and grandeurs of such gunner moves are evoked in, e.g., P. J. Campbell, *In the Cannon's Mouth* (Hamish Hamilton, London, 1979), pp. 60–4, at the start of the Passchendaele battle. Even in the infantry it was accepted that only fresh troops could make the fifth bound: see Fifth Army *Memorandum on Attacks*, 5 October 1916, in PRO WO158 344.
40 OH 1917, vol. 2, p. 206, has an estimate of German artillery strength at 558 guns on 19 August.
41 M. Farndale, *History of the Royal Regiment of Artillery: Western Front 1914–18* (RA Institution, Woolwich, 1986), p. 218, and OH 1917, vol. 3, p.47, for the German artillery strength at Cambrai.
42 The impossibility of achieving surprise had been noted by the Chief of the General Staff before the battle – OH 1917, vol. 1, p. 178.
43 Ibid., pp. 241, 275 and Sketch 11 opposite p. 241.
44 OH 1917, vol. 2, pp. 294–5; and John Terraine, *The Road to Passchendaele* (Leo Cooper, London, 1977), pp. 273, 279–382.
45 In recent times the historiographical debate has been continued in highly partisan terms by such popular writers as John Terraine for Haig, or Winter for the 'Easterners'. The truth, however, surely lies in some more subtle middle ground between the two.
46 OH 1918, vols 4 and 5, passim. See also the GHQ 1917 discussion of frontages and manpower in files 131–2, 139, in PRO WO158 20. For a recent vindication of French willingness to help in March 1918, see Tim Travers, *How the War Was Won, Command and Technology in the British Army on the Western Front, 1917–18* (Routledge, London, 1992), pp.

65–70.
47 E.g. Robert Graves's experiments with coordinated rifle volleys against machine guns at night, *Goodbye to All That*, pp. 142–3.
48 E.g. Dunn, *The War the Infantry Knew*, pp. 81–2 (2/RWF stand at La Cordonnerie Farm, La Bassée-Armentières sector, 24 October 1914); Charles E. Carrington, *Soldier From the Wars Returning* (Hutchinson, London, 1965), p. 122 (repulse of counterattack at Leipzig Redoubt, Ovillers 18 August 1916). The ability of rifles alone to beat attacks, even without wire, is still boasted about in the basic infantry manual SS 143, February 1917, p. 12.
49 S. L. A. Marshall, *Men Against Fire* (New York, 1947). For the pre-1914 French combat psychologists such as de Grandmaison and de Maud'huy, see my *Forward into Battle* (2nd ed., Crowood, Ramsbury, Wilts, 1990), pp. 84–94.
50 John Ellis, *The Social History of the Machine Gun* (London, 1975). Note that Hollywood likes its weapons to be 'man sized' for macho choreography: even the Vulcan minigun, originally designed specifically for use on aircraft, is now available in a man-portable version!
51 Each machine gun was assessed in some British manuals as being worth anything between three and ninety rifles, with the consensus resting at around thirty; e.g. SS 197, *The Tactical Employment of the Lewis Gun* (GHQ, January 1918), p. 1.
52 Ewing, *The History of the 9th (Scottish) Division*, p. 57, has German defence based mainly on machine guns already at Loos, September 1915. For the same thing on the Somme, see Martel's analysis of July 1916 (intended to define the rôle of tanks) in PRO WO158 834; Lord Moyne, *Staff Officer*, p. 104 (August 1916).
53 Ludendorff's complaint at the poor standard of his men's musketry is cited in John A. English, *On Infantry* (Praeger, New York, 1981), p. 18; cf. R. H. Roy, ed., *The Journal of Private Fraser* (Sond Nis Press, Victoria BC, 1985), pp. 204–6, for very effective German rifle fire against the Canadian at Pozières-Courcelette, 15 September 1916.
54 The easy temptation to overestimate the number of enemy machine guns was noted by the French Captain Laffargue, *Impressions and Reflections of a French Company Commander Regarding the Attack* (HMSO, London, January 1916), p. 10.
55 General Maxse, a specialist on all aspects of battalion organisation, believed the Lewis section should have a minimum strength of nine men; *Organisation of a Platoon*, November 1917, in Maxse papers, Box 69–53–10, File 42. Admittedly the same number of servants was sometimes stretched to operating two guns rather than one, and by the time of his Corps Commander's Conference of 17–20 February 1918 even Maxse had accepted that four men might do the job at a pinch; *Questions and Answers*, p. 1, in same box, File 44.
56 SS 197, *The Tactical Employment of the Lewis Gun*, p. 2.
57 Compare the successful night attack of 14 July 1916 with the many failures of attempts to repeat the trick in subsequent weeks; Prior and Wilson, *Command on the Western Front*, pp. 197, 211.
58 Alan Thomas, *A Life Apart* (Gollancz, London, 1968), p. 97. Compare the account in OH 1917, vol. 1, pp. 264–5, which says the concentration of this shelling was 'such as few observers had ever witnessed'.
59 Dunn, *The War the Infantry Knew*, p. 269.
60 Experiences of a forward artillery observer at Ypres 1917 are in, e.g., P. J. Campbell, *In the Cannon's Mouth* (Hamish Hamilton, London, 1979), pp. 50–1, 74–81, 118–24. The intermittent manning of observation posts was recognised by Maxse, in *Organisation of Infantry Intelligence* (8th Division, 21 December 1917), Appendix D; in Maxse papers, Box 69–53–12, File 58.
61 Campbell, *In the Cannon's Mouth*, pp.

170–3, saw but was not allowed to engage a densely packed mass of Germans on 21 March 1918, near Chapel Hill. He wryly reflected that he won his MC for this failure.

62 E.g. Dunn, *The War the Infantry Knew*, pp. 65, 93; Crozier, *The Men I Killed*, p. 102.

63 J. B. A. Bailey, *Field Artillery and Firepower* (The Military Press, Oxford, 1989), pp. 132, 146–7, 187.

64 In the Cambrai counterattack it took over an hour for the first SOS flares to be seen, OH 1917, vol. 3, p. 176. On 21 March 1918 the effects of the mist were notorious; see Campbell, *In the Cannon's Mouth*, p. 170; and Appendix 2 on artillery in Maxse's account of the battle, Maxse papers, Box 69–53–10, File 45.

65 Despite mist in the initial British attack at Cambrai, visibility was as much as 200 yards – except where it was reduced by smoke; OH 1917, vol. 3, p. 50. During the Hundred Days mist was often still more useful, e.g. at Amiens on 8 August and on the Hindenburg Line on 28 September.

66 J. Ewing, *The History of the 9th (Scottish) Division*, pp. 98-9. Battle casualties seem to have been roughly split at about 58 per cent to shells, 37 per cent to small arms and the remaining 5 per cent to all other causes; John Terraine, *The Smoke and the Fire* (Sidgwick and Jackson, London, 1980), p. 127 (quoting OH).

67 Statistics extrapolated from Terraine's *The Smoke and the Fire*, and his *White Heat: the New Warfare 1914–18* (Sidgwick and Jackson, London, 1980). For the Gas Brigade, see chapter 6, note 56, below.

Chapter 3: Infantry during the first two years of the war

1 The predominant view, from Bidwell-Graham and Farndale (following Anstey), confirmed recently and vividly by Prior and Wilson, is that success depended almost entirely on the presence of artillery in sufficient density to cope with the size of the target to be attacked. See chapter 8, below, for a discussion.

2 Good coverage in Shelford Bidwell and Dominick Graham, *Fire-power* (Allen and Unwin, London, 1982), pp. 22–37; Tim Travers, *The Killing Ground* (Allen and Unwin, London, 1987), pp. 37–82, and his articles.

3 First published in the United Service Magazine, but subsequently reprinted several times, most recently (for the use of the post-Vietnam US army!) by the US Government Printing Office, 1985. In 1909 Swinton published *The Green Curve* as 'Ole Luk'Oie' (Blackwoods, Edinburgh), and soon after the outbreak of hostilities he became the official war correspondent in France for a time, as 'Eye-Witness'. Promoted Colonel, he then made his name most prominently as a leading figure in the development of the tank, and eventually rose to Major General. Short biographical sketch in Travers, *The Killing Ground*, p. 292.

Duffer's Drift won a new lease of life in 1930 when its format was stolen for H. E. Graham's *The Defence of Bowler Bridge* (Clowes, London); and again in 1940 for Colonel G. A. Wade's *The Defence of Bloodford Village* (Gale and Polden, Aldershot).

4 *The 'Soft Spot': An Example of Minor Tactics* (January 1919), in Maxse papers, Box 69–53–12, File 59, no. 13, along with very significant correspondence praising it. This particular manual must surely be seen as the ultimate high point of BEF tactics, and it is therefore particularly unfortunate that it appeared only after the armistice.

5 (Haig's) 1909 Field Service Regulations (FSR) explained in John Baynes, *Morale* (Cassell, London, 1967), p. 48. Compare discussion of this, the 1914 *Infantry Training* and (Haking's) 1914 *Company Training*, in Travers, *The Killing Ground*, pp. 70, 92 etc.; and Bidwell and Graham, *Fire-power*, pp. 24–5, 49–53.

6 Discussion in *The Killing Ground*, p. 71. The text of the maxim itself can

Notes

be found in the May 1916 divisional training instructions for the Somme, OH 1916, Appendix 17, p. 127; and in the contemporaneous Fourth Army tactical notes, ibid., Appendix 18, p. 135.

7   Ibid., p. 136.
8   Most commentators, drunk with unlimited hindsight, have tended to condemn the cult of the offensive as either an unmitigated moral evil or some sort of mass socio-psychological aberration. However, I believe that in the conditions of the nineteenth-century debate – and even of the twentieth-century debate before 1915 – it was a perfectly reasonable position for any thinking soldier with combat experience to adopt...quite apart from the more rarefied breed of military intellectuals. See 'The Empty Battlefield', chapter 3 of my *Forward into Battle* (2nd ed., Crowood, Ramsbury, Wilts, 1990); and note that the methods of the Russian General Skobeloff, the hero of Plevna, were specifically cited in the Fourth Army tactical notes, OH 1916, Appendix 18, p. 136.
9   Bidwell and Graham, *Fire-power*, p. 27.
10 The renown of the Old Contemptibles was firmly established from the very start of the war, both in popular propaganda and in the nostalgia of the regular officers who had commanded them in 1914 and who continued to run the remainder of the war.
11 E.g. Major R. M. Lucock, a key staff officer in preparing the Somme battle, believed the discipline, leadership and tactical training of the New Armies was 'not what obtained in our troops of a year ago'; *Preparations by Fourth Army,* p. 19, in PRO WO158 321.
12 J. C. Dunn, *The War the Infantry Knew* (new ed., Jane's, London, 1987), p. 284.
13 E.g. Charles E. Carrington, *Soldier from the Wars Returning* (Hutchinson, London, 1965), p. 121.
14 Bidwell and Graham, *Fire-power,* pp. 27–8, for the theoretical basis of the 'mad minute', as developed at Hythe.
15 It is worth noting, however, that when the Germans at Mons supposedly mistook the British rifle fire for machine guns they were in fact being opposed by machine guns as well as by rifles; Lieutenant Chilty's letter of 1/50th RWK to the Historical Section, in PRO CAB45 194; see also findings of the GHQ machine gun school in G. M. Lindsay's lecture in Chelmsford Corn Exchange, 10 August 1915, pp. 65ff., in Lindsay papers, Machine Gun file A–22.

Against this, there seems little basis beyond blind Teutophilia for M. Samuels's view, in *Doctrine and Dogma: German and British Infantry Tactics in the First World War* (Greenwood, Westport, Conn., 1992), p. 164, that the Germans were actually better at high-volume rifle fire and fieldcraft than the BEF of 1914.
16 The cohesion and determination of the Coloniale do not appear to have served it at all well when almost a whole division was destroyed at Rossignol on 25 August 1914.
17 E.g. General T. Capper warned of the dangers of training conducted without 'bullets or nerves' during a Hythe course, 29 May 1914, in Lindsay papers, Machine Gun file 1-R.
18 Baynes, *Morale,* p. 92, echoes Wavell in saying that cohesion is far more important than 'Tactics', and complaining that historians too often get this the wrong way round. For an unexpected endorsement of this view, albeit from a prewar regular battalion which miraculously survived intact for longer than it had a right to expect (i.e. 2/RWF), see Robert Graves, *Goodbye to All That* (new ed., Penguin, London, 1982), p. 156. His battalion's good fortune during the early months of the war is related on pp. 105–6.
19 Well covered in Peter Simkins, *Kitchener's Armies* (Manchester University Press, 1988).
20 Ibid., pp. 279, 287, 290–2, 308. Cf. poor musketry standards in the

American Civil War derived from almost identical causes.

21 See, e.g., *Object and Conditions of Combined Offensive Action,* translated from the French, June 1915, in PRO WO33 717; *Tactical Notes compiled by the General Staff from both British and French Fronts,* 1915, in PRO WO33 725; and especially *Impressions and Reflections of a French Company Commander, Regarding the Attack* by Captain Laffargue, 25 August 1915, translated and issued by HMSO, January 1916, in Liddell Hart Archives 8/65. Apparently this was a reprint from a run of 20,000 issued in December 1915 as 'CDS 333', and should not be confused with the BEF's own separate translation and printing of 1,500 copies at around the same time (but no 'SS' number quoted). See entry in Peter T. Scott, 'The CDS/SS Series', in *The Great War,* vol. 2, no 3, May 1990, p. 108.

Compare this 1915 production of translations with the continuing monitoring of French practice, e.g. in documents of 27 September 1916 and 5 November 1917, in Maxse papers, Box 69–53–12, File 57.

22 Robin Prior and Trevor Wilson, *Command on the Western Front* (Blackwells, Oxford, 1992), especially pp. 25–30, 85, 190.

23 J. Ewing, *The History of the 9th (Scottish) Division* (Murray, London, 1921), p. 60.

24 J. Buchan, ed., *The Long Road to Victory* (Nelson, London, 1920), pp. 125–7. Ewing, *The History of the 9th (Scottish) Division,* p. 251, even claimed the BEF was using infiltration tactics in 1914.

25 Dunn, *The War the Infantry Knew,* p. 156, for rushes of alternate platoons in the 2/RWF at Loos.

26 Ewing, *The History of the 9th (Scottish) Division,* p. 23.

27 Ibid., p. 105; and OH 1916, vol. 2, pp. 69–70.

28 Laffargue, *Impressions and Reflections,* p.4. This insight is as modern as that of Maxse referred to in note 6 above: compare the equally modern ring of much French pre-1914 combat psychology: see my *Forward into Battle,* pp. 84–94.

29 Samuels, *Doctrine and Dogma,* p. 52, quoting Helmuth Gruss, *Die deutsche Sturmbataillone im Weltkrieg* (Berlin, 1939).

30 Laffargue, *Impressions and Reflections,* p. 15. For the German interpretation of the Boers see B. I. Gudmundsson, *Stormtroop Tactics* (Praeger, New York, 1989), p. 24.

31 Gudmundsson, *Stormtroop Tactics* pp.193–4. He suggests that Laffargue's influence on the Germans became known to the British only from German sources and only around 1939; but in view of the assiduous British analysis and dissemination of all captured tactical pamphlets throughout the war itself, the true date is likely to have been very much earlier.

32 E.g., Major General D. Campbell, *Training Notes For Young Officers* (Camberley, March 1918, currently located in Maxse papers, Box 69–53–12, File 58), p. 3.

33 For an account of the many very contradictory impressions of Laffargue's work and influence see Samuels, *Doctrine and Dogma,* pp. 53–4. On p. 54 he shows how a fatal mix-up in this entire debate was provoked by a confusion between Laffargue's *Impressions and Reflections* and a competing pamphlet entitled *L'Etude sur l'Attaque,* which was written by Commandant Lachèvre, a battalion commander in the 74th Infantry Regiment, two days after his attack on 25 September 1915. This was apparently published by the French army in November 1915; by the Paris publishers Plon in the early summer of 1916; and in translation by the British GHQ (as SS 113, *Notes on the Attack*) in June 1916 (see Edmonds, in OH 1917, vol. 2, p. 62, n. 1). Unfortunately the only version I have seen is an early French extract, in Maxse papers, Box 69–53–12, File 57. This does not appear to be the full text; but at least it repeats Laffargue's stress on all arms being held organically within the company,

to fight their own way forward. However, the author offers only a rather diluted version of Laffargue's general views because, contrary to Edmonds's assertion, he was certainly *not* Laffargue.

Related to this is another source of deep confusion which also seems to have been started by Edmonds. This is the completely (i.e. in all respects) erroneous belief that (a) none of this had any influence whatsoever upon either British or French tactical practice...but that (b) it certainly marked a decisive eye-opener for the Germans, when they eventually captured Laffargue's (or as it may perhaps have been, Lachèvre's) work. The modern Teutophile/Anglophobe school seems to have swallowed (a), while loudly but uncritically denying (b). It is thus demonstrably at least half wrong, and demonstrably a chronic case of intellectual confusion in relation to the real Captain Laffargue of May 1915!

34 OH 1916, Appendix 17, pp. 126–7.
35 OH 1916, vol. 1, p. 290. Compare Liddell Hart's equal scorn for the British and respect for the Germans in *The Future of Infantry* (Faber, London, 1933), pp. 26–7.
36 Liddell Hart, *Memoirs* (Cassell, London, 1965), vol. 1, p. 21, suspected that all available copies of the Red Book were burned by GHQ in order to hide the shameful obviousness of these tactics; yet his own furtively preserved copy in the Liddell Hart archives (8–66) appears to be textually identical to the one published openly and prominently in the official history. So much for the 'conspiracy' theory of GHQ cover-ups!
37 OH 1916, Appendix 18, pp. 134–5.
38 This was implied by the Fourth Army 'Red Book' and reinforced by the experience of raiding. Immediately after 1 July the authorities would lay additional, and more specific, stress on the 'mopping up' system; see *Preliminary Notes on the Tactical Lessons of the Recent Operations* (SS 110, elaborating the earlier SS 109), issued by GHQ, July 1916, p. 2.
39 Ibid., p. 1.
40 OH 1916, Appendix 18, p. 142.
41 Ibid., Appendixes 21 and 22, pp. 152–83.
42 Lindsay papers, Machine Gun file A-22.
43 OH 1916, Appendix 18, pp. 131–2.
44 E.g. 2nd Division's *Report on Operations, 17th February 1917*, in PRO WO95 1295.
45 Secret Memorandum by Army Commander, Reserve Army, 3 August 1916, in PRO WO158 344.
46 Prior and Wilson, *Command on the Western Front*, pp. 203–27, and especially their 'Summing up the Somme' in *History Today*, November 1991, pp. 37–43.
47 'Charles Edmonds', *A Subaltern's War* (Davies, London, 1929), pp. 59–107; and compare the same author's *Soldier from the Wars Returning* (this time as Charles Carrington), p. 121. Unfortunately the other Edmonds, in OH 1916, vol. 2, p. 101, notes the success without recognising either the achievement or its true author.
48 The entire focus of Samuels's thesis in *Doctrine and Dogma*, is that British defensive tactics on 21 March 1918 were less good than German defensive tactics during much of the rest of the war. This can surely be taken as read, simply because the Germans were perpetually on the defensive and the British very rarely were – and were in any case actively discouraged from defensive attitudes by their commanders. However, the real question, which Samuels very studiously avoids, is whether or not the British had learned more about attack techniques than had the Germans. The incontrovertible conclusion seems to be that they had indeed done so, and that is the reason why they were ultimately able to defeat the German defences.
49 Colonel C. W. Compton's letter of 26 February 1918 to the Historical Section, in PRO CAB45 194.
50 Report from Major Hilditch, passed on by Harrington, 10 July 1918, in Maxse papers, Box 69-53-11, File 53.

51 For the German concepts originating in raids, see Gudmundsson, *Stormtroop Tactics*, pp. 28–34, 51 and Appendix A; and Samuels, *Doctrine and Dogma*, p. 51.
52 Letter of 28 October 1915 covering Haking's report on 21 Division, in PRO WO158 272. See also general discussion in Travers, *The Killing Ground*, passim.
53 Disapproving but extensive commentary in Tony Ashworth, *Trench Warfare 1914–1918: The Live and Let Live System* (Macmillan, London, 1980), pp. 73–82, 178–98.
54 2nd Division's war diary for 24 January 1917, in PRO WO95 1294. Robert Graves became blasé and depressed after too many raids; *Goodbye to All That* (Penguin ed., London, 1982), p. 142.
55 *Training of Divisions for Offensive Action*, OH 1916, Appendix 17, p. 125.
56 Two examples of the arrangements for trench raids are in OH 1916, Appendices 6 and 7, pp. 42–62. They do not appear to be technically any less sophisticated than the German raids cited in Gudmundsson's appendices.
57 Wyn Griffith, *Up to Mametz* (new ed., Severn House, London, 1981), pp. 116–17, for all this paraphernalia in an unfortunately failed raid near Laventie in May 1916. The Canadians wore face veils for their raid of 16–17 November 1915 at le Petit Douve, OH 1916, Appendix 6, p. 44.
58 Ewing, *The History of the 9th (Scottish) Division*, p. 217, uses the expression for 9th Division action in April 1917; c.f. W. D. Croft for his 27th Brigade, also of 9th Division, in September 1917 – *Three Years with the Ninth (Scottish) Division* (Murray, London, 1919), p. 263. The Australians appear to have used the phrase only in 1918; see Peter Firkins, *The Australians in Nine Wars: Waikato to Long Tan* (Robert Hale, London, 1972), p. 126.
59 The Hamel attack is described in detail in *Notes on Recent Fighting*, no. 19, 5 August 1918, which was circulated throughout the BEF (in PRO WO158 70). Ironically, the Tank Corps traditionally appears to have seen its rôle more clearly in terms of raids than even the infantry, ranging from the 'tank raid' at Cambrai in 1917 to those at Arras in May 1940 and at Sidi Barrani later in the same year.
60 For the whole of this paragraph, see C. H. Foulkes, *Gas! The Story of the Special Brigade RE* (Blackwood, Edinburgh and London, 1934), passim.
61 For the pipe pusher, or 'mole', see chapter 6 below.
62 It should be added that the Germans would find exactly the same problem with their hastily trained 'stormtroop divisions' in the spring of 1918, which often degenerated into a shambling mass of closely packed targets that suffered very heavy casualties.

Chapter 4: The lessons of the Somme

1 *Preliminary Notes on the Tactical Lessons of the Recent Operations* (SS 110), p. 1.
2 PRO WO158 272.
3 Wyn Griffith, *Up to Mametz* (new ed., Severn House, London, 1981), pp. 78–86, 194–224. Artillery cooperation had been stressed in Fourth Army's Tactical Notes of May 1916 (OH 1916, Appendix 18), and would be again in SS 110 of July.
4 Griffith, *Up to Mametz*, p. 237.
5 Charlotte Maxwell, ed., *Frank Maxwell VC* (Murray, London, 1921), p. 146.
6 Chammier, *Report on Operations 1st October as Seen From The Air*, in PRO WO158 344. In fact Le Sars did not fall until 7 October, and the action described here was for the 'Flers switch line': OH 1916, vol. 2, pp. 430–2, 437. For further discussion of creeping barrages and other aspects of artillery support, see chapter 8, below.
7 *Memorandum on Trench to Trench Attacks* by a battalion commander in Fifth Army, 31 October 1916, p. 11, in PRO WO158 344. This 29-page

document was immensely important to official thinking, not merely in Fifth (or 'Reserve') Army, but also at GHQ. Alas, I have not been able to identify the author, who must be counted as one of the ten most influential British tacticians of the war.
8 *Preliminary Notes on the Tactical Lessons of the Recent Operations* (SS 110), p. 2.
9 J. Ewing, *The History of the 9th (Scottish) Division* (Murray, London, 1921), p. 13.
10 W. H. Watson, *A Company of Tanks* (Blackwood, Edinburgh, 1920), p. 87.
11 C. Carrington, *Soldier from the Wars Returning* (Hutchinson, London, 1965), p. 175.
12 Untitled Fifth Army circular giving 'notes on a recent attack', 16 September 1916, in PRO WO158 344.
13 W. D. Croft, *Three Years with the Ninth (Scottish) Division* (Murray, London, 1919), p. 106, at Arras, April 1917; cf. J. C. Dunn, *The War the Infantry Knew* (new ed., Jane's, London, 1987), p. 244, for how men of 2/RWF at High Wood felt helpless without bombs.
14 E.g. *Memorandum on Trench to Trench Attacks*: this document argues for the use of the bayonet alone, yet has two groups of company bombers in the first wave, and an élite force of regimental bombers in the fourth as a reserve. Something similar would be retained in SS 143, *Instructions for the Training of Platoons for Offensive Action*, February 1917.
15 Carrington, *Soldier From the Wars Returning*, pp. 25, 84, 121–2, 174–6, saw bombs as a fashion provoked by trench warfare and taught by the bull rings; but 'the Germans were more attached to them than us' and the rifle came back into favour with front-line British troops early in the Somme battle. Cf. Liddell Hart's wrong belief that the return to rifles and bayonets occurred only in 1918, in *H. G. Wells as a False Prophet*, Liddell Hart papers, 7–1918–25b.
16 Dunn, *The War the Infantry Knew*, p. 355.
17 *Memorandum on Trench to Trench Attacks*, p. 16.
18 Ibid., p. 11.
19 Laffargue, *Impressions and Reflections of a French Company Commander, Regarding the Attack* (HMSO, London, January 1916), pp. 15–16. Compare the opposite tendency of Maxwell, *Frank Maxwell, VC*, p. 155, at Trônes Wood, 14 July 1916, when he told *all* his men to open fire in the advance, as a deliberately devious psychological ploy to maintain them in a deployed line, and so prevent them from lapsing into a vulnerable single file among the shrubbery.
20 Kellett, 28 November 1916, *Lessons learnt during operations 13–17 November 1916*, in 2nd Division War Diary, PRO WO95 1294. Early in the following year this division would start a special school for riflemen and the rifle ethos.
21 Including Wellington and Marshal Ney: see my *Forward into Battle* (2nd ed., Crowood, Ramsbury, Wilts, 1990), chapter 2.
22 *Infantry and Tank Co-operation and Training*, Issued to tank brigades by GS Tank Corps (continuing SS 135, chap. 16), p. 3; in PRO WO158 803.
23 Note that Kiggell was actually a much cleverer tactical commentator, in his own right, than has often been acknowledged. He was not merely Haig's mouthpiece, and was surely very well aware of trench conditions before he 'sent men into them'.
24 Siegfried Sassoon, *Memoirs of an Infantry Officer* (new ed., Faber, London, 1965), p. 11. Compare the equally vivid description in Robert Graves, *Goodbye to All That* (Penguin ed., London, 1982), p. 195.
25 Edmund Blunden, *Undertones of War* (new ed., Collins, London, 1978), p. 80; 'Mark VII', *A Subaltern on the Somme in 1916* (Dent, London, 1927), p. 121.
26 The Rev. R. F. Callaway (a private soldier) cited in John Laffin, *Letters from the Front, 1914–18* (Dent, London, 1973), p. 67; John Terraine, ed., *General Jack's Diary, 1914–18*

(Eyre and Spottiswoode, London, 1964), p. 227; F. Mitchell, *Tank Warfare, the story of the tanks in the Great War* (Nelson, London, n.d.), p. 71.
27   Liddell Hart, *Memoirs* (Cassell, London, 1965), vol. 1, p. 19; Graham H. Greenwell, *An Infant in Arms* (new ed., Allen Lane, London, 1972), p. 142.
28   *Variants of the Offensive*, 3 December 1915, in PRO WO158 831.
29   Maxwell, *Frank Maxwell, VC*, p. 176. The same sentiment is repeated on p. 199.
30   Ibid., pp. 189, 203.
31   *Miscellaneous notes from Divisions, from recent fighting*, 17 August 1916, in PRO WO158 344.
32   G. D. Sheffield and G. I. S. Inglis, eds., *From Vimy Ridge to the Rhine: The Great War Letters of Christopher Stone, DSO, MC* (Crowood, Ramsbury, Wilts, 1989), p. 131.
33   Maxse to Wigram, in Maxse papers, Box 69–53–12, File 58.
34   E.g. in his *Hints on Training issued by XVIII Corps* (HMSO, August 1918), pp. 8, 15.
35   Croft, *Three Years with the Ninth (Scottish) Division*, pp. 18, 59; cf. Ewing, *The History of the 9th (Scottish) Division*, pp. 275, 278, for the maximum musketry ranges being 50–200 yards.
36   Croft, p. 62.
37   Ibid., pp. 71-3.
38   Ibid., pp. 83, 92. To the modern armchair conservationist, however, it is by no means immediately clear whether the tiger or the anarchist would normally have won...
39   Graves, *Goodbye to All That*, p. 114; 'Mark VII', *A Subaltern on the Somme in 1916*, pp. 187–90. Compare Sassoon, *Memoirs of an Infantry Officer*, p. 9, for the 'sporting' instruction in sniping he received from a former big game hunter.
40   Tony Ashworth, *Trench Warfare 1914–18: The Live and Let Live System* (Macmillan, London, 1980), pp. 57–8.
41   See Maxse's exhaustive lists of battalion organisation in, e.g., Maxse papers, Box 69–53–10, File 42; or PRO WO158 90, *Establishments 27th May 1918*. In 1914 a battalion's theoretical establishment of specialist snipers appears to have been 4; by 1916 it was 32, but by 1917 it had been cut back to 16.
42   The SS 143 manual, *Instructions for the Training of Platoons for Offensive Action, 1917* (14 February 1917), envisaged one sniper and one scout per section.
43   At the time of writing I have unfortunately not obtained sight of Hesketh-Pritchard's famous account of the central sniping school, *Sniping in France*.
44   In PRO WO158 344, p. 14.
45   Both the Germans and the French favoured the use of direct-fire infantry cannons or 'assault cannon' during much of the war: see Bruce I. Gudmundsson, *Stormtroop Tactics: Innovation in the German Army 1914–18* (Praeger, New York, 1989), pp. 29, 48, 77ff., for the German 37 mm and captured Russian 76.2 weapons. The French 37 mm, also used by the Americans, was a lighter and handier gun.
46   Dunn, *The War the Infantry Knew*, p. 339.
47   Ibid., p. 447.
48   E.g. Brigadier Cumming's dismissal by General Shoubridge, 3rd Division, for refusing to renew his attack on Bullecourt prematurely, 12 May 1917: H. R. Cumming, *A Brigadier in France, 1917–18* (Cape, London, 1922), pp. 71–87.
49   See chapter 3, above, for Prior and Wilson's interpretation of the Somme.
50   E.g. Currie's debrief of the Canadian Somme experience, 8 October, came to this conclusion; A. Hyatt, *General Sir Arthur Currie* (National Museums of Canada, Toronto, 1987), pp. 60–1. Jacob also circulated the details of Cumming's meticulously prepared 11 January attack on Munich trench (PRO WO95 1295, 2nd Division 'Diary' file; and compare Cumming's autobiography, *A Brigadier in France, 1917–18*, pp. 27-33).
51   II Corps GS to division commanders,

19 September, in PRO WO158 344; and II Corps GS, *Notes on the Attack*, 12 September 1916, in PRO WO 158–344. It is not quite clear, however, whether his first line still consisted of several waves. Probably it did, since the 99th Brigade's provisional order of attack for 9 February 1917 specified four waves forty yards apart, 'so that as many of the assaulting troops as possible may cross No Man's Land before the hostile barrage closes down'; in PRO WO95 1294.

52 II Corps to division commanders. In avoiding dawn attacks Jacob is at one with the anonymous battalion commander's *Memorandum on Trench to Trench Attacks*, p. 20; with Croft, *Three Years with the Ninth (Scottish) Division*, pp. 100–2; and with Currie in his 8 October 'wash up' on the Somme, described in A. Hyatt's biography, *General Sir Arthur Currie* (National Museums of Canada, Toronto, 1987), pp. 60–1.

53 Fifth Army *Memorandum on Attacks*, 5 October 1916, in PRO WO158 344.

54 II Corps directive of 18 September 1916, in PRO WO158 344. Gough's modernity also extended to making personal reconnaissances from aircraft: Liddell Hart papers, 7–1916–36B (information derived from a conversation with John Buchan, from GHQ).

55 For a recent questioning of the rigid German counterattack habit, see Archer Jones, *The Art of War in the Western World* (Oxford UP, 1989), pp. 456, 458, 463; and compare J. Terraine, *The Smoke and the Fire* (Sidgwick and Jackson, London, 1980), pp. 120–123.

56 II Corps to division commanders.

57 Notes on some Bombing Operations in the Reserve Army, 8 October 1916, in PRO WO158 344.

58 Alas, my lack of familiarity with the Great War sources has prevented me from attributing any of these key documents to particular officers, although I feel sure that this will make a fruitful target for subsequent researchers.

59 9th Division claimed in early 1916 to have invented the idea of section-based tactics (Croft, *Three Years with the Ninth (Scottish) Division*, p. 36); cf. John A. English, *On Infantry* (Praeger, New York, 1981), p. 20, who claims Currie was converted to 'sections' rather than 'platoons' following his visit to the Verdun battlefield in January 1917. For Maxse's fluctuating views on this subject, see chapter 5, above.

60 SS 143, p. 14.

61 Ibid., p. 7: a point that is reinforced in SS 197, *The Tactical Employment of Lewis Guns*, p. 1, and especially p. 17.

62 Introductory speech to senior officers' conference 17 February 1918, in Maxse papers, Box 69–53–10, File 44.

63 Carrington, *Soldier From The Wars Returning*, p. 174 (over bombs); cf. Croft, *Three Years with the Ninth (Scottish) Division*, pp. 72–3, who stressed that his battalion's musketry training should be under realistic battle conditions.

64 Carrington, *Soldier From the Wars Returning*, pp. 177–8, says that 'good divisions' had worked out correct tactics by the end of the Somme, i.e. at the same time as SS 143 itself was being elaborated. Misleadingly, however, Carrington believed all this had to wait for Liddell Hart in 1920 before it could be properly codified.

65 Graves, *Goodbye To All That*, p. 145, for the historical reasons why the first and second battalions seemed so different in character.

66 Ibid., p. 152.

67 Graves specifies the 2nd, 7th, 29th and Guards Divisions, although he might just as well have added the 1st, 3rd and 4th, which were also regular.

68 Croft, *Three Years with the Ninth (Scottish) Division*, p. 59.

69 Ibid., p. 87, for the 9th Division's 'tank virginity', although it should be added that in fact four tanks did offer abortive help on the first day of Arras (Ewing, *The History of the 9th (Scottish) Division*, p. 196). We should however remember that a 9th Division officer actually helped to *invent* the tank – i.e. Captain A. I. R.

Glasfurd of 27th Brigade in 1915; see OH 1916, vol. 2, p. 247.
70 Maxse's comments on the Cambrai inquiry, pp. 4–6 (Maxse papers, Box 69–53–11, File 55[1]), points to the advantages of maintaining the same divisions permanently under the same corps commander, staff and training regime.
71 There were two ANZAC corps in 1916 and 1917, until the five Australian divisions amalgamated as 'The Australian Corps' on 1 January 1918.
72 See S. Allinson, *The Bantams* (H. Baker, London, 1981) for the whole pathetic Bantam phenomenon.
73 I.e. the 29th and 51st Divisions, although even then Harper of the 51st wilfully refused to take advice on how his infantry should cooperate with tanks.
74 By way of comparison, Liddell Hart's notes on a conversation with John Buchan in late 1916 record what today looks a rather eccentric selection of Haig and Gough as the true geniuses of 1916. They are followed by Lieutenant Colonel Freyburg (of 63rd Division, who would rise to the doomed command of Crete in 1941) and Generals Horne, Jacob, Pulteney (who commanded III Corps on the Somme), Hunter Weston (who brought VIII Corps from Gallipoli to the Somme). Rawlinson, A. A. Massingberd (4th Army's CoS and CIGS in the 1930s), Congreve and Plumer also win honourable mentions. See Liddell Hart papers, 7– 1916–36b.

Chapter 5: The last eighteen months

1 See Tim Travers, *The Killing Ground* (Allen and Unwin, London, 1987), almost passim.
2 A recurrent shorthand for victory in the Anglo-Saxon Chronicle of a millennium before.
3 E.g. Plumer's fireplan for Passchendaele, 29 August 1917, in PRO WO158 208. The expression 'deep battle' is actually a neologism; but it accurately explains the great extension of range which occurred at about this time.
4 E.g. in OH 1917, Appendices 12, 15 (Canadian plans for Vimy Ridge); PRO WO106 402 (ditto); WO158 215 (2nd Army plans for Messines); and *Machine Guns – Tactical and Technical Lessons Learned*, IX Corps, June 1917, in PRO WO158 418 (also Messines); WO158 208 and 209 (Second Army operations at Third Ypres).
5 Ernst Jünger, *The Storm of Steel* (R. H. Mottram, ed., Chatto and Windus, London, 1929), pp. 107–8, 110 etc.
6 M. Farndale, *History of the Royal Regiment of Artillery: Western Front 1914–18* (RA Institution, Woolwich, 1986), p. 165.
7 Some of the tunnels may still be visited today. I am extremely grateful to the late Dennis Gillard for explaining them – and the whole subject of mine warfare – to me.
8 The total casualties in the first three days were some 13,000: OH 1917, vol. 1, p. 273.
9 J. Ewing, *The History of the 9th (Scottish) Division* (Murray, London, 1921), p. 197.
10 OH 1917, vol. 2, pp. 32–95, and additional details in Farndale, *History of the Royal Regiment*, pp. 184–91. Note that one of the relatively few technical successes of the British over the Germans during the first half of the war had been in mine warfare, in which they had come to dominate the field entirely. The spectacular Messines mines were dug in this period, although they would be fired only in mid-1917. During the second half of the war, by contrast, neither side bothered much with digging new mines, but turned to other technologies in which the British almost always seemed to hold the edge.
11 OH 1917, vol. 1, pp. 241–3.
12 PRO WO158 20, item 128.
13 'Harooshing' had already been identified as a reckless habit before the war, e.g. in General Capper's address to the Hythe course in May 1914 (Lindsay papers, Machine Gun file

1–R). It was applied to Gough, e.g. by Rawlinson in 1917, quoted in Robin Prior and Trevor Wilson, *Command on the Western Front* (Blackwell, Oxford, 1992), p. 270.
14. Lord Moyne, *Staff Officer: The Diaries of Lord Moyne 1914–18* (Leo Cooper, London, 1987), p. 162.
15. For the course of the battle see OH 1917, vol. 2, passim; John Terraine, *The Road to Passchendaele* (Leo Cooper, London, 1977), supplemented for gunner aspects by Farndale, *History of the Royal Regiment*.
16. OH 1917, vol. 2, p. x. See also his sketch 28, which purports to prove by science that even in November the area was mostly dry and well drained!
17. Maxse rather unkindly called them 'Semi-Immobile' Operations, in a cover note to 18th Division's tactical plan of 28 September; Maxse papers, Box 69–53–11, File 54. Wavell called the Somme 'open warfare at the halt', quoted in Shelford Bidwell and Dominick Graham, *Fire-power: British Army Weapons and Theories of War, 1904–45* (Allen and Unwin, London, 1982), p. 99.
18. See *Second Army's Notes on Training and Preparation for Offensive Operations*, 31 August 1917, in OH 1917, vol. 2, Appendix XXV, pp. 459–64.
19. OH 1917, vol. 2, pp. 177, 279, 291.
20. OH 1917, vol. 2, pp. xi, 303.
21. Comments extracted from various Fifth Army reports between 12 and 30 October, in PRO WO158 250 and 251. For the futility of tanks at Ypres see especially F. Mitchell, *Tank Warfare: The Story of the Tanks in the Great War* (Nelson, London, n.d.), pp. 99–126.
22. See Edmonds's special pleading, e.g. OH 1917, vol. 2, pp. 360–362.
23. For the Americans, see Du Cane's extremely critical *Notes on American Offensive Operations*, in PRO WO 106 528 (but compare my own book, *America Invades*, Mallard, New York, 1991, chapter 2).
24. All of this was foreseen in Major Carr's lecture on defence for XVIII Corps, 18 February 1918, at which both Gough and Maxse were present. It was alleged, however, that good training would be enough to solve all problems. Maxse papers, Box 69–53–10, File 44.
25. OH 1918, vol. 1, passim; Martin Middlebrook, *The Kaiser's Battle* (Allen Lane, London, 1978), supported by many more local accounts – e.g. Ewing's *The History of the 9th (Scottish) Division*, pp. 253–93, and many references in the Maxse papers. I am especially grateful to Dr Steven Badsey for first bringing this battle to my attention, in a particularly memorable way, in 1982.
26. Correlli Barnett, *The Swordbearers* (Penguin ed., London, 1966), chapter 4, for the March offensives and the creation of Foch as supreme commander, and compare the discussion in Tim Travers, *How the War Was Won: Command and Technology in the British Army on the Western Front, 1917–18* (Routledge, London, 1992), passim.
27. This was more a French than a British failure. The British contributed only IX Corps to the fighting, and it seems to have objected strongly to French complacency. See Sidney Rogerson, *The Last of the Ebb* (Barker, London, 1937), pp. 6–26.
28. R. E. Priestley's signals history, *The Signal Service in the European War of 1914 to 1918 (France)* (official RE history, Chatham, 1921), pp. 258–76.
29. The BEF was represented at Soissons only by XXII Corps.
30. E.g. H. Guderian, *Achtung Panzer* (English translation, Arms and Armour Press, London, 1991), pp. 96–110ff.
31. Compare note 20, above.
32. Cf. Travers, in *How the War Was Won*, pp. 174, 181, who sounds a note of caution that the losses of the Hundred Days were still, for all the fanfaronade, extremely heavy in real terms. Prior and Wilson, however, point out that Fourth Army suffered only 60,000 battle casualties, although its losses to influenza were higher.
33. Prior and Wilson offer several useful

'operational art' explanations for the success; *Command on the Western Front,* pp. 302–90.

34 Maxse papers, Box 69-53-12, File 57, document of 1 November 1917, quoting from XVIII Corps' platoon commander's course in the previous summer.

35 *Hints on Training issued by XVIII Corps,* p. 8.

36 XVIII Corps letter, 5 February 1917, in Maxse papers, Box 69-53-10, File 42.

37 *Organisation of a Platoon,* November 1917, in Maxse papers, Box 69-53-10, File 42. A similar proposal was put forward from GHQ by Bonham Carter, 5 December 1917, in Maxse papers, Box 69-53-11, File 55(1).

38 Harper's *Guiding Principles in Tactics and Training,* 23 October 1918, p. 2, in Maxse papers, Box 69-53-12, File 58.

39 Dugan's report on 31st Division, 29 August 1918, in Maxse papers, Box 69-53-12, File 60.

40 *Opening Remarks by IGT,* 23 July 1918, in Maxse papers, Box 69-53-11, File 53.

41 Lecture to Corps Commander's Conference, 17 February 1918, in Maxse papers, Box 69-53-10, File 44.

42 SS 143 as elaborated in SS 144, *The Normal Formation for the Attack;* see Maxse papers, Box 69-53-10, File 42.

43 Ibid., Notes (on 24th Division's use of lines in their attack of 7 June 1917) for Maxse's approval; cf. lecture entitled *The Attack,* for his training inspectorate (n.d., July 1918?), in Box 69-53-11, File 53.

44 Maxse papers, Box 69-53-12, File 57, report from 8th French infantry regiment for GOC British I Corps, 5 November 1917.

45 E.g., despite earlier experiments with other arrangements, the 9th Division at Meteren, 10 August 1918, reverted to a skirmish line followed by worms; Ewing, *The History of the 9th (Scottish) Division,* p. 330.

46 Ibid., p. 226. The idea was for the leading 'worms' to push on until they were stopped, which amounted to a type of infiltration.

47 Maxse papers, Box 69-53-10, File 42: *The Counter Stroke,* pamphlet, pp. 2–6.

48 Maxse papers, Box 69-53-11, File 53: account of experience at Amiens by Major General MacDonnell, 15 August 1918.

49 Major Carr's lecture on defence, 18 February 1918, Maxse papers, Box 69-53-10, File 44. In fairness he does talk more about consolidating counterattacks than about defence proper.

50 Brigadier White's report on the fighting of 21–22 March, Maxse papers, Box 69-53-11, File 53. Note, however, that in *Hints on Training issued by XVIII Corps,* p. 11, Maxse retained his earlier penchant for section blobs unedited.

51 Liddell Hart papers, 7/1917/112; *Notes on Platoon Attack Exercise,* returned with corrections by Colonel H. A. N. May, Tidworth, 25 November 1917.

52 *Hints on Training issued by XVIII Corps,* p. 23. An early version of the 'Hints' was his cyclostyled *Notes on Training Infantry Companies,* June 1917, later referred to as the 'Brown Book'. He urges this on his subordinates in, e.g., a lecture to 57th Divisions, 21 November 1917: both in Maxse papers, Box 69-53-10, File 42.

53 XV Corps school, September 1918; Maxse papers Box 69-53-12, File 58. De Lisle had already favoured diamonds in his April 1917 training pamphlet for 29th Division, cited in S. Gillon's *The Story of the 29th Division* (Nelson, London, 1925), pp. 100–2.

54 Maxse papers, Box 69-53-12, File 59.

55 Maxse papers, Box 69-53-10, File 42.

56 Major Bulloch, in Maxse papers, Box 69-53-11, File 54.

57 *Hints on Training issued by XVIII Corps,* August 1918, p. 10.

58 Formally propagated in Inspector General of Training's pamphlet no.4, *Attack Formations of Small Units,*

October 1918 (in Maxse papers, Box 69–53–12, File 59).
59 1st Canadian Division report, section VI, *Observations and Lessons Learned*, p. 2, in Maxse papers, Box 69–53–10, File 46.
60 PRO WO158 422. Edmonds repeats this in OH 1918, vol. 5, p. 575.
61 Inspector General of Training's leaflet no. 13, January 1919, *The 'Soft Spot'. An example of minor tactics* (in Maxse papers, Box 69–53–12, File 59).
62 De Lisle's training pamphlet for 29th Division around April 1917, Gillon, *The Story of the 29th Division*, pp. 100–1.
63 Ewing, *The History of the 9th (Scottish) Division*, pp. 233–4, also p. 241 for the later Lekkerboterbeek attack.
64 Canadian report on Amiens, 15 August 1918, in Maxse papers, Box 69–53–11, File 53.
65 *General Notes* on operations of V, IV and VI Corps in the Hundred Days, in Maxse papers, Box 69–53–12, File 58.
66 OH 1918, vol. 3, p. 255.
67 *Hints on Training issued by XVIII Corps*, p. 5.
68 Ibid., p. 20, for a drill in which three men are supposed to be able to dig a trench six feet deep, adequate to protect them all, within an hour.
69 See my *Battle Tactics of the Civil War* (Yale University Press, 1989). A baroque variant of drill was already being used as a circus act before that war (pp. 101–2 for Ellsworth's Zouaves); but the first serious infantry manual to stress tactical analysis – rather than pure drill – appeared only after it (pp. 103–4 for Upton's 1867 manual).
70 *Hints on Training issued by XVIII Corps*, p. 10.
71 Ibid., p. 6.
72 The first Liddell Hart reference I could find to 'a form of attack... reduced to a drill' is in his *Platoon Attack Exercise* issued by Cambridge University Press, 24 January 1918. The actual phrase 'Battle Drill, or Attack Formations Simplified' seems to have been reserved until October of the same year (also Cambridge University Press): See Liddell Hart papers, 7/1918/2 and 7/1918/7. Yet in his *Memoirs* (2 vols, Cassell, London, 1965), vol. 1, pp. 31–2, he seems to imply that most of the army, apart specifically from the fuddy-duddy ceremonialist Guards (among whom, n.b., Maxse himself must be numbered), was already following his wonderful new inspiration from around the middle of 1918.
73 According to Brian Bond's biography, *The Captain Who Taught Generals* (Cassell, London, 1977), p. 34 n., before 1921 the name 'Hart' stood in exactly the same relation to the more ancient and aristocratic-sounding 'Liddell Hart' as the name 'Anthony Wedgwood-Benn' used to stand in relation to the more modern and popular-sounding 'Tony Benn'. The lust for social credibility is indeed a multifaceted phenomenon!
74 A reference to the 'drill-like practice of tactical exercises' is cited from De Lisle's April 1917 training pamphlet for 29th Division, in Gillon's divisional history, *The Story of the 29th Division*, p. 100.
75 The Liddell Hart archives for 1917–18 (Refs. 7/1916/1–38; 7/1917/1–29; 7/1918/1–28) show that most of the material he published at that time in England was simply a popularisation and elaboration of manual SS 143, which had originated in the BEF overseas. Yet in his *Memoirs*, vol. 1, p. 37, Hart does not shrink from the claim that 'the Army's training manuals of pre-1914 vintage were out of date, while the official training pamphlets that the War Office had produced during the war to supplement them were focussed on trench warfare and did not fit the problem'.

Chapter 6: The search for new weapons

1 Siegfried Sassoon, *Memoirs of an Infantry Officer* (new ed., Faber, London, 1965), p. 38.
2 Edmund Blunden, *Undertones of War*

(new ed., Collins, London 1978), p. 97. The operating manual was issued as CDS 20 in May 1915. Gamages was known to the author in the early 1950s for its exciting conjuring tricks and semi-explosive novelties, so obviously it had changed little in forty years.

3 Alan Thomas, *A Life Apart* (Gollancz, London, 1968), p. 82. Note that major revolutions in both ceramics and plastics would be needed, combined with a zippy high-tech descriptive upgrading into a 'flak jacket', before the humble 'body shield' could ever hope to win general approval. This would occur only in the 1960s.

4 J. C. Dunn, *The War the Infantry Knew* (new ed., Jane's, London, 1987), p. 194.

5 E.g. correspondence of Lieutenant Colonel McMullen, of GHQ, concerning Bacon's schemes and those of his naval constructor, Mr. Lilicrap, 8 January and 15 June 1917, in PRO WO158 20, items 112 and 125. See also OH 1917, vol. 2, p. 117.

6 G. Hartcup, *The War of Invention* (Brasseys, London, 1988), p. 189, shows that less than one invention out of every 200 submitted to the Munitions Invention Department was adopted.

7 Peter Simkins, *Kitchener's Armies* (Manchester UP, 1988), pp. 127-33, 256-95 for an invaluable guide to the procurement chaos which obtained during the first year of the war.

8 For this paragraph see Peter T. Scott, 'Mr Stokes and his Educated Drainpipe', in *The Great War*, vol. 2, no. 3, May 1990, pp. 80-94. This short article is a model of 'procurement historiography'. A shorter version of the story is in Hartcup, *The War of Invention*, pp. 65-7.

9 A major problem was the design of the bombs. One tester reported that 'The ammunition for the Stokes gun is infinitely more difficult to produce than the guns themselves': quoted in C. H. Foulkes, *Gas! The Story of the Special Brigade RE* (Blackwood, Edinburgh and London, 1934), p. 111.

10 Churchill was so impressed by the demonstration that he considered the Stokes gun as a breakthrough weapon potentially equivalent in power to the tank: *Variants of the Offensive*, 3 December 1915, in PRO WO 158-831. See also Foulkes, *Gas!*, pp. 50-1, for his own input into the procurement of the Stokes for gas-throwing.

11 *Bomb Throwing Gun or Trench Gun*, in PRO MUN 7 273.

12 For this paragraph and the next, see PRO WO140 15, *Reports on Trials, September 1915 to November 1916*. For further details of Hythe, see Shelford Bidwell and Dominick Graham, *Firepower, British Army Weapons and Theories of War, 1904-45* (Allen and Unwin, London, 1982), pp. 27-32.

13 This report was filed on 7 October 1915. Rifle batteries were nevertheless being officially manufactured by 6 November 1915: see *Weekly Reports: Rifles, Machine Guns &c*, in PRO MUN 2 29. Clearly the verdict of one testing authority was far from automatically the last word on any given subject.

14 Hemming, *My Story*, IWM PP MCR 155, pp. 74-5.

15 PRO MUN 2 29, passim. There is a short account of the difficulties in obtaining good optical glass in Hartcup, *The War of Invention*, pp. 181-4.

16 Admittedly it has been only in the 1980s that an optical sight has been issued as standard on the general service rifle.

17 Compare the tanks and aeroplanes offered to the French war minister in the 1830s: see my *Military Thought in the French Army, 1815-51* (Manchester UP, 1989), pp. 77-8.

18 The awakening of public interest is exemplified by, e.g., the highly insecure complaint made in open Parliament in July 1915 that the supply of top secret chlorine gas cylinders to France was suffering delays which might prejudice the imminent top secret surprise attack with them... (Foulkes, *Gas!* p. 55).

19 Hartcup, *The War of Invention*, pp. 4–9.
20 Ibid., pp. 22–3. The Royal Society's 'advisory council' only gradually developed into the Department of Scientific and Industrial Research.
21 David R. Woodward, *Lloyd George and the Generals* (Associated University Presses, East Brunswick, New Jersey, 1983), pp. 48–54, for the goatish Welsh wizard's exploitation of the shell shortage.
22 Hartcup, *The War of Invention*, pp. 26–9. Compare Simkins, *Kitchener's Armies*, pp. 146, 153, for Lloyd George's seizure of Ordnance reponsibilities from the War Office.
23 Scott's 'Mr Stokes', p. 80.
24 John Terraine, *White Heat: The New Warfare 1914–18* (Sidgwick and Jackson, London, 1982), p. 230. For the committee's desperate losing battle against Mrs Ayrton's entirely useless gas fan, see Foulkes, *Gas!*, p. 101.
25 The most recent has been Tim Travers, in *How the War Was Won: Command and Technology in the British Army on the Western Front, 1917–18* (Routledge, London, 1992).
26 OH 1917, vol. 1, p. 360, and W. H. Watson, *A Company of Tanks* (Blackwood, Edinburgh and London, 1920), pp. 44–6.
27 Foulkes, *Gas!*, p. 102.
28 See my *Forward into Battle* (2nd ed., Crowood, Ramsbury, Wilts, 1990), chapters 3 and 5, for my own thoughts on the somewhat similar French 'pseudo sciences' of the mid-nineteenth century, and on the many conflicting US tactical interest groups in Vietnam.
29 Private communication to the author by Dr Paul Harris.
30 For the Germans, see Bruce I. Gudmundsson, *Stormtroop Tactics: Innovation in the German Army 1914–18* (Praeger, New York, 1989); and M. Samuels, *Doctrine and Dogma: German and British Infantry Tactics in the First World War* (Greenwood, Westport, Conn., 1992); and T. T. Lupfer, *The Dynamics of Doctrine* (US Staff College, Kansas, 1981).
31 A. O. Pollard, *Fire Eater: the Memoirs of a VC* (Hutchinson, London, 1932), p. 24, for the bayonet; p. 93 for the bomb.
32 Ibid., p. 96. The present author's knowledge of bombs in 1976 proved to be sufficiently sketchy for him to summon needlessly (but utterly memorably) a whole squad of Paris police all the way up to a seventh floor *chambre de bonne,* upon the discovery of just such a souvenir among his late father-in-law's stored artefacts.
33 Pollard, *Fire Eater,* pp. 218–23
34 Robert Graves, *Goodbye to All That* (Penguin ed., London, 1982), p. 159. The incident occurred in early 1916, soon after the Mills bomb's time fuse had in fact been perfected. Also see J. Ewing, *The History of the 9th (Scottish) Division* (Murray, London, 1921), pp. 13–14, for the dangers of percussion stick bombs. The development of the Mills bomb is described in Hartcup, *The War of Invention,* pp. 61–3. Pollard, *Fire Eater,* p. 221, for how the Mills is heavier but less accurate than the German stick bomb.
35 Simkins, *Kitchener's Armies,* p. 286.
36 W. D. Croft, *Three Years with the Ninth (Scottish) Division* (Murray, London, 1919), p. 43.
37 Tony Ashworth, *Trench Warfare 1914–18* (Macmillan, London, 1980), p. 69. At first the manuals for such bombing squads had to be improvised locally, e.g. the composer 2/Lt George Dyson, 22nd Royal Fusiliers, wrote his own hand-grenade manual in 1915, 'since there was no official manual on the subject': G. Dyson, *Fiddling While Rome Burns* (OUP, London, 1954), p. 16.
38 *Lessons from Recent Operations,* July 1916. For 'Fumite' see OH 1916, vol. 2, p. 262.
39 Repeated cries in PRO WO95 1294. Cf. the anonymous battalion commander's *Memorandum on Trench to Trench Attacks,* p. 19, which says that 'P' bombs probably will not burn dug-outs, but 'Fumite' ones will. He also recommends that Verey flares be used in this rôle.
40 *Notes on Some Bombing Operations in*

the Reserve Army, p. 3, in PRO WO158 344. Note that this practice would seem to run counter to the general feeling that the Mills was too heavy.
41 Blunden, Undertones of War, p. 23.
42 The Newton Pippin model was one of the designs in use, but the Hales model was more widespread. For rifle-grenades see also Ewing, The History of the 9th (Scottish) Division, p. 14.
43 Notes on Some Bombing Operations in the Reserve Army, p. 4.
44 See Pollard, Fire Eater, p. 142, for a rack of 6 rifle-grenades fired as 'an engine of fearfulness' (Brigade HQ soon told him to use fewer than 200 rounds per day). Blunden, Undertones of War, p. 64, for another 6-shot rifle-grenade battery.
45 E.g. many incidents in 1916 in Notes on Some Bombing Operations in the Reserve Army and many more from 1917–18 are cited in OH: thus men of the 52nd Division bombed their way forward when they had lost their barrage on 27 September 1918, OH 1918, vol. 5, p. 34; on 30 September the 3rd Australian Division at Bony used a combination of bombs, bayonets and Lewis guns, ibid., p. 136.
46 E.g. by II Corps on the Somme, 17 August 1916, Miscellaneous notes from the recent fighting, in PRO WO158 344.
47 By 1918 each division had a TM battalion and two TM batteries RA. See M. Farndale, History of the Royal Regiment of Artillery: Western Front 1914–18 (RA Institution, Woolwich, 1986), p. 132 and Annex G, p. 367, for the organisation of TMs and the rejection of the suggested Trench Mortar Corps (equivalent to the MGC) in May 1916.
48 This point is fully explained in Ashworth, Trench Warfare 1914–18, pp. 63–6. Ewing, The History of the 9th (Scottish) Division, p. 70, gives the 9th Division 'retaliation tariff' as a salvo of 18-pounder shells for each trench mortar bomb and two salvoes, plus two of 4.5-inch shells, for an aerial torpedo. Note, however, that the TM's much shorter range than artillery made it correspondingly more difficult for large numbers of tubes to concentrate their fire upon a single point.
49 At Passchendaele Gough wanted one TM to be attached permanently to each battalion, for immediate responsiveness when needed — Fifth Army conference, 30 October, in PRO WO158 251, item 17.
50 Many examples of very close-range use during the Somme battle in Notes on Some Bombing Operations in the Reserve Army. Compare frequent references to similar action in OH for the Hundred Days, e.g. vol. 4, pp. 48, 269 and even, on p. 207, an example of a Stokes gun capturing an enemy tank!
51 Reserve Army Report on a Brigade Action (on 15 September), 2 October 1916, in PRO WO158 344.
52 Haig, 22 November 1918, Increase to the establishment of Medium Trench Mortar batteries, in PRO WO106 432. These measures had first been mooted in July 1918, but their implementation came too late to help the war effort.
53 Foulkes, Gas!, p. 51ff., for the value of the 4-inch Stokes; see also Ashworth, Trench Warfare 1914–18, pp. 63–7; Ewing, The History of the 9th (Scottish) Division, p. 81; and Scott's 'Mr Stokes'.
54 Foulkes, Gas!, pp. 165–181.
55 Frontispiece to Gas!.
56 He claims it was forty times as effective as infantry formations of comparable size, having lost only 5,384 casualties in the whole war — few of them to gas — out of an establishment of about 6,000, but having inflicted around 200,000 casualties on the enemy, one of whom was no less a personage than Corporal Adolf Hitler (Foulkes, Gas!, pp. 302, 327, 338). It should be added, however, that total BEF gas casualties were themselves around 200,000, mainly to mustard.
57 The first German gas attack occurred on 22 April; but British retaliation could not begin to be planned for

Notes

58 about a month thereafter, mainly owing to Kitchener's vacillation: Foulkes, *Gas!*, pp. 17–20.
58 With no pun intended, the French had used this substance before the war in their dyeing industry. Once Foulkes had got his hands on it, however, it was found to constitute just about the most toxic and penetrative military gas of the era, with a particularly nasty delayed action effect. The French themselves, however, preferred to persevere with their ferocious-sounding but actually innocuous nerve agent, Prussic Acid (Foulkes, *Gas!*, pp. 49, 52–3, 104–8).
59 The Germans quickly moved away from cloud gas to the less efficient artillery method of delivering gas. See Foulkes, *Gas!*, pp. 156, 182.
60 Ibid., pp. 242ff.
61 Ibid., p. 238.
62 Ibid., pp. 263–70.
63 Ibid., p. 249.
64 Ibid., p. 320.
65 Ibid., p. 250.
66 Ibid., pp. 250–3. Our Russian readers will be fascinated to learn that this 'most effective chemical weapon ever devised' was never used against the Germans at all, but was liberally blown all over the Bolshevists at Archangel in 1919, apparently to the complete satisfaction of General Rawlinson.
67 Ibid., p. 169.
68 This was the best of three designs: see ibid., p. 111, where Foulkes himself adds that 'I was standing near him [i.e. the inventor of another flamethrower, not Vincent's] when the apparatus exploded and set him on fire, and I put out the flames with his overcoat'.
69 *Note on the Position of Landships,* 8 January 1916, reported that a 'trench warfare ship with a flamethrower' would be ready in six weeks' time, PRO WO158 831.
70 Reserve Army, *Report on Action at — by 'A' Infantry Brigade* [*sic*; the action took place on 15 September], 20 September 1916, p. 5, in PRO WO158 344.
71 OH 1916, vol. 2, pp. 187, 195.
72 R. E. Priestley's signals history, *The Signal Service in the European War of 1914 to 1918 (France)* (official RE history, Chatham, 1921), p. 127; and see also pp. 127–8 for (unsuccessful) attempts to employ trench-digging tanks and excavators.
73 OH 1916, vol. 2, p. 128. This was an Australian attack at Fromelles (i.e. not on the Somme itself).
74 Such a gas attack was made at Arras on 18 March 1917 using seven pipes, although no 'results' were ever recovered from the enemy: Foulkes, *Gas!*, p. 196. It should also be noted that a *definitely* successful push-pipe HE attack was made on 21 August near Mouquet Farm, where a German barricade was demolished: OH 1916, vol. 2, p. 224.
75 The 'obvious' was perhaps less so at the time, and the combined flame/HE push-pipe combination was repeated on 3 September 1916, when it won results remarkably similar to those of 12 August: ibid., pp. 255, 269. Only after that was it finally set aside.
76 Foulkes, *Gas!*, pp. 166–7, 169–73.

Chapter 7: Automatic weapons

1 Lindsay papers, Machine Gun file A1–J and A1–L. The same phrase is remorselessly reiterated in almost every one of his essays and lectures from the Great War period, e.g. in his *Fire in Battle,* in Machine Gun file A1–V.
   Lindsay himself has received little modern attention beyond a brief mention in Tim Travers, *The Killing Ground* (Allen and Unwin, London, 1987), p. 69.
2 Lindsay papers, Machine Gun file A3.
3 Ibid., file A1–i. Note that German achievements in these areas do not appear to have improved during the war; Tim Travers, *How the War Was Won* (Routledge, London, 1992), pp. 86–8.
4 Lecture at the Chelmsford Corn Exchange, 10 August 1915, in Lindsay papers, Machine Gun file

A22, p. 10.
5 Ibid., file A10.
6 Ibid., file A11. The Army Council made this official policy on 14 March 1915.
7 Some press cuttings boosting the machine gun are to be found e.g., ibid., files A17, A21. Lloyd George is quoted on the necessity for more machine guns in, e.g., John Terraine, *The Smoke and the Fire* (Sidgwick and Jackson, London, 1980), p. 133, and cf. David R. Woodward, *Lloyd George and the Generals* (Associated University Presses, East Brunswick, New Jersey, 1983), p. 54.
8 Lindsay Papers, Machine Gun file A16: letters from Haig, 15 April, and Smith-Dorrien, 13 April 1915.
9 Correspondence from June 1915 is to be found ibid., files A19 and A20. The questionnaires themselves are in A29, and their findings encapsulated in *Notes on the Employment of Machine Guns* (GHQ, July 1915, in PRO WO33 718). Unfortunately at the time of writing I have been unable to obtain sight of Baker-Carr's celebrated memoirs, *From Chauffeur to Brigadier*; but see the discussion in Terraine, *The Smoke and the Fire*, pp. 130–42.
10 Lindsay, working from England, appears to have continued to believe in the pastoral greenery of the battlefield — complete with standing corn, haystacks and upper storeys to farmhouses — for at least a year after mobilisation, before he learned wisdom. Ironically the final weeks of the war, during which he would personally take the field as a combat commander for the first and only time, happened to coincide with the restoration of precisely that happy condition.
11 SS 109, *Training of Divisions for Offensive Action*, in Appendix 17 to OH 1916, p. 129. Lindsay's tactical thought at this time is most clearly represented by his Chelmsford Corn Exchange speech in file A22. A suggestion for infiltrating machine guns forward was apparently made by J. F. C. Fuller in 1911, cited in G. S. Hutchinson, *Machine Guns* (Macmillan, London, 1938), p.113.
12 Lindsay's Chelmsford Corn Exchange speech, p. 57.
13 Hutchinson, *Machine Guns*, p. 133.
14 J. Ewing, *The History of the 9th (Scottish) Division* (Murray, London, 1921), p. 26.
15 Hutchinson, *Machine Guns*, p. 122.
16 For the 32-gun barrage demonstrations on to the beach at Camiers on 18 or 20 August 1917, attended by all the leading officers of the BEF, see ibid., p. 193; and Lindsay papers, Machine Gun file C36 (unfortunately these two sources seem to disagree over the precise date of the demonstration). Hutchinson, *Machine Guns*, p. 194, goes on to show that the Germans began to look into barrage fire, at Spandau, only in November 1917. Colonel N. K. Charteris believed they used it only in the spring of 1918; see his *Narrative of MG Operations, IV Army, April–November 1918*, in PRO WO158 332, chapter 9, p. 4.
17 SS 110 *(Lessons of the Recent Operations)*, July 1916, p. 7, referring to overhead fire designed to protect friendly troops consolidating an enemy position that they have just captured.
18 Angry letter from Brutinel to Lindsay, 2 September 1917, claiming credit for the development of barrage fire, in Lindsay papers, Machine Gun file C37. It is unfortunate that we lack Lindsay's own papers for the period between the spring of 1916 and the autumn of 1917, during which he was side-tracked into an infantry brigade at the front.
    Hutchinson, *Machine Guns*, pp. 116–17, suggests that the key mathematics were worked out at Hythe, notably by Captain R. G. Clarke of the Queen's Regiment.
19 E.g. 2nd Division at Miraumont in early February 1917 had a barrage scheme for 48 guns, in *Correspondence on Operations to be Carried Out*, in PRO WO95 1295.
20 Brutinel's claim to exclusivity at Vimy cannot be entirely substantiat-

ed, not only because of the British machine gun barrages or near-barrages fired in 1915 and 1916 which have already been mentioned, but also by the fact of barrages fired at other parts of the line on 9 April 1917 itself. 9 Scottish Division, for example, accompanied its artillery creeper with barrages from 54 trench mortars and 20 machine guns: Ewing, *The History of the 9th (Scottish) Division,* p. 188.

21 Lindsay papers, Machine Gun file B1: essay on *The Strategic and Tactical Value of Machine Guns,* dated 9 November 1915. Lindsay continued to include copies of this very important essay in the information packs he distributed throughout all the remainder of the war. Note that ibid., file B2, Lindsay told Baker-Carr that the 'Strategic' essay was originally written for Swinton at the Ministry of Munitions.

22 See chapter 2, notes 51–56 (p.227, above).

23 Brutinel's characterisation of Lindsay in his 2 September 1917 letter.

24 *The Strategic and Tactical Value of Machine Guns.* Official doctrine nevertheless set the value of one MG as only around 30 rifles, and Shelford Bidwell and Dominick Graham, *Firepower, British Army Weapons and Theories of War, 1904–45* (Allen and Unwin, London, 1982), p. 28, have estimates that a Maxim was worth 25 rifles and a Vickers worth 40; cf. Hutchinson, *Machine Guns,* p. 57, for a prewar German estimate that each gun was worth 120 rifles.

25 *First Army Troops: Motor Brigade, MGC,* in PRO WO95 246. Further correspondence is in PRO WO158 288.

26 *First Army Troops: Motor Brigade, MGC,* in PRO WO95 246, for all three examples. Note that where motor machine guns were not available in the Hundred Days, pack animals, half limbers and other improvised vehicles were often employed to enhance the mobility of ordinary machine guns — see Charteris in PRO WO158 332, pp.

24, 34 and conclusion on p. 7. Also see Hutchinson, *Machine Guns,* p. 315, for a eulogy of the Canadian Motor MGs at St Quentin in May 1918.

27 F. C. Hitchcock, *Stand To: A Diary of the Trenches* (new ed., Gliddon, Norwich, 1988), p. 290.

28 Private communication from Dr Paul Harris.

29 Lindsay papers, Machine Gun file B8. Unfortunately I have been unable to establish the identity of the General Staff luminary in question.

30 Ibid., file C25. In file C29 there is even a hint that GHQ is ready to accept the concept of a 'machine gun force', during the heady tactical dawn of May 1916.

31 Ibid., files A27, A28, B2, and especially B11.

32 Some of the administrative obstructions encountered even in 1918 are explained in Travers, *How the War Was Won,* pp. 46–8. In late 1917 Lindsay had returned to the charge as Lieutenant Colonel commanding the Camiers school and supervising the writing of the official machine gun manual, SS 192.

33 These issues are interestingly but perhaps inconclusively discussed in the Lindsay papers, machine gun files D3ff. Uncharacteristically even Hutchinson, *Machine Guns,* pp. 222-3, 265, was critical of the MGC contribution to the battle, thereby conceding points to its fierce opponents: e.g. Hollond (for Maxse), *Lessons No 2,* pp. 3–4, in Maxse papers, Box 69–53–10, File 45.

34 'Tactical Handling of Machine Guns in Defensive Operations', in *Notes on Recent Fighting, issued by the General Staff,* no. 9, 3 May 1918, in PRO WO158 70.

35 E.g. Hutchinson, *Machine Guns,* pp. 232, 269–89.

36 Charteris, *Narrative of MG Operations,* p. 6. Cf. OH 1918, vol. 5, p. 308, for a 50th Division MG barrage with 96 guns on 17 October.

37 Examples cited in Lindsay papers, Machine Gun file E15-B, supported by analysis by Major Atkinson from

Camiers, in file E5-B.
38 Lindsay papers, Machine Gun files D37, D40, E1 and especially E2A.
39 Ibid., files C35 and especially D18. This idea was widespread in all arms during the manpower shortages of 1918, and should not be used as a stick to beat the 'bad old' infantry and cavalry, as perhaps Tim Travers is sometimes tempted to do in his *How the War Was Won*.

By this time Lindsay had been promoted to be Horne's First Army machine gun officer, although he would receive staff training only in 1920.

40 Parts of this section are taken from my article entitled 'The Lewis Gun Made Easy', in *The Great War,* vol.3, No. 4, September 1991. I am especially grateful to Peter T. Scott, the editor, for his assistance.

The details of small arms may be found in, e.g., W. H. B. Smith and J. E. Smith, *Small Arms of the World* (Stackpole, Harrisburg, PA, 1943). See also G. M. Chinn, *The Machine Gun* (3 vols, Department of the Navy, Washington, DC, 1951) and M. M. Johnson and C. T. Haven, *Automatic Arms: their history, development and use* (Morrow, New York, 1941).

41 Although most reputable arms manufacturers had already invented their own types of AR well before 1914, only the Danish Army was truly senior to the British in practical adoption, with their quirky 22 lb 1902 Madsen being used by the Russian cavalry in Manchuria two years later and widely sold to other countries — albeit not usually in great numbers until after 1918. Cf. Peter T. Scott, 'Britain and the Madsen Machine Gun 1914–18' in *The Great War,* vol. 3, No. 4, September 1991.

See Bidwell and Graham, *Firepower,* p. 50ff., on prewar political funding for MGs, and on Lewis procurement.

42 *Weekly Reports,* in PRO MUN 2 29.
43 'Charles Edmonds' (Carrington), *A Subaltern's War* (Davies, London, 1929), p. 219.
44 John Terraine, ed., *General Jack's Diary, 1914–18* (Eyre and Spottiswoode, London, 1964), p. 170.
45 SS 143, p. 6.
46 General Staff training manual *The Tactical Employment of Lewis Guns* (SS 197, January 1918), pp. 1–3, assumes sixteen gun sections per battalion, and demands that this level should be maintained 'at all costs'. Normally there should be one gun per platoon — but the manual is careful to say that in some circumstances this may be increased to two or even three (although in other cases the gun may be removed altogether).
47 The Vickers weighed only a few pounds more than the Lewis, at 33 lbs excluding 8 lbs of water, 44 lbs of tripod and 15 lbs per ammunition belt. This compares with 34–44 lbs for the German MG 1908 Maxim, with more for water and a massive 65 lbs for a tripod or 83 lbs for the sled. Note also that *per round* Vickers belts were 40 per cent lighter than Lewis drums — a tactical disadvantage underlined in SS 197, *The Tactical Employment of Lewis Guns,* pp. 2ff. — and indeed the Vickers had originally been designated a 'light' machine gun, to distinguish it from the heavy brass prewar Maxim that it superseded.

Most other ARs were considerably lighter than the Lewis, e.g. the 18 lb French M1907 Chauchat (available to the troops only from 1916) and the 12 lb M1917 R.S.C.; the 16–19 lb American M1918 Browning, or the 15 lb Italian 1915 Vilar Perosa. By comparison the 26.5 lb French Benet Mercier Hotchkiss Model 1908 (US M1909, and French 1914 variant) was of comparable vintage and weight to the Lewis, whereas the relatively rare 39 lb German M1915 Dreyse MG13 was much heavier.

48 Charles E. Carrington, *Soldier from the Wars Returning* (Hutchinson, London, 1965), p. 177.
49 *General Jack's Diary,* p. 183; see also 'Edmonds' (Carrington), *A Subaltern's War,* p. 33. Colonel W. D. Croft,

Three Years with the Ninth (Scottish) Division (Murray, London, 1919), p. 49, complained that the wheels on the gun carriage were too small for mobility in mud.
50 Ian V. Hogg, *Illustrated Encyclopaedia of Firearms* (Newnes, London, 1978).
51 By June 1918 the testing of automatic rifles was taking mud very seriously. In an extensive comparative test at Bisley one Lewis gun was immersed in a pool of mud so deep that it was lost for a long time! — Maxse papers, Box 63–67–12, File 64.
52 F. C. Hitchcock, *Stand To: A Diary of the Trenches, 1915–1918* (new ed., Gliddon, Norwich, 1988), p. 133.
53 Croft, *Three Years with the Ninth (Scottish) Division*, pp. 124, 167 for MG barrages not being as good as claimed.
54 J. C. Dunn, *The War the Infantry Knew* (new ed., Jane's, London, 1987), p. 183. Cf. a succinct summary of the intended rôle of the Brigade Machine Gun Officer in 'Simplex', *Instruction on the Lewis Automatic Machine Gun* (London, 1916), pp. 110–14.
55 OH 1918, vol. 2, p. 471, quoted in Hutchinson, *Machine Guns*, p. 305.
56 By August 1918 the infantry had actually won back an MMG section for each battalion: Dunn, *The War the Infantry Knew*, p. 183.
57 *General Jack's Diary*, p. 183. Compare the suggested 1918 layout for a horse-drawn Company Lewis Gun GS wagon (to carry 4 guns with 176 magazines) in Appendix II of SS 197, *The Tactical Employment of Lewis Guns*, p. 22.
58 F. P. Crozier was the victor in the battle of Sokoto which won all the Hausa states for the crown. In his *The Men I Killed* (Joseph, London, 1937), p. 146, he makes the acid comment that 'Thus has so-called barbarism been opposed by so-called civilization'.
59 Sometimes the Lewis did not even have a bipod mount. See W. Moore, *A Wood Called Bourlon* (Leo Cooper, London, 1988), p. 112, for one being fired while rested only on a wounded man (the Americans actually removed the bipods from their Browning ARs). There were nevertheless a number of different efforts to provide tripods for Lewis guns — ranging from a very light 'field mount', through a number of Anti Aircraft frames, to the very same heavy tripod that was used for the Vickers MMG. None of these seem to have caught on very widely.
60 E.g. SS 197, *The Tactical Employment of Lewis Guns*, p. 13, stresses the distinction by recommending that Lewis guns should cover small gaps in the longer range defensive fireplan of the MMGs. It also suggests that the superior mobility of the Lewis qualifies it far more than the Vickers for moving forward with attacking troops. On p. 9 this includes the idea that consolidating captured ground against a counterattack is the Lewis gun's 'greatest opportunity'.
61 C. E. Crutchley, *Machine Gunner, 1914–18* (MG Corps Old Comrades Association, Northampton, 1973), p. 40. There are numerous stories of short-range use of the Vickers in Hutchinson, *Machine Guns*, including an appalling '6 yard field of fire' in March 1918 (p. 302).
62 Dunn, *The War the Infantry Knew*, p. 210.
63 Edmund Blunden, *Undertones of War* (new ed., Collins, London, 1978), p. 142.
64 'Mark VII', *A Subaltern on the Somme* (Dent, London, 1927), p. 75, for the Somme, August 1916; William Moore, *A Wood Called Bourlon*, p. 154, for the Guards counterattack on 30 November 1917.
65 Siegfried Sassoon, *Memoirs of an Infantry Officer* (new ed., Faber, London, 1965), p. 65, for the Somme, July 1916.
66 See note 11, above; but note that SS 110, 16 July, says that Lewis guns were actually effectively used ahead of attacks in the initial Somme fighting. The idea certainly seems to have been widespread; e.g. a related suggestion recurs in *General Jack's Diary*, p. 218, when he wants to patrol No

Man's Land with Lewis guns as roving snipers, in ordinary trench warfare, spring 1917.

67 Carrington, *Soldier from the Wars Returning*, pp. 177–8. See also his *A Subaltern's War*, pp. 143, 147, where at Ypres in mid-1917 he hints that he used true fire and movement with Lewis guns.

The first Lewis gun school was set up in June 1915, according to Tony Ashworth, *Trench Warfare 1914–18: The Live and Let Live System* (Macmillan, London, 1980), p. 61, although it is unlikely that tactical usage was stressed in the syllabus for well over a year thereafter.

68 Army and Navy Register, quoted in Chinn, *The Machine Gun*, p. 289.
69 Cited in Hutchinson, *Machine Guns*, p. 217.
70 E.g. Graham H. Greenwell, *An Infant in Arms* (new ed., Allen Lane, London, 1972), p. 172, used Lewis guns to destroy enemy machine guns in his successful attack on 28 March 1917. Lewis guns were decisive in VI Corps, 28 September 1918: OH 1918, vol. 5, p. 49. In 46th Division on 3 October 1918 they were able to suppress field guns: ibid., p. 162.
71 Criticisms retailed, respectively, in Robin Prior and Trevor Wilson, *Command on the Western Front* (Blackwell, Oxford, 1992), p. 161; Bidwell and Graham, *Fire-power*, p. 55.
72 SS 197, *The Tactical Employment of Lewis Guns*, pp. 9, 12, and Appendix I on p. 20. The manual makes it clear that firing from the hip is highly inaccurate and unduly exposes the firer to the enemy, and should be used only in special circumstances.
73 Moore, *A Wood Called Bourlon*, p. 70.
74 Ibid., p. 88. Compare Bidwell and Graham, *Fire-power*, p. 28, for the 1908 trials which achieved 18 per cent hits with a hip-fired Vickers!
75 *General Jack's Diary*, p. 226.
76 Carrington, *Soldier from the Wars Returning*, p. 177.
77 Bruce I. Gudmundsson, *Stormtroop Tactics: Innovation in the German Army 1914–18* (Praeger, New York, 1989),

p. 98ff. Note that the British would adopt a belt-fed GPMG of their own only in the 1960s, when it had been made much lighter than the 08/15 Maxim.

Chapter 8: Artillery

1 See Guy Hartcup, *The War of Invention* (Brassey's, London, 1988), pp. 8–10, 48–54; Shelford Bidwell and Dominick Graham, *Fire-power: British Army Weapons and Theories of War, 1904–45* (Allen and Unwin, London, 1982), p. 98; also I. V. Hogg, *The Guns 1914–18* (Pan/Ballantine, London, 1973), pp. 7, 17–19.
2 See chapter 4, note 35, above, for BEF musketry on the Somme; and compare my *Rally Once Again* (Crowood, Ramsbury, Wilts, 1987) for the American Civil War. Note that even in terms of purely technical capability, as opposed to actual battlefield usage, the American Civil War rifle was already pretty accurate to 1,000 yards whereas the Great War SMLE service rifle was pretty accurate to 2,000 yards.
3 Billy Congreve's diary, 1 November 1914, in L. H. Thornton and P Fraser, eds, *The Congreves, Father and Son* (Murray, London, 1930), p. 247. The expression 'shell shock' was nevertheless applied quite indiscriminately to any front line soldier suffering any type of nervous breakdown, regardless of whether or not he had been exposed to shelling. It remained a highly ambiguous and controversial term.
4 D. Porch, *The March to the Marne* (Cambridge UP, 1981) for an unnecessarily anti-French account of this debate. Hogg, *The Guns 1914–18*, p. 100, for the jealously guarded secret of the 75.
5 Excellent general introductions to these subjects in Bidwell and Graham, *Fire-power*, pp. 7–21, and in J. B. A. Bailey, *Field Artillery and Firepower* (The Military Press, Oxford, 1989), pp. 127–51.
6 A good selection of the new genera-

tion of maps may be found in Peter Chasseaud, *Topography of Armageddon* (Mapbooks, London, 1991). These maps also inspired the beautiful maps which so liberally illustrate all volumes of the official history. Bidwell and Graham, *Fire-power*, pp. 104–8 for survey; pp. 108–11 for location.

7 Note that accurate survey, although now using such innovations as satellite imagery or computerised terrain-matching, has been as crucial to the modern precision weapons as it was in 1917.

8 Bailey, *Field Artillery and Firepower*, p. 150.

9 The Germans' *Stollen*, or deep bunkers, made a particularly important contribution to their fighting efficiency on the Somme. See R. E. Priestley's signals history, *The Signal Service in the European War of 1914 to 1918 (France)* (official RE history, Chatham, 1921), pp. 77–8, 115 for the successive deepening of cable buries from 2 ft 6 in to 6 ft.

10 I. V. Hogg, *The Guns 1914–18*, p. 25, for an explanation of how complicated even the shrapnel shell had become by 1914. Cf. *Handbook of Shell charged with chemical, incendiary and smoke mixtures at present in use with the field armies*, September 1918, in PRO WO33 933.

11 M. Farndale, *History of the Royal Regiment of Artillery: Western Front 1914–18* (RA Institution, Woolwich, 1986), p. 73.

12 Bidwell and Graham, *Fire-power*, p. 98.

13 *Abstract of Statistics of the military effort of the British Empire* (HMSO, London, 1922), p. 486: approx 4,500,000 shells were given to Britain's allies during the war — mostly to Russia — but this amounts to less than one month's production!

14 Unsuccessful Hythe tests to cut wire with MGs, on 5 June 1916, are mentioned in PRO WO140 15; standing corn appears to have been even more resistant than wire, according to tests held in France on 15 July 1918. Whereas between 41 per cent and 66 per cent of shots could hit a target at 50 yards' range through corn, only around 6 per cent could penetrate 150 yards. The critical depth of corn appeared to be between 60 and 70 yards, at which distance the bullets were deflected and ricocheted in all directions. Charteris, *Narrative of Machine Gun Operations, Fourth Army*, conclusion, p. 10, in PRO WO158 332.

15 J. Ewing, *The History of the 9th (Scottish) Division* (Murray, London, 1921), p. 78, reports a Bangalore torpedo raid supported by dummies worked by strings (!) on 1 March 1916; J. C. Dunn, *The War the Infantry Knew* (new ed., Jane's, London, 1987), p. 192, reports another on 9 April 1916 in which the torpedo failed to explode. Frank Richards, *Old Soldiers Never Die* (new ed., Mott, London, 1983), p. 145, has yet another (presumably different one?) that did not explode.

16 *Lessons from Messines, Part 1, Tanks*, p. 10, in PRO WO158 298.

17 The South African brigade at Arras in early April 1917 deliberately made craters in No Man's Land, to act as cover for their forming up to attack; Ewing, *The History of the 9th (Scottish) Division*, p.193.

18 W. D. Croft, *Three Years with the Ninth (Scottish) Division* (Murray, London, 1919), p. 244.

19 Ibid., p. 87.

20 Bidwell and Graham, *Fire-power*, p. 98, places the introduction of the 106 percussion fuse in spring 1917, following the demand for a replacement for the delayed action fuse 100, and the technical failure of its successor, the 101 percussion fuse.

21 Cf. chapter 6, above, for gas; and OH 1916, vol. 2, p. 203, for the shortfall on ordered smoke shell supplies during September 1916. Ibid., p. 569, shows smoke shell became available for artillery only in November 1916, although it was already being fired from mortars in 1915. OH 1918, vol. 5, p. 606, for the first appearance of Lachrymator gas shell in September 1916 and of Lethal gas shell in May 1917; and p. 95 for the first use of

Mustard gas ('BB') shell on 29 September 1918.
22 E.g. Thermit shells fired to mark divisional boundaries, 14 October 1918; OH 1918, vol. 5, p. 274.
23 After one of its GOCs had been broken in spirit by preparing for the battle of Loos, and his successor killed there, the 9th Division was lucky to be taken over by the keen gunner W. T. Furse, who was promoted to be Master General of Ordnance at the end of 1916 — a position which he used to expedite the procurement of 'tactical' munitions such as smoke shell. His 9th Division BGRA was the equally 'tactical' H. H. Tudor, who designed fiendish fireplans on the Somme and throughout 1917, most notably for Cambrai, and who eventually took command of the division following the March 1918 retreat.
24 Ewing, *The History of the 9th (Scottish) Division*, pp. 187, 194–8, for 9th Division's attack at the start of the battle of Arras, when it used up Third Army's entire supply of smoke shell.
25 Farndale, *History of the Royal Regiment*, pp. 123, 147; cf. Robin Prior and Trevor Wilson, *Command on the Western Front* (Blackwell, Oxford, 1992), p. 164, who say that this was only a lifting barrage that came 'very near' to being a creeper. A classification of the various types of lifting barrages is in Bailey, *Field Artillery and Firepower*, pp. 132–4. See also Charles E. Carrington, *Soldier from the Wars Returning* (Hutchinson, London, 1965), p. 192, for the terminology of 'standing', 'creeping' and 'searching' barrages.
26 Farndale, *History of the Royal Regiment*, pp. 143–7; Bailey, *Field Artillery and Firepower*, pp. 136–7. Apparently the Germans adopted the creeper only relatively late and without the full sophistication of the allies.
27 E.g. in Bidwell and Graham, *Firepower*, pp. 112-14.
28 Classic barrage plans include Vimy Ridge, 9 April 1917, e.g. in PRO WO106 402; and Messines, 7 June 1917, e.g. in PRO WO158 413. See also Second Army at Ypres, August–September 1917, especially *General Principles on which the Artillery Plan will be drawn*, 29 August 1917, in PRO WO158 208.
29 Second Army *Notes on Training and Preparations for Offensive Operations*, 31 August 1917, in OH 1917, vol. 2, Appendix XXV, p. 462.
30 The Germans and their Colnel Bruchmüller are often credited with having perfected this type of 'orchestration'; but in reality the British knew as much about it, if not more.
31 A striking recent witness to the nervousness with which the secrecy of zero can become charged was the bitter reaction against hints dropped on the BBC World Service emanating from the otherwise highly unruffled 2nd battalion of the Parachute Regiment, just before its entirely successful attack on Goose Green, with 'H' Hour on 28 May, 1982.
32 E. K. G. Sixsmith, *British Generalship in the Twentieth Century* (Arms and Armour, London, 1970), pp. 96–8, shows that Maxse wanted his infantry to start moving forward a few moments before the creeper started, in order to keep well up with it.
33 E.g. Birch's letter of 23 September 1918, listing the possible ways to suppress anti-tank guns, in PRO WO158 832. The same dossier also has several documents which show a growing nervousness about mines, enemy tanks and 'friendly fire' due to faulty recognition — since in this phase of the war a majority of active German tanks were actually captured allied machines.
34 See, e.g., Appendices A1, A2, A3 to *Appendix 16* of OH 1916, pp. 108–20.
35 The 9th Division frequently made 'Chinese Attacks' with creeping barrages without an infantry assault, in order to alarm the enemy into prematurely manning his parapets (see Ewing, *The History of the 9th (Scottish) Division*, pp. 154, 193). However, when itself receiving an enemy barrage unsupported by infantry, this division could apparently distinguish

Notes 251

that fact without difficulty (ibid., p. 235).

Compare Edmund Blunden's 'trench theatricalities', with 190 dummy soldiers and a smoke barrage, in a Chinese attack on the Somme (in which all the dummies were 'killed' by the enemy's fire), *Undertones of War* (new ed., Collins, London, 1978), p. 96. OH 1918, vol. 5, p. 248, shows that 'dummy tanks and infantry operated from shell holes' were still being used as late as October 1918.

36 For Tudor's rôle in planning Cambrai see Ewing, *The History of the 9th (Scottish) Division* (Murray, London, 1921), pp. 221–3, and OH 1917, vol. 3, pp. 6–7. It would appear that German preparatory barrages in 1918 always lasted at least two hours, because they had never achieved fully predicted fire; e.g. they were starting to calibrate their pieces only at the start of 1918, whereas the British had at least a six-month start (Farndale, *History of the Royal Regiment*, p. 217).

37 The main exception was the attack on the Hindenburg Line, where the defences were so formidable that an extensive preliminary bombardment was thought necessary, with a start of zero: Prior and Wilson, *Command on the Western Front*, p. 363 and associated discussion.

38 Statistics derived from John Terraine, *The Smoke and the Fire* (Sidgwick and Jackson, London, 1980), pp. 219, 307. 50,000 members of the Royal Artillery lost their lives in the war, which was about as many as in one or two infantry divisions: Farndale, *History of the Royal Regiment*, p. 317.

39 I. V. Hogg, *The Guns 1914–18*, p. 83, referring specifically to anti-aircraft gunnery.

40 Ibid., and OH 1914, vol. 1, pp. 416–28. Note that this figure refers to 1 cavalry division in 1914 and 3 in 1918, as well as 5 and 61 infantry divisions, respectively. Bailey, *Field Artillery and Firepower*, p. 127, gives 6.3 guns per 1,000 infantry in 1914, rising to 13 in 1918.

41 Even among the 18-pounders, many batteries found they were 'army' rather than 'division' troops, and hence nobody's children in particular. See H. Siepmann, *The Echo of the Guns* (Hale, London, 1987) for the travails and frustrations of just such a battery.

42 OH 1918, vol. 5, p. 595.

43 E.g. the BEF had no less than 26 million shells stockpiled at the start of March 1918: *Abstract of Statistics*, p. 480.

44 Prior and Wilson, *Command on the Western Front*, pp. 30, 33, 84–5, especially 167–9, and indeed passim.

45 E.g. both Prior and Wilson, and Farndale.

46 See the next chapter and Priestley's *The Signals Service* for the almost uncanny matching of the BEF's signalling efficiency with the general efficiency of each of its battles.

47 See discussion in G. Blaxland, *Amiens 1918* (Muller, London, 1968); and the French General Fayolle, 1916, quoted (from OH 1917, vol. 3, p.278) in Moore's *A Wood Called Bourlon*, p. 120: 'The third days of battles are never worth anything'.

48 Farndale, *History of the Royal Regiment*, pp. 159, 184–5.

49 E.g., ibid., p. 208, for Plumer at Passchendaele.

50 Ibid., pp. 2, 86, 120, 128.

51 Ibid., p. 146.

52 Bidwell and Graham, *Fire-power*, p. 100.

53 Farndale, *History of the Royal Regiment*, pp. 156–8, 185.

54 Lieutenant Colonel H. H. Hemming, *My Story* (microfilm PP MCR 155, Imperial War Museum, c.1976), passim. The rôle of Major (later Colonel) Winterbotham, RE, seems to have been absolutely crucial in overcoming the artillery's resistance and setting up the whole counterbattery circus, and in being the patron of both Hemming and Bragg. He was surely a key mover in the whole BEF war effort.

55 Ibid., p. 85. Fawcett finally disappeared on one of his Amazonian expeditions soon after the war.

56 Ibid. Thus it is clear that Colonel J. F.

C. Fuller was not the only practising spiritualist who rose to high places in the BEF! Note that Hemming's subsequent terrain analysis revealed that none of Fawcett's 'ouija board' shoots actually hit anything apart from mud.
57 Ibid., pp. 86–100. The flash-buzz system was nevertheless prudently retained as an option until the end of the Second World War.
58 This and the next two paragraphs are derived from Bragg's paper 'Sound Ranging in France 1914–1918', reprinted in Farndale, *History of the Royal Regiment*, Appendix L, pp. 374ff. I am also grateful to Professor Gordon Lorimer — who like Hemming is a scientific Canadian Anglophile — for pointing out Bragg's former house in Manchester.
59 Something very similar seems to have happened within the US armed forces in Vietnam. Before the key watershed of 1968 the troops were enthusiastic but the technology was only 'promising'; whereas after that date the technology worked well but the troops had tired of war.
60 For this and the next paragraph see Chasseaud, *Topography of Armageddon*, pp. 8ff. See also Bidwell and Graham, *Fire-power*, pp. 101, 106ff., for further gunner details.
61 For this and the next three paragraphs I have relied mainly upon the Air Historical Branch's *A Short History of the Royal Air Force* (London, 1929).
62 Ibid., p. 197, and cf. J. F. C. Fuller's suggestion that ground attack aircraft should be armoured — 'in fact they must become flying tanks' — in *Tank Programme*, p. 2, in PRO WO158 865. Exactly the same idea has often been pinned on to the attack helicopter during the past quarter century, which flies twice as fast as the fighter-bombers of 1917 and carries a heavier payload.
63 The rise of German interdiction bombing made a big impression on the BEF; e.g. Gough intoned that 'This form of warfare is certain to increase', Fifth Army circular on *Future Policy*, 1 November 1917, in PRO WO158 251, document 19.

64 OH 1918, vol. 4, p. 84. Note that the *Short History of the RAF* tastefully fails to mention the distressing matter of RFC/RAF losses: some 6,166 killed and 10,457 wounded during the war. Some 55,000 aircraft were procured during the war, of which a large majority was lost. The RAF's front-line strength on 11 November 1918 was 1,799 aircraft; but it was not unusual to lose at least 50 per cent of these in every month's combat flying (OH 1918, vol. 5, pp. 591, 597, 601).
65 E.g. Blunden, *Undertones of War*, p. 177; and Blunden's introduction to A. Wade, *The War of the Guns* (Batsford, London 1936), p. xi.
66 'Charles Edmonds' (Carrington), *A Subaltern's War* (Davies, London, 1929), p. 21.
67 J. C. Dunn, *The War the Infantry Knew* (new ed., Jane's, London, 1987), p. 167.
68 Wade, p. 17.

Chapter 9: Controlling the mobile battle

1 Siegfried Sassoon, *Memoirs of an Infantry Officer* (new ed., Faber, London, 1965), p. 8, on a Fourth Army school towards the end of 1916. When Robert Graves became an instructor at the Harfleur bull ring earlier in the year, he had taught strictly 'trench' subjects; *Goodbye to All That* (Penguin ed., London, 1982), p. 151.
2 The official historian Edmonds, an engineer, took the lead in characterising the cavalry as obsolete. Infantrymen such as Rawlinson and Plumer appear to have shared this view, but expressed themselves more diplomatically to the cavalryman Haig.
3 Stephen D. Badsey's unpublished thesis (Cambridge University, 1981), 'Fire and the Sword', and his private communications to the author.
4 W. D. Croft, *Three Years with the Ninth (Scottish) Division* (Murray, London, 1919), p. 59, referring to Longueval 1916, Arras 1917, and

Broodeseinde 1917.
5   Graham H. Greenwell, *An Infant in Arms* (new ed., Allen Lane, London, 1972), p. 114. Note that this same phrase would become a recurrent slogan of the much-admired Field Marshal Montgomery during the Second World War.
6   John Terraine, ed., *General Jack's Diary* (Eyre and Spottiswoode, London, 1964), pp. 229, 276.
7   W. H. Watson, *A Company of Tanks* (Blackwood, Edinburgh and London, 1920), pp. 182–3, 195.
8   Ibid., p. 245. F. Mitchell, *Tank Warfare: The Story of the Tanks in the Great War* (Nelson, London, n.d.), pp. 249, 251, for the heavy losses to officers walking in front of tanks at Amiens in August 1918.
9   E.g. the paper by 3rd Cavalry Division circulated by GHQ, 24 August 1916, in PRO WO158 186, document 293.
10 Badsey, 'Fire and the Sword'; and for High Wood see T. Norman, *The Hell They Called High Wood* (Kimber, London, 1984), pp. 81–2, 96–101.
11 OH 1918, vol. 5, pp. 219, 527–30, for discussion of vanguards and cavalry operations. See also discussion of Brutinel's and Lindsay's motor machine gun forces in chapter 7, above; and Greenwell, *An Infant in Arms*, p. 167 for his flying column in early 1917 during the German retreat, including armoured cars and cavalry.
12 E.g. Fuller's *Tank Programme*, p. 2, 12 March 1918, in PRO WO158 865. Note that the term 'cruiser tank' did not emerge until the 1930s; but in essence the 20 mph Medium Ds projected for 1919 were ultimately intended to fill an analogous rôle. Unfortunately it is extremely doubtful that anything like adequate numbers of these machines could have been manufactured in time for a spring campaign in 1919, and even then they would surely have been mechanically very unreliable.
13 At Cambrai the artillery fired lifting concentrations rather than a true creeper — probably to give more room for manoeuvre to the tanks, but conceivably owing to the artillery's doubts about the accuracy of predicted fire.
14 Not least in H. Guderian's *Achtung Panzer!* (English translation, Arms and Armour Press, London, 1991).
15 For asphyxiation in Mk V 'Star' and other tanks, see Mitchell, *Tank Warfare*, pp. 247, 254, 264; and Watson, *A Company of Tanks*, p. 245.
16 I am grateful to Dr Paul Harris for pointing out the importance of this factor in British tank doctrine during two world wars. For the stress on gunnery on the move in 1917 training curricula, see *Gunnery School*, first document, pp. 2–3, and second document, p. 3, in *Tank Corps Schools*, in PRO WO158 802.
17 File on *Tank Corps Centre*, p. 1, in PRO WO158 802.
18 Birch's GHQ memorandum, 23 September 1918, in PRO WO158 832.
19 Edmonds admits the pro-tank bias of his official histories in OH 1918, vol. 5, p. v.
20 Kiggell's rather lukewarm report on the tanks' achievements at Flers, written 5 October 1916, is document 310 in WO158 186, and is also reproduced in WO158 832. See also the revealing map of each tank's progress in OH 1916, vol. 2, opposite p. 319.
21 Watson, *A Company of Tanks*, pp. 41–66.
22 *Lessons from Messines, Part 1, Tanks*, in PRO WO158 298.
23 E.g. actions at St Julien, 28 August. and Poelcapelle Road, 9 October, in Watson, *A Company of Tanks*, pp. 133–158; Mitchell, *Tank Warfare*, p. 124.
24 Lord Moyne, *Staff Officer: The Diaries of Lord Moyne 1914–18* (ed. Brian Bond and Simon Robbins, Leo Cooper, London, 1987), p. 131.
25 William Moore, *A Wood Called Bourlon: The Cover-up after Cambrai, 1917* (Leo Cooper, London, 1988), pp. 125–7, attempts — not altogether conclusively — to get to the bottom of the bell-ringing epidemic.
26 See chapter 8, above.
27 Mitchell, *Tank Warfare*, p. 132.

28 OH 1917, vol. 3, p. 90, elaborated in John Terraine, *The Smoke and the Fire* (Sidgwick and Jackson, London, 1980), p. 154.
29 OH 1917, vol. 3, p. 288.
30 R. E. Priestley's *The Signal Service in the European War of 1914 to 1918 (France)* (official RE history, Chatham, 1921), p. 238. The irony is that the wires had not been buried, which would have saved them from the tank tracks, in order to allow the tanks to achieve surprise.
31 Fuller's *Infantry and Tank Co-operation and Training*, issued to brigades by GS Tank Corps, 27 January 1918, in PRO WO158 803.
32 Swinton was already worried that the tanks were not 'field gun proof' as early as 6 July 1916 — letter to Director of Staff Duties at the War Office, in PRO WO158 833; as late as August 1918 the artillery was seen as the best recourse against anti-tank guns, in SS 214, *Tanks and their employment in co-operation with other arms*, p. 13, in PRO WO158 832.
33 Ibid., p. 4.
34 The full curriculum is itemised in PRO WO158 802.
35 OH 1918, vol. 4, pp. 384–5.
36 Note that the official statistics are more complete than mine as presented in Table 12 (p.167, above), since they tend to include reserves held just behind the fighting, and other incidents not listed in OH.
37 Some individuals nevertheless went into action sixteen times in the last three months: Mitchell, *Tank Warfare*, p. 271.
38 Ibid., pp. 268–70.
39 OH 1918, vol. 4, p. 24. Of course they were not all in the front line of the attack; e.g. the Whippets were reserved for the exploitation echelons. In addition to the fighting tanks there were also 120 supply tanks, making a grand total of 534.
40 *Infantry and Tank Co-operation and Training* suggested there should be one tank to every 1–200 yards of front but grouped in units of three tanks, presumably with one group every 500 yards. This would give three groups (nine tanks) to each divisional frontage of 1,500 yards.
41 OH 1917, vol. 3, pp. 27–8. There were also 54 supply tanks and 44 other auxiliary tanks for obstacle crossing or signals, making a total of 476 in all.
42 SS 214, *Tanks and their employment in co-operation with other arms*, p. 6.
43 For this and many other aspects of tanks in the Great War, see D. J. Fletcher, 'The Origins of Armour', in *Armoured Warfare*, ed. J. P. Harris and F. H. Toase (Batsford, London, 1990), pp. 5–26.
44 Even then, it is hard to see how Japan's army, fleet and air force could have been more seriously depleted before the bomb was dropped than they actually already had been by conventional means. In 1917 the German armies in the West had not been reduced to a comparable level, and would therefore have been more resistant and reactive to a massed tank surprise.
45 The official formula was that 'It is unwise, however, to place too much reliance upon mechanical contrivances', SS 214, August 1918, p. 3. It is unfortunate that some misguided ultra-modernists have chosen to ridicule this eminently commonsense attitude as excessively reactionary, when it really was not.
46 For this and much of what follows, see Priestley's excellent *The Signal Service*.
47 Ibid., p. 59.
48 Ibid., pp. 60–3.
49 Ibid., p. 65.
50 Bragg in Appendix L to M. Farndale, *History of the Royal Regiment of Artillery: Western Front 1914–18* (RA Institution, Woolwich, 1986), p. 377.
51 Priestley, *The Signal Service*, pp. 119–23.
52 Ibid., pp. 196-201.
53 Edmund Blunden, *Undertones of War* (new ed., Collins, London, 1978), p. 99, and cf. Priestley, *The Signal Service*, pp. 98–106.
54 Priestley, *The Signal Service*, p. 105.
55 E.g. Frank Richards, *Old Soldiers Never Die* (new ed., Mott, London,

1983), pp. 196, 223.
56 For use in aircraft, the 20 lb 'Stirling' set had provided a lightweight and ideal solution since 1915; but for ground troops the uneven terrain always posed many additional problems.
57 Priestley, *The Signal Service*, pp. 273–5.
58 Ibid., pp. 180ff., for SS 148: *Intercommunication in the Field*, later superseded by SS 191 (with the same title).
59 Priestley, *The Signal Service*, pp. 231-6.
60 Ibid., pp. 258–67.
61 Charlotte Maxwell, ed., *Frank Maxwell, VC* (Murray, London, 1921), p. 159. The culprit was exactly the same General Shoubridge who would later stellenbosch Brigadier Cumming for knowing too much about the true situation in the front line at Bullecourt in May 1917; H. R. Cumming, *A Brigadier in France* (Cape, London, 1922), pp. 79–87.
62 In XVII Corps' *Lessons Learnt During the Operations, 21st August–7th September 1918*, p. 1, in PRO WO158 422.

Chapter 10: Doctrine and training

1 E. K. G. Sixsmith, *British Generalship in the Twentieth Century* (Arms and Armour, London, 1970), pp. 110–30 etc., echoes Maxse's condemnation of his predecessors on much of this; and see especially the bitter letter from Deputy CIGS to CIGS, 11 July 1918, calling for the creation of a British military doctrine under a 'thinking branch', Maxse papers, Box 69–53-11, File 53.
2 Dossier 'f' — *Army Printing and Stationery Depot* — in PRO WO95 4189.
3 Ibid., entry for 9 October 1914.
4 Note for 10 August 1917 in the Army Printing and Stationery Service's war diary, PRO WO95 81.
5 Ibid., for all the P and SS transactions from 6 December 1914 onwards.
6 Ibid., 11 February 1918; on 16 May he found 535 machines in First Army. If we accept that higher HQs and the Line of Communication absorbed the majority of typewriters, Partridge's figures would suggest an average of only around twenty typewriters to each fighting division — perhaps one per 800 men — although the complete census he claimed does not seem to have been taken.
7 The scale of issue was set on 17 November 1914 as two per man per week, or one per Indian; but on 2 July 1916 there was a poignantly hasty double issue to Fourth Army, and a hurried ruling that in future, additional postcards would be allowed to all troops engaged in heavy fighting.
8 There was a problem with propaganda intended for enemy consumption, since under French law it was illegal to print anything in the German language — see appendix of 13 October 1915 in PRO WO95 81.
9 An impressively researched modern listing up to SS 659 is Peter T. Scott's unfinished ten-part 'The CDS/SS Series of Manuals and Instructions', in *The Great War*, vol. 1, no. 2 to vol. 3, no. 4.
10 Entry for 6 December 1914 in *Army Printing and Stationery Depot,* in PRO WO95 4189; and for 8 January and 9–10 February 1915 in PRO WO95 81.
11 E.g. ibid., entry for 2 January 1917 complaining about — and suppressing — pamphlets that are no more than extracts from routine orders. On 13 October 1917 this was followed by a complaint that 'armies are too ready to rush into print'.
12 SS 58 was secret; SS 390 was not. New 'distinctive' editions of both were demanded on 14 February 1916; but by 8 April 1918 both were being deliberately held out of circulation in order to reduce demand.
13 The main British manuals tended to have SS numbers in the low hundreds, whereas most of the German material was in the 400s and 500s. The 300s were largely taken up with line of communication or administra-

tive matters such as courts martial, the abolition of flies, how a soldier should get his pay, etc; see Scott, 'The CDS/SS Series'. A uniquely complete listing of all SS print jobs — not just 'SS' numbered items — during the first four months of 1916, including number of copies in each case, is in PRO WO95 4189, dossier (a).

14 The order of battles series was referenced as 'OB/17'. That for 11 November 1918 was reprinted in 1989 by the IWM. Note also the parallel series of SS 407, *Composition of Headquarters*, from 11 September 1917, of which ten different editions are in Maxse papers, Box 69–53–13, File 67.

15 Entries for 26 September, 7 and 19 November 1916, in PRO WO95 81; and discussion in Shelford Bidwell and Dominick Graham, *Fire-power: British Army Weapons and Theories of War, 1904–45* (Allen and Unwin, London, 1982), p. 103.

16 Some of the most significant 'classics' are:- SS 125, *Instructions for Training* (various editions from June 1917 onwards); SS 126, *Training and Employment of Bombers* (September 1916, replacing SS 398 of March 1916); SS 135, *Training of Divisions for Offensive Action* (December 1916); SS 143, *Training of Platoons for Offensive Action* (February 1917); SS 191, *Intercommunication in Battle* (November 1917, based on SS 148 of March 1917); SS 192, *Employment of Machine Guns* and SS 197, *The Tactical Employment of Lewis Guns* (both of January 1918, superseding CDS 36 of June 1915, SS 106 of March 1916 and SS 122 of September 1916); SS 210, *The Division in Defence* (May 1918); SS 214, *Tank-infantry Co-operation* (August 1918); SS 388, *Defensive Measures Against Gas Attacks* (January 1916, replaced by SS 419 of May 1916).

17 Collected in Maxse papers, Box 69–53–12, File 59.

18 *The Song*, p.1: typescript approvingly annotated in Maxse's own handwriting, in Maxse papers, Box 69–53–11, File 53.

19 For Solly Flood's post, see OH 1916, vol. 2, p. 571.

20 Sixsmith, *British Generalship*, p. 139.

21 Letter of 11 July 1918, in Maxse papers, Box 69–53–11, File 53.

22 Sixsmith, *British Generalship*, pp. 96, 98; compare discussion of such matters in chapters 4 and 8, above.

23 Charles E. Carrington, *Soldier from the Wars Returning* (Hutchinson, London, 1965), p. 196, for the use of such pro-formas at Ypres in 1917. Cf. 24th Division's overprinted maps, designed for a similar purpose in the summer of 1917, in Maxse papers, Box 69–53–10, File 42.

24 E.g. John Terraine, ed., *General Jack's Diary, 1914–18* (Eyre and Spottiswoode, London, 1964), p. 139.

25 A. Hyatt, *General Sir Arthur Currie* (National Museums of Canada, Toronto, 1987), pp. 60–1.

26 *Lessons from Messines*, in PRO WO158 298.

27 F. Mitchell, *Tank Warfare: The Story of the Tanks in the Great War* (Nelson, London, n.d.), p. 64.

28 J. C. Dunn, *The War the Infantry Knew* (new ed., Jane's, London, 1987), p. 341.

29 E.g. Partridge was constantly complaining that manuals were not standardised between the War Office in Britain and the BEF in France: PRO WO95 81, passim.

30 E.g. the made-in-Australia *Syllabus of Training for Light Horse, Infantry and Machine Gun Reinforcements* (Australian Imperial Force, Melbourne, 1917).

31 See especially his *Hints on Training, Issued by XVIII Corps* ('The Brown Book', HMSO, August 1918), but also the thrust of all his training exhortations.

32 *General Jack's Diary*, p. 137.

33 Brigadier C. Bonham Carter, *Training in France*, 25 June 1918, p. 3, in Maxse papers, Box 69–53–11, File 53.

34 For details of schools and training the reader is referred to forthcoming work by Simon Robbins of the IWM. While not conducting any sys-

tematic analysis of his own, the present author has certainly been struck by the apparent ubiquity of schools throughout the BEF, see e.g. Maxse papers, Box 69–53–11, File 54; Box 69-53-12, File 58; the Lindsay papers, passim; and, for tanks, PRO WO158 802.
35 For Lindsay's career, see chapter 7 above.
36 W. D. Croft, *Three Years with the Ninth (Scottish) Division* (Murray, London, 1919), p. 96.
37 H. R. Cumming, *A Brigadier in France, 1917–18* (Cape, London, 1922), pp. 88–90.
38 E.g. G. Ashurst, *My Bit: a Lancashire Fusilier at War* (Crowood, Ramsbury, Wilts, 1987), p. 110; A. French, *Gone for a Soldier* (Roundwood, Kineton, Warks, 1972), p. 26. However, the present author has absolutely no intention of becoming involved in the 'Monocled Mutineer' controversy at this point....
39 Edmund Blunden, *Undertones of War* (new ed., Collins, London, 1978), p. 200. The present author knows all about the very real problem of teaching theoretical subjects in warm classrooms to buckish young officers who have spent the previous night digging trenches, standing sentry or carousing. No military system can expect good learning from such officers until they have been fully rested for 48 hours.

Chapter 11: Conclusion

1 It is of course a pure coincidence that General Norman Schwarzkopf, who expunged the Vietnam failures by his Gulf campaign, has a Germanic name.
2 Not even Robin Prior and Trevor Wilson, *Command on the Western Front* (Blackwell, Oxford, 1992), despite their excellent critique of Rawlinson's many failures in 1916, are able to offer a complete explanation for his impressive successes in 1918.
3 E.g. Shelford Bidwell and Dominick Graham, *Fire-power: British Army Weapons and Theories of War, 1904–45* (Allen and Unwin, London, 1982), pp. 112, 116, 118 etc., are in error when they assume British infantry failed to reform itself or failed to understand the Lewis gun. They seem to be looking at the matter too much from a gunner's — and perhaps also a Canadian — viewpoint.
4 OH 1917, vol. 2, pp. 294–5, 318, 337.
5 Note that I do not hold to the 'stab in the back' theory about Vietnam, since many of the wider political problems in 1968 were clearly of the US Army's own making — just as one of (Douglas!) Haig's main failings had been his inability to express himself clearly to the politicians in London. The US Army in Vietnam should certainly have predicted its PR defeat, in precisely the same way that Haig should have predicted the rain at Ypres or the Lloyd George manpower freeze.
6 See chapter 3, above.
7 E.g. Currie in A. Hyatt, *General Sir Arthur Currie* (National Museums of Canada, Toronto, 1987), p. 121.
8 W. D. Croft, *Three Years with the Ninth (Scottish) Division* (Murray, London, 1919), p. 232.
9 Ibid., p. 87.
10 Robin Prior and Trevor Wilson, *Command on the Western Front;* M. Farndale, *History of the Royal Regiment of Artillery: Western Front 1914–18* (RA Institution, Woolwich, 1986); J. B. A. Bailey, *Field Artillery and Firepower* (The Military Press, Oxford, 1989).

Notes to Appendix 2

1 Correlli Barnett, *The Swordbearers* (Penguin ed., London, 1966), p. 362.
2 Bulfin's opinion, quoted in Tim Travers, *The Killing Ground: The British Army, the Western Front and the Emergence of Modern Warfare 1900–1918* (Allen and Unwin, London, 1987), p. 12, n. 33.

Bibliography

This listing does not purport to be either systematic or exhaustive — it is simply a note of the books and documents that I happen to have read. They constitute far less than ten per cent of the relevant material commonly available; but I hope they are at least adequate to provide a revealing *sondage* of its main points.

Note on the official histories

As befits a gigantic war fought within the era of mass literacy, the military literature of 1914–18 is indeed prodigious. Personal impressions and popular illustrated descriptions of the fighting were already appearing within days of the outbreak — and even well before that, if one counts the many 'future histories' that prefigured the event. Throughout the war there was also a steady flow of more or less authoritative historical treatments, which grew into a flood after the armistice. One prominent branch dealt with higher strategic questions and the contribution made by individual generals; but less controversially there was also a sustained effort of professional tactical analysis. This type of work was heavily overshadowed by the much better-known outgrowth of memoirs and novels in the 1920s; but it can nevertheless still tell us a very great deal about how the war was fought. In more recent times it has been supplemented by many additional collections of memoirs and diaries, and the opening to students of a mass of correspondence, both public and private.

Of particular significance in this process has been the publication of the official histories, most notably *Military Operations, France and Belgium*. This fourteen-volume work was planned, overseen and to a considerable extent written by Brigadier Sir James Edmonds — a somewhat unstable and insecure engineer officer who had done well with Kitchener in South Africa, had helped to set up MI5, and had even written a book about the Americans that was disarmingly entitled *Fighting Fools*. However, he ultimately failed to be appointed as a field commander in the Great War. He was undoubtedly conscientious and intelligent, but also devious, over-political and too deeply opinionated. His 'Official' History thus turns out to be every bit as much an idiosyncratic personal statement as were the many dissenting or individual personal opinions which he often made it his business to censor out of the record. He was surely acting only on behalf of his own rather unsystematic set of *private* conspiracies, rather than for some sinister centrally controlled hit list. For example, few of Edmonds's implied criticisms of our allies, the French, can have been officially welcomed by the Foreign Office (at least in the volumes which appeared before 1940!); nor can his notoriously distressing and cheating habit of massaging the reputations of individual generals possibly be seen as a 'pure' historiographical method that would be officially recommended.

Once we have recognised these important traits in Edmonds's character, however, we must also concede that he only relatively rarely allowed his (often highly colourful and totally misleading) private gossip and prejudices to distort the *ex cathedra* statements which he finally made in his published official history. Perhaps contrary to normal practice, he was often far more honest in his published texts than he was in his private scuttlebutt. Edmonds actually succeeded in placing a great deal more of the true record of the BEF's achievement in front of the public than did many other competing agencies — and this is far more true in the field of tactics than in those of operational, strategic or political affairs. It seems especially unfortunate that Edmonds has been so universally damned by the 'war and society' school, since he actually has so very much to offer to the 'pure military history' school.

The general layout of Edmonds's OH offers us an important clue to the general drift of his thinking. Whereas nine of his volumes (or 64 per cent) deal with the last 27 months of the war in which the overwhelming majority of the fighting took place (see Table 1, p. 18, above), the almost minuscule operations of 1914 and 1915 could still be strung out to a total of four volumes (or 29 per cent of the total, for at most 20 per cent of the fighting). This should be contrasted with the paltry two volumes given to the exceptionally heavy action of 1917 before 20 November at Cambrai, particularly to the massive (but controversial) battle of Passchendaele. This disproportionate coverage in a sense represents deeply hidden propaganda for the Old Contemptibles — written by one of them — which seems to have influenced rather too many later students and to some extent elbowed aside the victories won in the 'real' war of the subsequent citizen armies. We should also remember that Edmonds was a committed opponent of the Machine Gun Corps, for which no formal official history was ever written, and that he gives remarkably little space to the war-winning artillery (see his excuses in *1918,* vol. 5, p. v), which has also been badly served by official historians. It is equally perhaps no accident that almost half of the OH volume in which the arrival of Kitchener's new armies is heralded (*1916,* vol. 1) is centred on their disastrous 1 July on the Somme. For Edmonds it is only the regulars and the colonials who really seem to count as 'proper' infantry.

This view is reinforced, accidentally or otherwise, by the emphasis laid on the tank corps, as a type of technological *deus ex machina* which arrives to rescue the Kitchener armies from the consequences of their supposed inefficiency. Edmonds gives it pride of place in a whole (disproportionately lengthy) volume on Cambrai, and constantly cites it separately in the other volumes from *1916, vol. 2,* onwards. Something similar may be said of the RFC and RAF, even though they were accorded an entirely separate six-volume series of Official Histories of their own.

Like every other historian who has ever lived, Edmonds was a pretty direct product of his own particular personal circumstances. The present author would, however, like to put on record that he has drawn not merely massive funds of detailed information from Edmonds's encyclopaedic work, but also a great historiographical inspiration from its transparently individualistic tone, its lucid general organisation and its extremely wide scope. Either despite or because of its often irritatingly close grain, it would not be too much to say that this British Official History for France and Belgium is positively the best book ever written, or ever likely to be written, about the Great War on the Western Front — and by quite a long margin.

Unpublished sources

Public Record Office, Kew

MUN series

2 / 29:	Offprints of Weekly Reports on the production of rifles, machine guns and small arms ammunition: text and statistics for September 1915–December 1917.
4 / 3590:	One carton out of an extensive collection of intelligence reports on enemy weapons and tactics during the last year of the war; includes examples of regular British intelligence summaries as well as the monthly gas and 'inventions' newsletters.
7 / 273:	Semi-automatic, rifled 'bomb throwing gun or trench gun' designed by Captain S. F. Stokes, RE: i.e. the *other* Stokes who designed trench mortars! This one was rejected by the Munitions Inventions Department on a number of occasions, 1915–18.

WO series

(i) Administration, weaponry and supply

33 / 667:	Final Report of the Committee on Automatic Rifles, 1914.
33 / 933:	1918 (September) *Handbook for Shell:* technical guidelines.
33 / 934:	1918 (August) *Handbook of Bombs and Grenades:* technical guidelines.
98 / 5 :	Awards of the VC to Divisions and Units 1914–18.
98 / 6 :	*VCs Awarded During the Campaign August 1914–April 1920.*

Bibliography 261

106 / 432:	Decision to abandon heavy trench-mortars but increase the mobility of medium mortars, November 1918.
158 / 154:	Notes on the construction of concrete block houses, June–August 1918.
158 / 802:	Notes on the schools set up in England for tank crew.
158 / 803:	Notes on the schools set up in France for tank crew; also *Infantry and Tank Co-operation and Training,* issued to tank brigades by Tank Corps General Staff, encapsulating the training for and lessons from Cambrai.
158 / 807:	Includes notes on Tank Corps organisation, 1918.
158 / 961:	Papers on the reorganisation of the Intelligence Branch, GHQ, 1917–18.

(ii) Operations and tactics before the battle of the Somme, 1916

33 / 717:	*Object and Conditions of Combined Offensive Action* (Translated by GHQ — from the 16 April 1915 French original — June 1915): Beautiful summary of Joffre's principles at this point in the war.
33 / 718:	*Notes on the Employment of Machine Guns and the Training of Machine Gunners* (July 1915) — a long, interesting but perhaps already outdated summary of the state of the art; the fruits of Lindsay and Baker-Carr's questionnaires.
33 / 721:	Tactical Notes (GHQ) 1915, n.d. Shows that the outline of modern tactical ideas is already available, but the detail is not very fully thought through.
33 / 725:	Tactical Notes (GHQ) 1915, n.d. Ditto, but concentrates on the Attack....
158 / 288:	Establishments of Motor Machine Gun Battalions, February–May 1915.
158 / 831:	Notes on tanks, 1915–16.
158 / 833:	Correspondence on tanks, early 1916.

(iii) Operations and tactics in the battle of the Somme, 1916

158 / 186:	First Army correspondence with GHQ July–November 1916, including the action at Fromelles and abortive plans for diversionary attacks on Vimy Ridge.
158 / 321:	Notes by R. M. Lucock on Fourth Army's planning and preparation for the battle of the Somme.
158 / 322:	Fourth Army *Summary of Operations* (log) for 1 to 13 July 1916.
158 / 344:	Notes and Lessons from operations, June to November 1916: assorted reflections on infantry tactics on the Somme, all very useful. See especially *Memorandum on Trench to Trench Attacks* by a Battalion Commander

	(Fifth Army, 31 October 1916).
158 / 410:	VIII Corps' plan for 1 July 1916 (the attack on Serre and Beaumont Hamel).
158 / 834:	Notes on the tactical employment of tanks, second half of 1916.

(iv) Operations and tactics in 1917

33 / 816:	*A Note on the Recent Cavalry Fighting,* up to 7 April 1917 (10 April 1917): a handbook for 'Open Warfare'.
106 / 402:	Canadian Corps report on operations against Vimy Ridge, 9–16 April 1917. Appendices missing.
158 / 20:	General Staff Notes on operations and manpower, January 1917–October 1918.
158 / 208:	Second Army orders and planning for 3rd Ypres, September–October 1917.
158 / 209:	Second Army orders and planning for 3rd Ypres, October–December 1917.
158 / 215:	Planning papers and orders for the battle of Messines, 1917.
158 / 250 and 251:	Fifth Army HQ conferences for Third Ypres, 31st August to 19th November 1917.
158 / 298:	*Lessons from Messines,* collected in response to a Sec-ond Army questionnaire concerning Tanks, Cavalry, Organising the area captured, and Machine Guns.
158 / 347:	*Training Memoranda in Connection with Schemes,* November 1917: III Corps arrangements for tank-infantry training for the battle of Cambrai.
158 / 413:	IX Corps instructions for the Messines offensive, Part 2, Artillery.
158 / 418:	*Machine Guns, Tactical and Technical Lessons Learnt* by IX Corps in the battle of Messines, 1917.
158 / 814:	Includes notes on tanks in the battle of Arras, 1917.
158 / 858:	Includes notes on tanks in the battle of Messines, 1917.

(v) Operations and tactics in 1918

158 / 70:	*Notes on Recent Fighting, Issued by the General Staff.* Useful printed series for distribution down to battalions: 20 issues appeared from 5 April to 6 November 1918.
158 / 90:	*Establishments,* May–September 1918: Maxse's attempt to persuade GHQ to raise the minimum authorised strength of infantry battalions.
158 / 332:	Colonel N. K. Charteris, *Narrative of Machine Gun Operations, Fourth Army April–November 1918.*
158 / 411:	VIII Corps' defence scheme NE of Ypres, January–February 1918.

158 / 422:   XVII Corps' narrative of operations 27 September–11 November 1918.
158 / 832:   Notes on the employment of tanks, 1918, including training manual SS 214, *Tanks and their Employment in Co-operation with Other Arms*: the final formal expression of three years' over-heated argument and speculation!
158 / 835:   Includes notes on tanks early in 1918, with ideas for the defensive.
158 / 855:   Includes notes (by J. F. C. Fuller?) on tank co-operation with other arms, immediately postwar.
158 / 865:   *Tank Programme 1919*: J. F. C. Fuller's ideas on the use of tanks, as at 12 March 1918.

Imperial War Museum

General Sir Ivor Maxse's Papers, in cartons 69–53–10, –11, –12, –13, –14. These include Maxse files 42–48, 53–67 and 69–71, covering mainly the last two years of the war. I have not looked into the earlier Maxse cartons, which cover 1901–1916. In 1916–18 Maxse collected a wide range of training, organisational and tactical material while in command of 18th Division, XVIII Corps, and finally the Inspectorate of Training. Of particular interest are his repeated insistence on regular platoon organisation, his detailed arrangements for defence in March 1918, and his series of IT training manuals issued during the last four months of the war.

Microfilm PP MCR 155: Lieutenant Colonel H. H. Hemming, *My Story* (c.1976). Entertaining memoirs of the young Canadian inventor of the 'flash and buzzer board' for flash spotting, who saw counterbattery reconnaissance from close quarters throughout the war.

Misc 23, Item 398: Infantry training programmes at 42nd Division School of Instruction, November 1918: just two sheets, revealing a very basic training course lasting two weeks.

91 15 1: Captain A. R. Pelly, *Some Notes on Infantry Drill — Special Hints for Volunteers* (published Chelmsford, March 1918): short notes for basic training in drill, the PH Helmet, Hotchkiss gun etc., for new recruits. His 1915 manuscript diary, including Gallipoli, shows how an interfering colonel could ruin a battalion's training efforts.

Other

The G. M. Lindsay archive, held at the Tank Museum, Bovington, Dorset. The Machine Gun series (covering mainly the Great War) includes cartons A to L, all full of essential material on this key figure in the tactical history of the war. Unfortunately the crucial year between the battle of the Somme and the start of Passchendaele appears to be unrepresented in these documents, since Lindsay was in the front line at the time.

The Liddell Hart Archive at King's College, The Strand, London: Hart's papers covering 1916–18, i.e. 7/1916/1–38; 7/1917/1–29; 7/1918/1–28. Also consulted in this collection were manuals 8/34 (SS 135, *Instructions for the Training of Divisions for Offensive Action*); 8/65 (Laffargue, *Impressions and Reflections of a French Company Commander, Regarding the Attack*); and 8/66 (Fourth Army *Tactical Notes*, or 'Red Book', of May 1916).

Stephen D. Badsey, 'Fire and the Sword' (unpublished doctoral thesis, Cambridge, 1981): a stimulating analysis of British cavalry in the late Victorian and Great War eras, and the way it reformed its tactics. This thesis casts considerable (and long overdue) doubt on the perceived uselessness of this much-maligned arm.

Matthew Buck, 'French Operational Doctrine and the Offensives of 1914' (unpublished PhD 'Registration Piece', Cambridge University, 1991): interesting modern analysis of the oft-overlooked Bataille des Frontières.

'*The Kirke Committee Report*', Memorandum by CIGS to PUS, March 1932 (PRO document WO32/3115 [a]) and *Report on the Staff Conference Held at the Staff College, Camberley*, 9–11 January 1933 (PRO document WO 32– 3116): interesting common-sense reflections on WW1 tactics, albeit expressed in officialese that would surely have been anathema to most trench fighters.

Official (and semi-official) publications

General staff training manuals

(Apart from those consulted in OH, PRO, Maxse or Liddell Hart papers)

SS 119, *Preliminary Notes on the Tactical Lessons of the Recent Operations* (July 1916, elaborating the earlier SS 109): fascinating update made during the initial Somme battles, particularly stressing artillery cooperation, reliance on the rifle rather than the bomb, the flow of information within divisions, and the special qualities of fighting in close country such as woods or villages.

SS 143, *Instructions for the Training of Platoons for Offensive Action, 1917* (14 February 1917): the epoch-making change from 'wave' tactics to small unit platoon initiatives relying on integral firepower.

SS 197, *The Tactical Employment of Lewis Guns* (January 1918): an extremely useful analysis of how the Lewis was finally understood towards the end of the war.

Other

Anon, *Abstract of Official Statistics of the Military Effort of the British Empire During the Great War, 1914–20* (HMSO, the War Office, March 1922): 'Devastating Statistics'!

Anon, *British Tactical Notes* (US Government Printing Office for Army War College, Washington DC, December 1918): a 52-page summary of British minor tactics and training from the last two years of the war.

Anon, *Order of Battle of the British Armies in France 'OB / 17'* (GHQ, 11 November 1918, reprinted by IWM, London, 1989): complete for units and lines of communication troops, but does not include the composition of HQs.

Anon, *A Short History of the Royal Air Force* (Air Historical Branch, London, 1929).

Anon, *Syllabuses of Training for Light Horse, Infantry and Machine Gun Recruits* (Australian Imperial Force, Melbourne, 1917).

R. V. K. Applin, *Machine Gun Tactics* (3rd ed., Rees, London, 1915): somewhat dated thoughts based mainly on the South African and Manchurian campaigns, but widely circulated at the start of WW1.

Leonard P. Ayres, *The War With Germany, a Statistical Summary* ('Issued by Authority', Washington DC, n.d.): official propaganda from immediately postwar, purporting to 'prove' that USA played a considerably greater rôle in the victory than was actually the case.

A. F. Becke, *Order of Battle of Divisions* (British official history, 8 vols, HMSO, London, 1937–45, reprinted by Sherwood and Westlake, 1986–9): vol. 1 covers the regular army, vols 2A and 2B the territorials, vols 3A and 3B the New Armies, vol. 4 the higher staffs, and vols 5–6 the colonials. Invaluable and detailed, although sadly reticent on corps and army troops.

G. Casserly, *Tactics for Beginners — for the use of the new armies and volunteers* (Hodder and Stoughton, London, 1915): a standard training text which looks naïvely backwards to many 19th-century lessons, but nevertheless exhibits a number of what would later be called 'stormtroop' tactics.

Peter Chasseaud, *Topography of Armageddon* (Mapbooks, London, 1991): full coverage of the BEF's frontage in facsimile trench maps, mainly in 1:10,000 scale, together with some interesting explanatory material about the survey.

J. E. Edmonds, general editor (but with many other contributors, notably G. C. Wynne for 1915, W. Miles for 1916, C. Falls for 1917 and A. F. Becke for maps), *History of the Great War, Based on Official Documents: Military Operations, France and Belgium* (i.e. British Official History, or OH). 1914–16 have two volumes for each year, in addition to maps and appendices. 1917 has three volumes and 1918 five volumes, plus maps and appendices. All published by Macmillan, London, 1922–49. Some volumes recently reprinted by IWM.

N. V. Lothian, *The Load Carried by the Soldier* (Army Hygiene Advisory Committee Report No.1, War Office, London 19 January 1922): an influential historical overview which shows that no foot soldiers since the Greek hoplites had been asked to carry as much dead weight as the infantry of 1918. This question much exercised Basil Hart.

I. Maxse, ed., *Hints on Training, Issued by XVIII Corps* ('The Brown Book'; HMSO, August 1918, compiled by IT in May 1918 from a variety of fragments much used by Maxse during the preceding twelve months): an interesting insight into the strongly held ideas about platoon training of the new BEF Inspector General of Training, who was nevertheless also the recently defeated spearhead Corps Commander of 21 March....

R. E. Priestley, *The Signal Service in the European War of 1914 to 1918 (France)* (official RE history, Mackay, Chatham, 1921): positively the most impenetrable book ever written on the war; but equally certainly one of the most revealing and informative.

'Simplex', *Instruction on the Lewis Automatic Machine Gun* (first published 1916; 3rd ed., Forster Groom, London, n.d.).

Anon, *Handbook of the .303" Lewis Machine Gun (part 1, provisional, 40/WO/ 3695)* (HMSO, London, 1915).

Histories of specific units, formations, corps or services

Sidney Allinson, *The Bantams: The Untold Story of World War I* (H. Baker, London, 1981): a highly journalistic account of a tragically botched recruiting initiative, upon which the 35th and 40th Divisions, and other units, would be based.

J. B. A. Bailey, *Field Artillery and Firepower* (The Military Press, Oxford, 1989): an excellent and wide-ranging technical overview of the whole history and future of combat gunnery.

John Baynes, *Morale: A Study of Men and Courage* (Cassell, London, 1967): seminal micro-history, including magnificent sociological and psychological analysis, of 2nd Scottish Rifles in the battle of Neuve Chapelle, 1915.

J. Ewing, *The History of the 9th (Scottish) Division* (Murray, London, 1921): an exemplary history of an exemplary formation. The 9th Division's gunners and tacticians were among the most innovative on either side, although by the fag-end of Passchendaele not even this division could continue to deliver the goods with any certainty, and the South Africans especially fell on hard times as a result of poor reinforcement arrangements.

M. Farndale, *History of the Royal Regiment of Artillery: Western Front 1914–18* (RA Institution, Woolwich, 1986): a fascinating modern version of Anstey's unpublished work on the subject — almost an official history — revealing the decisive importance of the artillery for the outcome of almost every attack. An exceptionally good book, despite its detractors.

Peter Firkins, *The Australians in Nine Wars: Waikato to Long Tan* (first published 1971; London ed., Robert Hale, 1972): entertaining ANZAC mythology, taking absolutely no prisoners from the effete Poms.

D. J. Fletcher, 'The origins of armour', in J. P. Harris and F. H. Toase, eds, *Armoured Warfare* (Batsford, London, 1990), pp. 5–26: a succinct, albeit 'tankie', summary of Great War tanks.

C. H. Foulkes, *Gas!, The Story of the Special Brigade RE* (Blackwood, Edinburgh and London, 1934): an important, very full and 'near official' history of this innovative and deadly unit.

S. Gillon, *The Story of the 29th Division: A Record of Gallant Deeds* (Nelson, London, 1925): an adequate if not brilliant account of an élite formation, especially interesting for General De Lisle's tactical ideas.

G. S. Hutchinson, *Machine Guns: Their History and Tactical Employment*

(Macmillan, London, 1938): the persuasive champion of the MGC, at his 'tens-of-thousands-slaying' best!
F. Mitchell, *Tank Warfare, The Story of the Tanks in the Great War* (Nelson, London, n.d., no. 15 in 'The Nelsonian Library', reprinted Donovan, 1987): a popular but very full account, by a participant in the first tank v. tank duel in history.
Desmond Morton, 'The Canadian Military Experience in the First World War, 1914–18', in *The Great War, 1914–18,* ed. R.J.Q. Adams (Macmillan, London, 1990) pp. 79–133: useful short overview of the Canadian experience, from 'Drill Hall Sam' Hughes onwards.
G. H. F. Nichols, *The 18th Division in the Great War* (Blackwood, Edinburgh and London, 1922): a somewhat disappointing account of an epic formation.
Bill Rawling, *Surviving Trench Warfare, Technology and the Canadian Corps, 1914–18* (University of Toronto Press, Toronto, 1992): an excellent analysis of the Canadians' tactical development and battlefield methods, which puts some of the wilder claims for their originality into a welcome perspective. A highly recommended work which expands Hyatt's biography of Currie, but which unfortunately I read only when it was too late for its findings to be incorporated in the present text.
G. D. Sheffield, 'British Military Police and their battlefield role 1914–18', in *The Sandhurst Journal of Military Studies,* issue I, 1990: an interesting modern analysis of the Corps of Military Police.

Biographical materials and eyewitness accounts

George Ashurst, *My Bit: A Lancashire Fusilier at War 1914–18* (edited by Richard Holmes, Crowood, Ramsbury, Wilts, 1987): memoirs from the gas attack at Second Ypres, Beaumont Hamel on 1 July 1916, and many other stricken fields.
Edmund Blunden, *Undertones of War* (first published 1928: new ed. Collins, London, 1978): a powerful poet who fought through the thick of the war in 11th Royal Sussex.
Malcolm Brown, *The Imperial War Museum Book of The First World War* (Sidgwick and Jackson, London, 1991): a miscellany of letters, diaries and memoirs from the IWM.
P. J. Campbell, *In the Cannon's Mouth* (Hamish Hamilton, London, 1979): a subaltern in the field artillery, from Third Ypres to the end of the war.
Charles E. Carrington, *Soldier from the Wars Returning* (Hutchinson, London, 1965): a more discursive and informative, but perhaps less immediate, latter day rewrite of 'Charles Edmonds's' subaltern's memoirs (see below).
George Coppard, *With a Machine Gun to Cambrai* (London, HMSO, 1969): excellent Vickers gunner's view of the war.
W. D. Croft, *Three Years with the Ninth (Scottish) Division* (Murray, London, 1919): lively account by an intelligently 'thrusting' brigadier, which goes deeper into tactical details than most. A dedicated opponent of the hand-grenade, Croft should be read in conjunction with A. O. Pollard's (see below) enthusiastic support for that particular weapon!

F. P. Crozier, *The Men I Killed* (Joseph, London, 1937): a notorious 'thruster' from 36th Division who was nevertheless innocent enough to believe — simultaneously — in both the deterrent value of shooting fugitives and in the inner purity of the military system. Astonishingly, it needed the depradations of his 'Auxiliaries' in Ireland to teach him otherwise....

C. E. Crutchley, ed., *Machine-Gunner 1914–18* (Machine Gun Corps Old Comrades Association, Northampton, 1973): variegated memoirs from the MGC, a body of devoted soldiers who were deprived of a postwar home.

H. R. Cumming, *A Brigadier in France, 1917–18* (Cape, London, 1922): interesting memoirs from a scientifically competent officer who did not appear to know how to blow his own trumpet in 7th Division, although he was later accepted in 21st Division.

J. C. Dunn, *The War the Infantry Knew* (first published 1938, new ed., with introduction by K. R. Simpson, Jane's, London, 1987): fascinating overview of the famous 2/RWF, from its long-serving (and long-suffering) battalion doctor.

'Charles Edmonds' (= Charles E .Carrington), *A Subaltern's War* (Davies, London, 1929): one of the all-time classics. The author was a triumphantly successful — and therefore apparently unduly enthusiastic — officer of 48th Division, at both the Somme and Third Ypres.

Anthony French, *Gone for a Soldier* (Roundwood, Kineton, Warwickshire, 1972): interesting short memoir by a literary corporal (and sniper) of 15th London Regiment, who was wounded near High Wood on 25 September 1916.

Robert Graves, *Goodbye to All That* (first published by J. Cape 1929; Penguin ed., London, 1982): this most celebrated and scandalising of all the war memoirs is written with great gusto, wit and acidity, but also with far more technical military expertise than many pacifists would wish to acknowledge.

Graham H. Greenwell, *An Infant in Arms* (first published 1935; new ed., with introduction by John Terraine, Allen Lane, London, 1972): a subaltern notorious as a 'war lover', but no less credible for that, from 48th Division.

Wyn Griffith, *Up to Mametz* (first published 1931; new ed., Severn House, London, 1981): a beautifully written 'literary' memoir from 115th Brigade, 38th Welsh Division on the Somme, containing some fascinating insights into the technicalities of brigade handling.

Donald Hankey, *A Student in Arms* (first published 1916; 12th ed., London, 1917): mawkish religiosity and unconvincing public school musings from the start of the war. Famous, even so.

B. H. ('Liddell') Hart, *Memoirs* (2 vols, Cassell, London, 1965), vol 1: contains a neatly compressed summary of the guru's earlier years, albeit obviously lacking the mordant critique that a J. J. Mearsheimer might have wished to add....

Ian Hay, *Carrying On: After The First Hundred Thousand* (Blackwood, Edinburgh & London, n.d., 1918?): amusing and interesting personal view of 9th Scottish Division from Loos through the Somme.

F. C. Hitchcock, *Stand To: a Diary of the Trenches, 1915–1918* (first published 1937; new ed., edited by Anthony Spagnoly, Gliddon, Norwich, 1988):

excellent and long analysis from a subaltern in the 2/Leinsters, especially for 1915–16.
A. Hyatt, *General Sir Arthur Currie* (National Museums of Canada, Toronto, 1987): a fascinating and revealing modern study, slightly let down in places by obscure phrasing and defective proofreading.
Ernst Jünger, *The Storm of Steel* (R. H. Mottram, ed., Chatto and Windus, London, 1929): memoirs of the celebrated and grittily unrepentant ideologue of German stormtroop tactics.
John Laffin, *Letters from the Front, 1914–18* (Dent, London, 1973): a well-selected anthology, although mostly from the first half of the war and written mostly by officers who failed to survive the experience.
General Lanrezac, *Le Plan de Campagne Français et le Premier Mois de la Guerre* (Payot, Paris, 1920): the controversial — and short-lived — anglophobe commander of Fifth Army in August 1914.
'Mark VII' (Max Plowman), *A Subaltern on the Somme in 1916* (Dent, London, 1927): a Scots batallion in 17th (Northern) Division, with some telling accounts of the autumn battles. 'Mk VII' is, of course, a reference to the improved rifle cartridge that was introduced just before the war began.
Charlotte Maxwell, ed., *Frank Maxwell, VC* (Murray, London, 1921): an inspiring professional colonial soldier comes to terms with the trenches in 1916, commanding the 12th Middlesex and then briefly the 27th Brigade (9th Division) before being sniped dead at Ypres, September 1917. His admiring successor was W. D. Croft of 11/RS (see above).
Lord Moran, *The Anatomy of Courage* (first published 1946; new ed., Constable, London, 1966): marvellous psychological analysis of battle stress, from a member of the 'old school' who went on to become Churchill's medical adviser in the Second World War.
Lord Moyne, *Staff Officer: The Diaries of Lord Moyne 1914–1918* (edited by Brian Bond and Simon Robbins, Leo Cooper, London, 1987): the revealing (and unusual) memoirs of a very 'upper crust' brigade major (and MP), first in 25th Division, then in the astonishing Major General H. K. Bethell's 66th Division, during the second half of the Hundred Days.
R. A. Nye, *The Origins of Crowd Psychology* (SAGE, London, 1975), chapter 6, pp.123–53, 'Gustave Le Bon and Crowd Theory in French Military Thought Prior to the First World War': an important insight into the predominant psychological theories of the day.
A. O. Pollard, *Fire Eater: The Memoirs of a VC* (Hutchinson, London, 1932): a 'mad bomber' in every sense of the term, who started in the 3rd Division, then went to 63rd Division but gravitated to the Military Police and finally the RAF. Essential reading.
Geoffrey Powell, *Plumer, the Soldier's General* (Leo Cooper, London, 1990): a judicious modern analysis of an excellent tactician who suffered from a lack of the right social connections until he helped in the relief of Mafeking — and even then he continued to be a political lightweight. The book is inevitably marred by Plumer's failure to leave an archive or, indeed, any of the exciting cattiness displayed by so many of his colleagues in their memoirs.
Robin Prior and Trevor Wilson, *Command on the Western Front* (Blackwell,

Oxford, 1992): misleadingly advertised as a sort of biography of Rawlinson, this is actually the first modern, serious and technical 'operational art' analysis of Haig's key battles in 1915–16, and as such hugely to be welcomed. A deeply researched and highly recommended work.

Robin Prior and Trevor Wilson, 'Summing Up The Somme', in *History Today*, November 1991, pp. 37–43: a neat digest of the authors' damning conclusions on Rawlinson on the Somme.

Brian Reid, *J. F. C. Fuller, Military Thinker* (Macmillan, London, 1987): straightforward modern hagiography of an extremely important character.

Frank Richards, *Old Soldiers Never Die* (first published 1933; new ed., Mott, London, 1983): a fascinating 'other ranks' story of signalling in 2/RWF, on the Somme and elsewhere.

Sidney Rogerson, *The Last of the Ebb* (with narrative of German operations by Major General A. D. von Unruh; Arthur Barker, London, 1937): the Battle of the Aisne, May 1918, by a subaltern at 23rd Brigade HQ, and a rather more senior German *Doppelgänger*.

E. Rommel, *Infantry Attacks* (first published Potsdam, 1937; trans. G. E. Kidde, published in 1944 by the Infantry Journal, Washington, DC; new ed., Greenhill, London, 1990): a classic account of how the German infantry was all supposed to work, but significantly little from the Western Front.

Reginald H. Roy, ed., *The Journal of Private Fraser, Canadian Expeditionary Force* (Sond Nis Press, Victoria, BC, 1985) pp. 198-215: details of the attack from Pozières to Courcelette, mid-September 1916.

Siegfried Sassoon, *Memoirs of an Infantry Officer* (first published 1930; new ed., Faber, London, 1965): the poet's classic progress from intrepid trench-raider with 1 and 2/RWF, to equally fearless antiwar protester (unfortunately I have not read the two other volumes of his memoirs).

G. Serle, *John Monash: A Biography* (Melbourne University Press, 1982, reprinted 1985): stronger on family history than on military technicalities.

G. D. Sheffield and G. I. S. Inglis, eds, *From Vimy Ridge to the Rhine: the Great War letters of Christopher Stone, DSO MC* (Crowood, Ramsbury, Wilts, 1989): the trench letters of a 2nd Division infantry subaltern who became first a staff officer, and then the original superstar postwar disc jockey!

Harry Siepmann, *Echo of the Guns: Recollections of an Artillery Officer 1914–18* (Hale, London, 1987): an officer whose laudable main interest seems to have been thieving materials with which either to build grandiose shell-proof structures or to give his men treble rations.

Keith R. Simpson, 'Capper and the Offensive Spirit', in *Journal of the Royal United Services Institution*, 1973, pp. 51–6: excellent summary of the shortage of staff officers at the start of the war, and how Capper chose to contribute to it suicidally in his own person, by 'leading from the front'.

Anthony Spagnoly, 'Captain R. B. Hodson, a Record of Service', in *Stand To!* spring 1985, pp. 27–31: the story of crossing one field at Third Ypres.

John Terraine, ed., *General Jack's Diary, 1914–18* (Eyre and Spottiswoode, London, 1964): a classic 'thruster' who came to command 19th Brigade in 9th (Scottish) Division. Jack was also a friend of the author Sidney Rogerson, whom he had commanded in 2nd West Yorkshires, 1916–17.

Alan Thomas, *A Life Apart* (Gollancz, London, 1968): the well-written memoirs of a self-doubting but extremely effective young officer of 6/RWK, who was luckier even than Carrington in finding himself in astonishingly successful advances; in this case on the first days of both Arras and Cambrai.

L. H. Thornton and Pamela Fraser, *The Congreves* (Murray, London, 1930): the lives of General Sir Walter Norris Congreve and his son, Brevett Major William La Touche Congreve, who both won the VC. Disappointingly short on tactics.

A. J. Trythall, *'Boney' Fuller: The Intellectual General* (Cassell, London, 1977): a smilingly 'onside' biography of this bizarre and highly unconventional staff officer.

Edward Campion Vaughan, *Some Desperate Glory: The diary of a young officer 1917* (first published 1981; Papermac ed., London, 1985): useful memoirs from 8th Royal Warwickshire Regt (in 48th Division — the same formation as Charles Carrington and Graham Greenwell) at Arras and especially 3rd Ypres, 1917.

Aubrey Wade, *The War of the Guns: Western Front, 1917 and 1918* (Batsford, London, 1936): the exciting story of a Forward Observation Officer at Third Ypres and subsequent battles.

W. H. Watson, *A Company of Tanks* (Blackwood, Edinburgh and London, 1920): a highly recommended 'sponson's eye view' of early armour, not without respect for the horsed cavalry.

Denis Winter, *Haig's Command: A Resassessment* (Viking, London, 1991): notorious 'conspiracy theory' not only that Haig showed unrelieved incompetence which was rewarded by the highest possible promotion, but also that he successfully covered his tracks at every stage, especially with the connivance of the official British historian. We are tempted to suggest that this book 'goes over the top' with all the aggressive spirit, social connections and lack of system that Haig himself would have approved in one of his more enthusiastic subalterns.

David R. Woodward, *Lloyd George and the Generals* (Associated University Presses, East Brunswick, NJ, 1983): dense insights into 'The Goat' and his remorseless thrust for power.

Secondary sources: general narrative history

Correlli Barnett, *The Swordbearers* (first published 1963; Penguin ed., London, 1966): stimulating general analysis of Moltke's Marne, Jellicoe's Jutland, Pétain's 1917 mutinies and Ludendorff's 1918 spring offensives.

Gregory Blaxland, *Amiens 1918* (Muller, London, 1968): good general coverage of a splendid little battle.

F. Gambier and M. Suire, *Histoire de la Première Guerre Mondiale* (2 vols, Fayard, Paris, 1968): one of the best general French coverages.

Alistair Horne, *The Price of Glory: Verdun, 1916* (Penguin ed., London, 1964): journalistic and even hackneyed, but nevertheless inspiring.

Martin Middlebrook, *The Kaiser's Battle* (Allen Lane, London, 1978): excellent analytical collection of firsthand accounts of a momentous day — 21 March 1918 — unfortunately lacking in analysis of subsequent days.

William Moore, *A Wood Called Bourlon: The Cover-up after Cambrai, 1917* (Leo Cooper, London, 1988): not a great work, but it touches on some interesting aspects.

Jonathan Nicholls, *Cheerful Sacrifice: The battle of Arras 1917* (Leo Cooper, London, 1990): excellent overview of a lamentably overlooked battle that is nevertheless very important indeed. Arras was the way the Somme was supposed to be, but couldn't quite manage.

R. G. Nobécourt, *Les Fantassins du Chemin des Dames* (Bertout, Luneray 76810, France, 1983): in effect a history of the entire war 'in the East', as seen by the *poilu*.

Terry Norman, *The Hell They Called High Wood* (Kimber, London, 1984): an excellent blow-by-blow account of some of the fights that were actually more central to the Somme battle than anything that happened on 1 July.

Barrie Pitt, *1918: The Last Act* (first published 1962, Papermac ed., London, 1984): a journalistic but lively account of an epic period.

John Terraine, *Mons: The Retreat to Victory* (Batsford, London, 1960): a good account which fails to conceal the ineptitude of British generalship in 1914.

John Terraine, *The Road to Passchendaele* (Leo Cooper, London, 1977): a bizarre but wide-ranging collection of quotations from the high command, marred by its failure to look lower than Army HQs.

John Terraine, *The First World War 1914–18* (first published 1965; Papermac ed., London, 1983): a workmanlike short summary.

Ian Uys, *Longueval* (Uys, Germiston, South Africa, 1986): unusual peep at the German side of the epic South African battle for Delville Wood, 1916.

Secondary sources: analysis and technical aspects

Anderson et al., 'An Experiment in Combat Simulation: The Battle of Cambrai 1917', in *Journal of Interdisciplinary History*, 1972, pp. 229–49: a rare attempt to use carefully controlled wargame techniques to gain historical insights.

Tony Ashworth, *Trench Warfare 1914–1918: The Live and Let Live System* (Macmillan, London, 1980): one of the very best Great War analyses to appear in recent times. It discusses the tacit understandings that predominated in 'quiet sectors', as opposed to attacks, raids, strafes or 'hates'.

Tony Ashworth, 'The Sociology of Trench Warfare', in *British Journal of Sociology*, 1968, pp. 406–21: an early and pregnant sketch for the author's eventual book.

W. Balck, *Tactics* (new ed., Posen, 1908; trans. W. Kreuger, Fort Leavenworth, Kansas, 1911; reprinted Greenwood, Westport, Conn., 1977): methodical and vibrant picture of the prewar state of the art.

W. Balck, *The Development of Tactics, World War* (trans. H. Bell, Fort Leavenworth, Kansas, 1922): a collection of various tactical experiments tried out in the war, somewhat lacking in convincing final conclusions.

Arthur Banks, *A Military Atlas of the First World War* (first published 1975; new ed., Leo Cooper, London, 1989): an informative reference work which

suffers from a chronic case of the widespread disease of fading away after 1 July 1916.
F. von Bernhardi, *Germany and the Next War* (new English ed., London, 1912): interesting insights from a militarist.
Geoffrey Best and Andrew Wheatcroft (eds.), *War, Economy and the Military Mind* (Croom Helm, London, 1976): excellent essays collected from 'war and society' experts.
Shelford Bidwell and Dominick Graham, *Fire-power: British Army Weapons and Theories of War, 1904–45* (Allen and Unwin, London, 1982): a magnificently authoritative, lengthy and almost definitive tactical history — which does nevertheless at times tend towards specifically 'gunner' history — and which this particular wretched individual hereby hopes to trump!
I. S. Bloch, *Modern Weapons and Modern War* (ed. W. T. Stead, London, 1899): statistical prediction of how the sheer volume of fire on the modern battlefield would cause tactics to bog down.
Gustave Le Bon, *Psychologie des Foules* (first published 1895; new ed., PUF, Paris, 1947): bestselling and highly influential psychology, from the era when that science was first founded.
Eugène Carrias, *La Pensée Militaire Française* (PUF, Paris, 1960): excellent overview.
George M. Chinn, *The Machine Gun* (3 vols, Bureau of Ordnance, Department of the Navy, Washington, DC, 1951): a long technical coverage.
Rose E. B. Coombs, *Before Endeavours Fade: A Guide to the Battlefields of the First World War* (first published 1976; new ed., After the Battle, London, 1986): essential guidebook to the battlefields, their monuments and cemeteries.
Elmar Dinter, *Hero or Coward* (Cass, London, 1985): interesting analysis of combat psychology.
John Ellis, *The Social History of the Machine Gun* (London, 1975): an elegant cultural analysis which might more properly have been entitled *The Machine Gun as Symbol and Myth*.
John Ellis, *Eye-deep in Hell: The Western Front 1914–18* (Croom Helm, London, 1976): a gruesome illustrated evocation of trench warfare.
John A. English, *On Infantry* (Praeger, New York, 1981): a useful overview of infantry tactics and organisation in the twentieth century, albeit unduly sympathetic to the claims of 'Liddell' Hart.
Paddy Griffith, *Forward into Battle* (2nd ed., Crowood, Ramsbury, Wilts, 1990): includes unconventional discussion of the prewar (and postwar) doctrinal background, but is totally superseded by the present volume for his understanding of the Great War itself.
H. Guderian, *Achtung Panzer* (English translation, Arms and Armour Press, London, 1991): the German tank maestro's 1937 impressions, mainly of tanks in the Great War.
Bruce I. Gudmundsson, *Stormtroop Tactics: Innovation in the German Army 1914–18* (Praeger, New York, 1989): a modern and 'attack-specific' manifestation of an honourable bloodline that runs back through Lupfer and English to Wynne. In common with these predecessors, Gudmundsson's thesis is nevertheless marred by his apparent ignorance of parallel tactical develop-

ments in the non-German armies, which were often far ahead of the Germans themselves.

B. H. (Liddell) Hart, *The Future of Infantry* (Faber, London, 1933): essential, albeit rather late, companion to the author's claim to have re-invented infantry tactics around 1920.

Guy Hartcup, *The War of Invention* (Brasseys, London, 1988): useful short guidebook overviewing the whole development of Great War technology, from nitrates to neurasthenia and from submarines to sound-ranging.

J. E. Hicks, *French Military Weapons, 1717–1939* (Flayderman, New Milford, Conn., 1964): just an illustrated list, but quite useful.

Ian V. Hogg, *The Guns 1914–18* (Pan/Ballantine, London ed., 1973): interesting and lively technical discussion of the weapons rather than the tactics.

Ian V. Hogg, *The Illustrated Encyclopaedia of Firearms* (Newnes, London, 1978).

Michael Howard, 'Men Against Fire: the Doctrine of the Offensive in 1914', in Peter Paret, ed., *Makers of Modern Strategy* (Princeton University Press, 1986): a disappointingly thin treatment of a potentially magnificent subject.

Melvin M. Johnson jr and Charles T. Haven, *Automatic Arms: Their history, development and use* (Morrow, New York, 1941).

Erich von Ludendorff, *The Nation at War* (trans. A. S. Rappaport, London, 1936): a bleak vision of total war, by someone who ought to know his subject well.

T. T. Lupfer, *The Dynamics of Doctrine: The Changes in German Tactical Doctrine during the First World War* (US Army Command and General Staff College, Fort Leavenworth, Kansas, 1981): an interesting updating of Wynne's analysis of German tactics, which should be read in conjunction with the work of Gudmundsson (see above) and Samuels (see below).

M. Samuels, *Doctrine and Dogma: German and British Infantry Tactics in the First World War* (Greenwood, Westport, Conn., 1992): a book about German stormtroops rather than about British tactics, but which is unfortunately blind to the many defects in the German 'system'. One should not attempt to contrast German theoretical principles with British battlefield experience.

Peter T. Scott, 'Mr Stokes and his Educated Drainpipe', in *The Great War*, vol. 2, no 3, May 1990, pp. 80–116: comprehensive technical description of how the Stokes mortar was born.

Peter T. Scott, 'The CDS / SS Series of Manuals and Instructions, a Numerical Checklist': invaluable series of ten short articles in *The Great War*, vol. 1, no. 2, February 1989, p. 50, through to vol. 3, no. 4, September 1991, p. 136.

Peter Simkins, *Kitchener's Armies* (Manchester University Press, 1988): exhaustive and well-written analysis of the raising of the New Armies.

Ian D. Skenneto and Robert Richardson, *British and Commonwealth Bayonets* (Skenneto, Margate, Australia, 1986).

E. K. G. Sixsmith, *British Generalship in the Twentieth Century* (Arms and Armour, London, 1970): something of a lost opportunity to make a convincing technical analysis, possibly due to excessive compression of a big story.

W. H. B. Smith and Joseph E. Smith, *Small Arms of the World* (Stackpole,

Harrisburg, Penn., first published 1943; 7th ed., 1962): classic technical specifications.
John Terraine, *Impacts of War 1914 and 1918* (Hutchinson, London, 1970): interesting comparative study of popular perceptions at the beginning and end of hostilities.
John Terraine, *The Smoke and the Fire* (Sidgwick and Jackson, London, 1980): provocative demolition of Great War myths, which generally succeeds despite itself.
John Terraine, *White Heat: the New Warfare 1914–18* (Sidgwick and Jackson, London, 1982): a lively rejigging of generally well-known themes, but alas failing to grapple with many of the key points of tactics.
P. A. Thompson, *Lions Led by Donkeys* (T. Werner Laurie, London, 1927): classic early attempt to 'grade' Great War generals, heavily based on the first few months of the conflict, to the near-exclusion of the last three years.
Richard J. Tindall, Edwin F. Harding et al., *Infantry in Battle* (The Infantry Journal, Washington, DC, 1934): George W. Marshall's uniquely revealing and pregnant set of low-level studies — but seemingly oddly ignorant of the British experience.
Tim Travers, 'The Offensive and the Problem of Innovation in British Military Thought, 1870–1915', in *Journal of Contemporary History*, vol. 13, no. 3, July 1978: an early and pregnant work by this scholar.
Tim Travers, *The Killing Ground: The British Army, the Western Front and the Emergence of Modern Warfare 1900–1918* (Allen and Unwin, London, 1987): disappointing for the tactical subjects promised by the title, and exceptionally thin for 1917–18; but a very interesting analysis of Haig's HQ, outlook and social class, and of Edmonds's official history of them.
Tim Travers, *How the War Was Won: Command and Technology in the British Army on the Western Front, 1917–18* (Routledge, London, 1992). Good for certain points such as Haig-Pétain relations and the German lack of tactics in their spring offensives; but generally a very disappointing book, considering the great mass of original documents that the author obviously has at his disposal. Apart from anything else, it really starts only in 1918.
J. David Truby, *The Lewis Gun* (Paladin, Boulder, Colorado, 1976).
Tom Wintringham and J. N. Blashford-Snell, *Weapons and Tactics* (Pelican ed., London, 1973): an informative general overview.
G. C. Wynne, *If Germany Attacks* (London, 1940): analysis of German defence tactics by one of the official historians, with oddly relatively little about methods of 'attack', and still less about what the British did right.

Poetry, fiction and literary criticism

S. Cooperman, *World War I and the American Novel* (Johns Hopkins, Baltimore, 1967): interesting trans-Atlantic view.
Cyril Falls, *War Books: a critical guide* (Davies, London, 1930): ideological bibliography opposing the 'antiwar' school that was so prevalent at that time, by one of the official historians.
Paul Fussell, *The Great War and Modern Memory* (Oxford UP, 1975): marvellously

stimulating overview of 'literary' trench literature which has been criticised too often, for its military inaccuracies, by 'pure military history' people.

C. S. Forester, *The General*: wonderful evocative summary of an early-war BEF 'Blimp'.

Brian Gardner, ed., *Up the Line to Death*: very useful anthology of war poems.

Holger Klein, ed., *The First World War in Fiction* (Macmillan, London, 1976): a fine and stimulating collection of critical essays on varied aspects of the subject.

R. H. Mottram, *The Spanish Farm* (first published 1924, new Penguin ed., 1941): a classic novel of the plight of the peasants in the zone occupied by armies of whatever colour.

Wilfred Owen, *Poems* (with 1931 'Memoir' by Edmund Blunden, new ed., Chatto and Windus, London, 1946, reprinted 1966): 'a seminal influence' makes far too weak a description of this work.

E. M. Remarque, *All Quiet on the Western Front* (Granada reprint, London, 1971). Ditto.

Index

The most important references to any subject are shown in **bold** type.

Abbeville, 170, 180
Accuracy, 137–8
Afghans, 132
Air photographs, 61, 111, 137, **153–8**, **183**, 188, 196
Air support for ground troops, 10, 22, 24–5, 41, 44, 50, 62, 66, 74, 86, 131, 137, **156–7**, 159, 189, 198, 200
Aircraft, air forces/RFC/RAF, 12, 25, 85–6, 88, 93, 108, 110–12, 126, 128, 130, **155–8**, 164–5, 182, 185, 187, 189, 193, 197–8, 260
Aire, 66
Alcohol, 28, 108
Alderson, Gen. Sir E. A. H., 216
All-arms co-operation, 6, 10, 111–12, 161, 193–4, 198
Allenby, F. M. Edmund ('The Bull'), 33, 35, 209–10, 215, 217
Ambulances, 129
America/US Army (in the Great War), 8, 19, 36, 56, **90**, 129, 134, 196, 219
American Civil War, 28, 30, 50, 53, 99, 131, 136, 161, **204–7**
Anderson, Gen. C. A., 215
Anarchists, 73
Anneaux, 134
Anti-tank artillery, 152, 165, 168
ANZACs, 7, 19, 62, 81–3, 110, 128, **163**, 192–3, 196, **214–18**
Appearance, 88, 110, 206
Armée de Chasse, 161, 209–10
Armentières, 68, 208
Armistice, 1918, 93, 108, 110, 199, 218
Armoured cars, 92, 129, 161, 166–7
Arnim, Gen. Sixt von, 8
Artillery, 13, 21, 24, 31, 34–5, **40–44**, 47, 54, 85, 88, 96, 103, 110–11, 114–19, 122, 124–5, 128, 131–2, **135–58**, 161, 169, 173–4, 181, 183, 185, 197–8, 212, 259
Artillery co-operation with infantry, 10, 24, 35, 47, 50, 53, 59, **61–7**, 69, 74, 76, 85, 87, 92–3, 95, 111, 115, **135–58**, 164, 189, 193, 199
Artillery direct fire, 49, 54, 63, 95, 145, 152, 165
Artillery, Directorate of, 105
Artillery location techniques, 107, 137, 152–5, 200
Ashworth, Dr Tony, and the 'Live and Let Live System', **15–16**, 40, 42, 68, 115
Assault cannon *see* Artillery direct fire
Atomic Bombs, 168
Attrition, 9, 13–14, **20–21**, 32–3, 84–5, 105, 157–8
Australian corps *see* ANZAC
Automatic rifles (ARs') *see* Lewis guns

Babel, Tower of, 184
Back-barrages, 143
Bacon, Adm. Sir Reginald, 36, 104

Bacteriological warfare, 116
Baden Powell of Gilwell, Lord R. S. S., Chief Scout, 47
Bailey, Lt. Col. J., 150
Baker-Carr, Major C. d'A., 123, 186
The Balkans, 125
Ball grenades, 113
Balloons, 137, 153, 174, 187
Bangalore torpedoes, 139
Bapaume, 31, 90, 146
Barbed wire/wire cutting, 3, 29–30, 34, 40, 60–61, 104, 107, 123, 125, **139–40**, 142, 145–6, 151, 158, 162, 164, 187, 198
Barrett hydraulic forcing jack *see* pipe pushers
Barrington Ward, journalist, 185
Battalion organisation, 73, 78–9, 95, 113, 122–9, 189
Battles
    Agincourt, 1415, 47
    Aisne, 1914, **50–51**, 170
    Aisne, 1918, 37, 92
    Albert, 1918, 16, 150
    Albuera, 1811, **14**, 47
    Amiens, 1918, 16, 34, 83, **92–3**, 128, 146–7, 149–50, 156, 161, 165–6, 169, 173, 198, 200, 208, 213
    Appomattox, 1865, 8
    Arras, 1917, 10, 16, 18, 26–8, 33, 35, 74, **85–8**, 91, 100, 113, 141, 144–5, 148, 150–51, 153, 155–6, 161, 163–4, 168, 173, 194, 210
    Arras, 1918, 93
    Aubers Ridge, 1915, 11, 53, 150, 156
    Austerlitz, 1805, 204
    Balaklava, 1853, 47
    Bullecourt, 1917, 16, 86, 91, **110**, 150, **163**, 168, 190
    Cambrai, 1917, 7, 10, 18, 34, 43, 60, 81–2, 86, 89–91, 134, 141, 146, 150, 158, **164–9**, 173, 195, 197–8, 200, 212, 259–60
    Cambrai, 1918, 89, 93
    Le Cateau, 1914, 5, 50, 208
    Crécy, 1346, 72
    Festubert, 1915, 53, 145
    Bataille des Frontières, 1914, 135, 205
    Gallipoli, 1915, 9, 36, 105, 209, 211, 214, 218
    German March offensive, 1918, 8–9, 18, 22, 43, 60, 63, 83, **90–92**, 97–8, 123, 146, 148, 150, 160, 173, 184, 192, 195, 199, 213, 218
    Gettysburg, 1863, 205
    Hamel, 1918, **62**, 128, 187
    Hastings, 1066, 47
    The Hundred Days, 1918, 9, 14, 16, 18, 34, 37, 43, 87, **93–5**, 98, 100, 116, 118, 128, 152, 160, 162, **166–7**, 171–4, 193, 195, 200, 219
    Loos, 1915, 11, 18, 39, **53–4**, **58**, **61**, 65, 67, 80, 91, 110, 113, 116, 119, 124, 142, 149–50, 156, 186, 193, 197, 209
    Lys, 1918, 18, 92–3, 126, 149, 214
    Mafeking, 1899–1900, 47
    Marne, 1914, 50, 208
    Messines, 1917, 10, 18, 85–8, 91, 93, 135, 148, 150–51, 156, 163–4, 173, 187, 194, 211
    Meteren, 1918, 37, 92, 142, 144
    Mons, 1914, 11, 50, 60, 81, 121, 135, 160, 170, 205, 208–9
    Mukden, 1905, 30
    Neerwinden, 1793, 204
    Neuve Chapelle, 1915, 11, 18, 31, 33, **53**, **67**, 124, 149–50, 155, 193
    New Orleans, 1815, 30
    Nieuport, 1917, **87**
    Nivelle offensive, 1917, 36
    Petersburg, 1864–5, 30
    Plevna, 1877, 50
    Port Arthur, 1904–5, 30
    Reims, 1918, 156
    Rorke's Drift, 1879, 47
    Salamanca, 1812, 75
    Sebastopol, 1853–4, 29–30
    The Somme, 1916, 4–5, 8, 10, 12, 15, 17–18, 31–3, 35, 40, 44, 54, 56, 58–9, **62–78**, 82–5, 87, 89–90, 100, 114, 118–19, 123–4, 130, 133, 140–45, 148–52, 154–7, 168, 171–4, 182–5, 187, 193–5, 199, 206, 208–12, 259
    Soissons, 1918, 92
    Torres Vedras, 1810, 30
    Verdun, 1916, 35, 60, 194
    Vimy Ridge, 1917, 16, 85–6, 124, 128, 150, 210, 212
    Waterloo, 1815, 23, 47, 204
    First battle of Ypres, 1914, 4, 18, 36, 208

Index 279

Second battle of Ypres, 1915, 4, 18, 29, 60, 208
Third battle of Ypres (Passchendaele), 1917, 4–6, 8–10, 18, 32–5, 44, 71, 82, **85–91**, 96, 100, 117, 124, 130, 134, 141, 144, 148, 150–51, 157–8, 161, 163–4, 168, 172, 194–5, 200, 204, 211–12, 259
Ypres, 1918, 4, 89, 93, 126
'Battle Drill', **99–100**
Bauer, Col. Max, 8
Bayonets, 49, **67–72**, 78, 104, 107, 113, 125–7, 130, 146, 158, 190, 194
Bazentin le Petit, 35
Beaulencourt, 146
Beaumetz, 126
Beaumont Hamel, 4, 82, 124, 150, 209
Becke, A. F., 18
British Army, *passim*
 and *see* BEF
BEF, formations and units
 Battalions
  5/Duke of Wellington's, 146
  6/King's Own Scottish Borderers, 43
  12/Middlesex, 66
  5/Royal Berkshire, 146
  11/RS, 28, 72, 79
  2/RWF, 14, 28, 41, 74, 79, 104, 131, 133, 187
  17/RWF, 66
  6/RWK, 26–7
  4/Seaforth Highlanders, 96
  1/5 Warwickshires, 59
 Brigades
  3rd, 216
  26th, 54–5
  27th, 28, 54–5
  98th, 28, 74
  99th, 124
  110th, 146
  115th, 66
 Divisions, **217–19**
  Guards, 216–18
  Lahore, 7, 19, 217
  Meerut, 7, 19, 216, 217
  New Zealand, 216, 218
  1st Regular, 215–17
  1st Australian, 142, 216, 218
  1st Canadian, 80, 96, 216–18
  2nd Regular, 70, 81–2, 124, 146, 209, 215–16, 218
  3rd Regular, 82, 142, 216, 218
  3rd 'Neutral' Australian, 83, 217–18
  4th Regular, 81–2, 86, 216, 218
  5th Regular, 215–18
  6th Regular, 142, 216–18
  7th Regular, 141, 209, 218
  9th (Scottish) 'K', 28, 54–5, 68, 80–83, 86, 96, 99, 124, 131, **141–2**, **144**, 146–7, 190, 198, 215, 218
  11th (Northern) 'K', 80, 82, 216, 218
  12th (Eastern) 'K', 142, 218
  14th (Light) 'K', 80–81, 218
  15th (Scottish) 'K', 53, 80, 142, 218
  16th (Irish) 'K', 81, 218
  17th (Northern) 'K', 82, 126, 218
  18th (Eastern) 'K', 80–83, 98, 141, 217–18
  19th (Western) 'K', 59, 80, 218
  20th (Light) 'K', 218
  21st 'K', 61, 65, 80–82, 146, 186, 218
  23rd 'K', 218
  24th 'K', 61, 186, 218
  25th 'K', 81, 87, 218
  27th Regular, 216
  28th Regular, 218
  29th Regular, 83, 216–18
  30th (Lancashire) 'K', 80, 218
  31st 'K', 216, 218
  33rd 'K', 28, 80, 218
  35th (Bantam) 'K', 81, 218
  36th (Ulster) 'K', 63, 80, 134, 218
  37th 'K', 182, 218
  38th (Welsh) 'K', 66, 217–18
  39th 'K', 218
  40th ('Mongrel') 'K', 81, 218
  41st 'K', 218
  42nd (East Lancashire) TF, 218
  46th (North Midland) TF, 80, 146, 218
  47th (2nd London) TF, 80, 218
  48th (1st South Midland) TF, 82, 218
  50th (Northumbrian) TF, 66, 218
  51st (Highland) TF, 80, 82, 216, 218
  52nd (Lowland) TF, 218
  55th (1st West Lancashire) TF, 80, 82–3, 218
  56th (1st London) TF, 80, 218
  62nd (2nd West Riding) TF, 134, 217, 218
  63rd (RN), 80, 82, 216, 218
  66th (2nd East Lancashire) TF, 28, 219

67th (2nd Home Counties) TF, 218
74th (Dismounted) Yeomanry, 218
Corps, **214–17**
   Cavalry Corps *see* Cavalry
   I Corps, 208–9, 215–16
   II Corps, 72, 82, 208, 215–17
   III Corps, 83, 146, 215–17
   IV Corps, 31, 146, 209, 215–16
   V Corps, 82, 208–9, 215–16
   VI Corps, 142, 152–3, 216
   VII Corps, 216
   VIII Corps, 216
   IX Corps, 216–17
   X Corps, 83, 216–17
   XI Corps, 215–16
   XIII Corps, 58, 82, 143, 146, 216–17
   XIV Corps, 216
   XV Corps, 83, 98, 143, 209, 216–17
   XVII Corps, 98, 142, 217
   XVIII Corps, 27, **96–8,** 179, 184, 217
   XIX Corps, 217
   XXII Corps, 99, 217
  Armies, **208–14**
   First Army, 16, 93, 128, 216
   Second Army, 16, 87–8, 93, 141, 215
   Third Army, 16, 28, 86, 91, 93, 141, 150, 152, 155
   Fourth Army, 16, 28, 35, 47, 49, 56, 93, 100, 123, 141, 152, 162, 219
   Fifth Army (Reserve Army), 16, 43, 60, 68, 77, 86–8, **90–92**, 98, 110, 150, 163, 218
Begbie Lamps, 104, 172
Béhagnies, 146
Belgian Army/Belgium, 17, 31, 36, 42, 104, 129, 131
Belts for machine guns, 31, 107, 127
Berlin, 205, 209–10
Bernafay Wood, **43**
Best, Prof. Geoffrey, 202
Bethell, Gen. J., 28
Béthune, 92
Bickford, 104, 113
Birch, Gen. Sir J. F. N., 83, 185
Birdwood, FM W. F., 83, 213, 215–16
Birmingham, and Britain's industrial heartland, **31,** 129–30
'Bite and hold' attacks, **32–3,** 86–8, 93, 159
Blacksmiths, 138
'Blitzkrieg', 203

'Blobs', 26–7, 56, 90, **96–7**
'Bloody Sunday', 15
'Blue Cross' gas, 117–18
Blunden, Prof. Edmund, 71, 114, 133, 172, 190, 192
Body shields, 104, 107
Boer War/Boer troops, 48, 51, 55, 74, 111, 120, 131
La Boisselle, 4, 59
Bombs/bombing, 21, 28, 51, 57, 61–2, **67–9,** 72–3, 77–8, 95, 103–4, 109, **112–19,** 125–6, 172, 182, 194, 199
Bourlon Wood, 161
Boyelles, 142
Bragg, Capt. W. L. (Prof. Sir Lawrence/ 'Willie'), 118, **153–4,** 170, 180
Braithwaite, Gen. Sir W. P., 216
Breakthrough operations, 9, 21, 32–5, 80, 86–7, 93, 117–18, 126, 128, **159–69,** 197–8
British Official History, 3, 11,18, 37, 47, 103, 131, 167, **259–61**
Broodseinde, 82, **88,** 150
Brüchmuller, Col. G., 8, 62, 200
Brutinel, Brig. R. E., 124–5, **128–9,** 192
Bull, Lucien, 153
'Bull rings', 80, 113, **187–91**
'Bumf', 180, 184, 186
Burnett Stuart, Gen. J. T., 127
Butler, Gen. Sir Richard, 216
Byng, FM Sir Julian, 83, 212, 215–16

Cable (telephone wire), 90, **170–74**
Cairo, 180
Calais, 116
Callaway, Capt. R. F., 71
Cambridge, 100, 196
Camiers, 127, 190
Campbell, Major Ronald B., DSO, **71–2,** 78, 190
Canadian Corps/Canadians, 8, 19, 80–83, 86, 89, 96, 98, 124, **128–9,** 142, 146, 187, 192, 196, 212, **218**
du Cane, Gen. Sir John, 83, 217
Carbon monoxide, 162
Carrington, Capt. Charles, **59–60,** 65, 68, 79, 130, 133
Casualties, 10, 14–16, 21, 27, 30, 32, 38–40, **43–4,** 51, 62–72, 80, 86–9, **92,** 97, 116–18, 143, 159, 163–7, 182, 192, 195, 218
Catapults, 104

Index

Cavalry/The Cavalry Corps/Cavalry Divisions, 5, 20, 25, 32–4, 41, 72, 78, 85–6, 92, 110, 126, 132, **159–69**, 175, 181, 187, 193, 198, **209–19**
Cavan, FM the Earl of, 215–16
Charteris, Brig. John, 84, 89
'Château generalship', 13, 23, 79, 92, 174–5, 192
Chatham, 121
Chelmsford, 124
Chemin des dames, 36
'Chinese' attacks', 58, 85, 144–5, 158
Chuignolles, 142
Churchill, Winston, 6, 36, **71–2**, 104–5, 107, 109, 111, 163
Cinematograph, 75, 153–4
Clapham Common, 105
'Clapham Junction', 88
Claremont Park, 106
Closing down batteries quickly, 10, 93, 151
Colt, Col. Samuel, 108
Column and line infantry assault formations, 51, **53–7**, 70, 75, 96, 195
Command and control, 60, 69, 75, 91–2, 127, 151–2, 159, **169–75**, 199
Computers, 180
Congreve, Gen. Sir Walter, 82–3, 216
Convoy system, 6
Cost of killing Germans, 118
Counterattacks, 32, 35, 57, 60, 68, **76**, 88–9, 113, 124, 168, 187, 189, 194, 200
Counter-battery fire, 21, 44, 85, 88, **136–58**, 171, 194, 200
Courcelette, 124
Creeping barrages, 22, 24, 53, 62–3, **65–7**, 69–70, 75, 77, **85**, 95, 97–8, 123, **140–50**, 151, 158–9, 162, 185, 189, 194, 199–200
Crimean War, 154
Croft, Brig. W. D., 28, 72, 79, 113, 131, 161, 190
Cromwell, Lord Protector Oliver, 38
Crozier, Brig. F. P., 28
Cumming, Brig. H. R., 190
Currie, Gen. Sir Arthur, 83, 187, 217
Cyclists, 126, 129, 161

Davidson, Gen. Sir J. H., 87–8
Delville Wood, 7, 124
Density of defences, 29–31, 33, 35, 87, 90, 93, 99, 162

Department stores, 104
'Depth battle' ('Deep battle'), 62, 85, 93, **153–8**, 200
Dervishes, 132
'Destructive' fire, 142–3
Diamond formation, **97–8**
Discipline, 52, 70, 99, 121, 188
'Doctrine', 7, 69, 75, 179–85, **186–91**, 196, 206
Douai, 93
Drocourt-Quéant Line, 142
Dropshorts, 24, **65–6**, 119, **143**, 198
Duffer's Drift, see Swinton
Duffour, Col., 96
Duffy, Dr Christopher, 202
Dunn, Dr, Capt. J. C., 74, 79, 104, 187

East Anglia, 180
Edmonds, Brig., Sir James, 7, 18, 22, 37, 56, 81, 88, 103, 131, 207, 213, **258–60**
Egypt, 82
Élite spearhead formations, 4, 43, 14, **79–83**, **103**, 121, **128**, 190, 192, 194, 198, 217
Engineers ('Pioneers'), 22, **25**, 48, 57, 62, **103**, 110, **112**, 116, 169, 185, 198
Epsom, 106
Esher, 106
Étaples, 127
Ewing, Col. J., 141, 144
Experimental Section, RE, 110

Fanshawe, Gen. Sir E. A., 82, 216
Fanshawe, Gen. H. D., 216
Farndale, Gen. M., 141, 150
Fawcett, Col., 152
Fergusson, Gen. Sir Charles B., 215, 217
Fieldcraft, **48–51**, 73–4, 121
Fireplans, 24, 34, 41, 53, 143, 152, 164, 200
Flamethrowers, 63, 114, **118–19**
Flares, 40–43, 58, 61, 117
Flers, 150, 163
'Flying Pigs', 115,
Foch, Marshal Ferdinand, 36, 92, 213
Fontaine les Croisilles, 74
Fortification, 8–9, 29–30, 48, 57, 60, 63, 92, 98, 140, 194–5
Foulkes, Brig. C. H., 103, 106, 110, **116–18**, 180, 186, 193
French Army/France, 4, 7, 11, 17, 31, 35–7, 42, **52–3**, **56**, 75, 80–82, 89, 92,

131, 137, 143, 148, 169, 181–2, 186, 188, 196, 208–15, 218–19, 260
Franco–Prussian War, 205
Frederick the Great, 38, 100
French, FM Sir John, 5, 29, 109–10, 208–9, 215
Frezenburg, 88, 99, 144
Friction, **26–7**, 34
Fuller, Col. J. F. C., 70, 83, 104, 110, 118, 120–21, 125–6, 132, 180, 183, 192, 204
'Fumite' bombs, 113
Furse, Gen. Sir W. T., MGO, 28, 83
Fuses, **136–58**

Gas/gas shells/gas brigade, 3–4, 25, **43**, 53, 58–9, 62, 85, 90, 103, 108, 110–11, **112–19**, 140–41, 145, 149, 169, 183, 186–9, 193, 197–8, 211
Gatling, Dr R. J., 108
Geneva Convention, 28, 131
German army/Germany/'Boche', 3–4, 7–10, 17, 22, 25–7, 31–5, 39, 41–4, 50, **52–63**, 67–9, 71–3, 75–6, 84–5, 87–93, 96–100, 112–19, 121, 123–9, 133–4, 145, 150, 153–8, 162, 164, 170–74, **179**, 182, 186–7, **192–200**, 203
Gheluvelt, 88
GHQ *see also* Haig, 8–9, 22, 26, 35–8, 68, 71, 73, 81–3, 87, 100, 106, 122–3, 127, 154, 163, 165, 169–70, 179–80, **182**, 184–8, 190, 193, 197, 211
GHQ Inventions Committee, 110
Godley, Gen. Sir A., 99, 216–17
Gomiecourt, 142
Gonnelieu, 146
Gough, Gen. Sir Hubert, 9, 20, 33, 35, 59–60, 68, 75, 86–8, 90–92, 98, **110–11**, 163, 168, 184–5, **209–15**
Graham, Prof. Dominick, 11
Grant, Gen. U. S., 111
Grantham, 127, 190
Graves, Capt. Robert, 14, 73, 80, 113
Greenwell, Lieut Graham H., 71
Grenades *see* bombs/bombing
Griffith, Capt. Wyn, **66**, 75, 159, 192
Guinness, Major, Walter (later Lord Moyne), 87, 135, 163
Gulf War, 1991, 93, 138, 158, 192–3, 197, 206
Gunners, 103, 128

Haig, FM Sir Douglas, **4–9**, 20, 31–3, 35–7, 61, 71, 77, 84, 87–93, **110–11**, 122, 169, 185–6, 208–16
Haldane, Gen. Sir Aylmer, 152, 216
Hales grenade, **114**
Haking, Gen. Sir R. C. B., 65, 215–16
Hamilton-Gordon, Gen. J., 216
Harfleur bull ring, 80, 113,
'Harooshing', 87
Harper, Gen. Sir G. M., 80, 216
Hart, Capt. Sir Basil ('Liddell'), 7, 71, **100**, 126, 196
Hausa, 132,
Le Havre, 180
Heath Robinson, cartoonist and inventor, **104**, 172
Hemming, Col. H. H., 107–8, **152–3**
Heriot-Maitland, Brig. J. D., 28
Hermies, 126
High Explosive (HE), 41–4, 88, 90, 112, 114–15, 135, **138–58**
High Wood, 118, 124, 161
Hindenburg Line, 16, 34, 74, 80, 84, 86, 93, 142, 146–7, 149, 160, 210–11, 213
Hippophilia, 153, 206
Hogg, I. V., 141
Holland, Gen. A. C. A., 215
Horne, Gen. H. S., 83, 103, 128, **209–13**, 216
Howard, Prof. Sir Michael, **202–3**
Hunter-Weston, Gen. Sir A. G., 105, 109, 216
Hutier, Gen. Oskar von, 8

Illustrations, **49**
Imperial College of Science and Technolgy, 109
Indian Corps/India, 7, 19, 83, 106, 124, 181, **215–18**
Infantry, *passim*
'Infiltration tactics', **53–7**, **59–64**, 90, 93, 97–100, 123, 194–6
Inspectorate General of Training, **96–100**, **183–6**
Intelligence, 21, 41–2, 73, 84, 92, 117, 152–3, 161, 197, 259
Interdiction, 85, **156–7**
Israeli Defence Force, 197
Italian Army/Italy, 6, 9, 36, 82, **89**, 218

Jack, Brig. J. L., 71, 130, 134, 161, 188
Jack, Major E. M., 154
Jacob, Gen. Sir Claud W., 72, 75–7, 80,

Index 283

82–3, 180, 185, 215–16
Jeudwine, Gen. Sir H. S., 81, 83
Junior officers, 7–9, 12, 15, 21–3, 41, 59, 61, 79, 95, 126, 174, 189–90, 196, 206, 214

Kavanagh, Gen. Sir Charles, 160, 215
Kellett, Brig. R. O., 70
Kemmel Hill, 154
Khyber Pass, 47
Kiggell, Gen. Sir Lancelot, **71**, 89, 183
Kipling, Rudyard, 47
Kitchener, FM Horatio Herbert, 5, 127, 180, 259
Kitchener armies, 11–13, 21–2, **51–3**, 61, **63–4**, 69, 79–82, 121, 186, 196, 206–7, 209, 217–18, 259–61
Kleptomania, 170
Kluck, Gen. Alexander von, 179

Laffargue, Capt. André, **54–7**, 69, 77, 195
Langemarck, 88
Lanrezac, Gen. Charles, 11, 180
Leonardo da Vinci, 108
Lewis guns, 13, 21, 52–3, 57, 73, 76–9, 95, 99, 115, 120–28, **128–34**, 162, 189, 194, 199
Light machine guns *see* Lewis Guns
Lindsay, Col. G. M., 58, 83, 104, **120–29**, 180, 186, 190, 193, 196
de Lisle, Gen. Sir H. de B., 83, 98, 100, 217
Livens, Capt. W. H., **118–19**, 193
Livens projectors, 85, **116–19**
Lloyd George, Prime Minister David, **5–7**, **9**, 36, 39, **89–91**, 105–6, 109, 111, 122, 125, 163
'Load of the soldier', 100, 125
Longueval, 35, **54**, 72, 141, 144, 150, 161
Lorries, 129, 180
Lossberg, Col. Fritz von, 8
Low explosives, 135
Low visibility, 35, **38–43**, 90, 123, 145, 147, 150, 157, 195
Ludendorff, Gen. Erich von, 8, 35, 53, 90, 92, 100, 195, 213
Ludwick, Capt., 124
Lusitania, the merchant vessel, 118

Machine guns/Machine Gun Corps, 6, 21, 25, 27, 31, 38–40, 43–4, 48, 54, 57–9, 61–2, 73, 76–7, 83, 90, 99, 106–7, 110, 112, 114–16, **120–34**, 139, 140, 143, 155, 161–4, 182–3, 185–7, 189–90, 193, 198–9, 206, 260
Machine gun barrages, 25, 39, 53, 62, 85, 103, **123–9**, 131, 144, 199
Madsen guns, 134
Mametz wood, 66, 75, 192
Manhattan Project, 108
Manpower, 6, 29–31, 37, 40, 47, **89–90**, 99, 173, 195, 205
Maps, 137, **154–8**, 183, 188, 196
'Mark VII' (Plowman, Capt. Max), 71, 73
Mathematicians, 147
Maxim guns *see* Machine guns
Maxse, Gen. Sir Ivor, 7, 27–8, 49–50, 72, 79–81, 83–4, **95–100**, 179–80, **184–6**, 188, 196, 217
Maxwell, Brig. Frank, VC, 28, 66, 72, 174
McClellan, Gen. George, 204
Meaulte, 142, 146
Mesopotamia, 9
Mess society, 117, 153
Military historiography, x–xii, 3–10, 13–14, 103, **201–3**, **259–61**
Military police, 127–8
Mills bomb, **113–14**, 130
Milward, Prof. A. S., 203
Mines, 85–6, 206, 211
Ministry of Munitions/Munitions Inventions Department, **105–10**, 122, 217
Miraumont-Irles, 124
de Miremont, Col. 'Count', 28
Mission Orders/'directive command', **58**, 62, 194
Mitchell, Col. F., 71
Mobilisation, 7, 9, 11–12, 14, 17, 51–3, 63, 108–9, 112, 180, 204–6
Monash, Gen. Sir John, 5, 83, 192, 217
Monchy le Preux, 26–7, **41**, 86
Monro, Gen. Sir Charles, 209, 215
Montauban, 141, 143
'Mopping up', 57, 72, 77, 114, 143, 159
Morale, 21, 32, 51–2, 61, 70–71, 74, 84, 90, 97, 128, 162, 198
Morland, Gen. T. L. N., 216
Morval, 150
Motor machine guns/motor-cycles /motorised troops, 126–7, **128–9**, **160–69**, 180–82
Mud, 3–4, 8, 32, 44, 67, 72, **88–9**, 105, 123, 144, 150, 156–7, 163, 186, 200, 211

Mules, 166
Munitions supply, 9, 11, 31–3, 35, 63, 67, 105–6, **108–11**, 122–5, 130, **138–40**, **147–9**, 157, 204
Murray, Gen. Sir Archibald, 109
Musketry, 21, 28, 38–40, **48–53**, 72–4, 120–22, 183, 188, 190
Mustard gas, **117–18**, 158

Nantes, 180
Napoleon, First Emperor of the French, 111, 204
National Physical Laboratory, 109
Nebelwerfer, 62, 116
Neuville St Vaast, 54, 69
New armies *see* Kitchener armies
New Weapons, 6, 9, 13, **103–19**, 140, 206
New Zealanders/New Zealand *see also* ANZACS, 81
Nissen Huts, 189
No Man's Land, 13, 26, 28, 38–41, 44, 53, 61, 75–6, 79, 110, 112, 118, 130, 133, 144, 174, 199
Noyon salient, 36
Numbers of troops in the fighting line, 11, 14–18, 29–31, **36–7**, 52, 85–7, 90, 125, 132, 147, **150–52**, 208–14

Observation posts, 24, **41**, 137, 151–3, 199
Oise River, 90
'Old Contemptibles', 11, 38, **48–52**, 60, 72, 74, 81, 260
'Open warfare'/'mobile warfare', 3, 9, 84–6, **88**, **92–3**, 112, 126, 147, **159–69**, 192–200, 209–10
Oppy, 124
Optical munitions, **108**
Ordnance board, 109
Ouija Boards, 153
Ovillers, **59**, 68, 172
Oxford, 201–3

Palestine, 218
'Panzer Armee', **110**
Paper recycling, 182
Partridge, Capt. S. G., **179–86**
Passage of lines/leapfrogging of units, 33, **54–7**, 86, 92, 174
Patrols, 40, 61, 74
Periscopes, 38, 104
Péronne, 156

Pershing, **90**
Pétain, Marshal Henri, 36
Phosgene Gas, **116–19**
Phosphorus ('P') bombs, 113
Pigeons, 12, 58, 172–3, 189
Pillboxes, 34, 60, 74, 76, 99, 116, 126, 133–4, 162
'Pimple' Hill, 72
Pinney, Gen. R. J., 27–8, 80
Pipe pushers, 63, **118–19**
Platoon tactics, 10, 22, 26–7, 51, 54, 56–9, 73, **76–9**, **95–100**, 113, 115, 130, 184, 188, 194, 196, 199
Plumer, FM The Lord Hubert, 21, 29, 33, 83, **87–9**, 143, 156, 187, **208–15**
Poelcappelle, 88
Poetry, 10–11, 13, 68–9, 71, 114
Pollard, Capt. A. O., VC, 112–13
Polygon Wood, 88
Polymaths, 201–3
Pom Pom, 131
Portuguese Army/Portugal, 19, 36, **218**
Postcards, **181**
Potije, 133
Preparation (or lack of), 9–10, 15, **59**, **61**, **74–9**, 85, 90
Press, 14, 49, 81, 100, 109, 122, 133, 163, 181, 185
Prior, Robin, 15, 59, 150
Prisoners of war, 27, 72
Propaganda, 181
Pulteney, Gen. Sir William, 215

'Quiet' sectors, **14–16**, 34, 36, 117, 119
Questionnaires, 123, 179, **186–7**, 196

RAF *see* Aircraft
RFC *see* Aircraft
Railways, 29, 31–2, 117, 162, 205
Rawlinson, Gen. Sir Henry, 15, 20, 31, 33, 35, 87, **209–15**
Red Dragon Crater, 133
Regimental system, 10, 24, 32–3, 47, **51–2**, 111, 128, 198
Remarque, Private E. M., 15
Reutel, 88
Richtofen, Baron Manfred von, 86, 155–6, 158
Rifles, 38–40, 48–51, **67–74**, 77, 95, 99, 104, 107, 111, 120–23, 125–6, 130, **132**, 135, 194, 199
Rifle-grenades, 13, 74, 76–8, 95, **113–19**,

Index 285

134, 162
Rifle rests, 106
Rimington, Gen. M. F., 106, 109, 215
Riqueval, 142
Robertson, Gen. Sir William, 109
Rockets, 111, 116, 172
Roeux, 86
Rotton, Brig. Johnny, 152
Rouen, 180
Royal Aircraft Factory, 109
Royal Military Academy, Woolwich, 48
Royal Navy, 36, 109, 211
Royal Society, 106
Royal United Services Institution, 48
Rumanian Army/Rumania, 6
Russian Army/Russia, 36, 99, **127**
Russian saps, 58, 76, 118
Russo-Japanese War (Manchurian War), 48, 50, 137

St Omer, 123
St Quentin, 146
Salonika, 9, 82, 180
Le Sars, 66
Sassoon, Capt. Siegfried, 20, 26, 104
Saxe, Marshal Maurice de, 108
School of Musketry, Hythe, **48**, **52**, 83, **106–7**, 109, 120, 122
Schwaben Redoubt, **63**
Scientists, 7
Second World War, 8–9, 24, 36, 39, 61, 76, 116, 118, 131, 157, 162, 203–5
Secrecy of Zero Hour, **34–5**, **42**, 58, 75, 83, 86, 90, 93, 117, 140, 142, 144–5, 151, 158, **172–3**
Security printing, 180–82
Selle, 146
Sentries, 38, **40**, **42**, 73
Serre, 4, 209
Shells, 40–43, **135–58**
Shrapnel shells, 41, 43, **138–58**
Shute, Gen. C. D., 216
Signals, 12, **23–5**, 41–2, 44, 48, 53, 66, 76, 90, 92, 100, 103, 109, 112, 127, 129, 143, 152, 157, 159, 161, **169–75**, 189, 199
'Silicon Valley', 31
Simkins, Peter, 11
Smith-Dorrien, Gen. Sir Horace, 5, 29, 122, 208, 215
Smoke, 28, 43, 53, 58, 62, 77, 85, **112–19**, **140–42**, 145, 148–9, 159, 195

Snipers, 21, 28, 38, 40, 50–51, 67, **73–4**, 77, 189, 196, 199
Snow, Gen. Sir T. d'O., 216
'Snowball' scheme, 121
Social control, 13
Soft spots, **97–8**, 195–6
Soixante Quinze, 137
Solly Flood, Brig. A., 83, **184–5**
'SOS' fire, 24, **41–4**, 75, 90, 115
South Africa/South Africans, 7, 19, 48, 80–81, 132, 137, 161, **218–19**, 258
Specialisation of rôles, 12, 25, 48, 77, 79, **95**, **111–12**, 115, 122, 127, 161, 189, 198–9
Staff College, Camberley, 48, 52, 111, 186
Staff officers/staffwork, 12, 17, 23, 25, 53, 74, 89, 92, 111, 126–7, 184–6, 189–90, 204, 217
'Stand to', 75, 145
Stationery, **179–86**
'Stellenbosching', 5, 29, 61, 83
Stephens, Gen. Sir Reginald, 217
Stokes mortars *see* Trench mortars
Stokes, Frederick William and Capt. S. F., **105–6**, 144
Stone, Capt. Christopher, DSO, 72
Stormtroops, 8, 25, 55, **57–62**, 68, 77, 79, 85, 100, **112**, 123, 168, **192–200**
Sturmabteilung Rohr, 60, 62
Survey, 103, **152–6**, 200
Sutton, Lieut. F. A., 105
Swinton, Gen. Sir Ernest, and *The Defence of Duffer's Drift*, **48–50**, 72, 77

'Tactics', **20–29**, **100**, **111**, and *passim*
Tank co-operation with Infantry, 70, 80, **165**, 183, 189
Tanks/Tank Corps, 6, 10, 12, 14, 25, 32–3, 62, 70, **80**, 83, 85–6, 89, 92–5, 103–4, 108–11, 118–19, 121, **125–6**, 128–9, 131, 140, 145–6, 152, 156, 159, **161–9**, 175, 185–7, 189, 192–3, 197–8, 212, 260
Terrain models, 26, **61**, 75, **188–9**
Territorial forces, 52, 196, 209, **217–18**
Tiananmen Square, 15
Thermit, 85, 113, 116, 134
Thiepval, 4, 72, 82, 150, 209
Thomas, Capt. Alan, **26–7**, 41
'Thrusters', 13, **29**
Tibetans, 132
Timescale of battles, **29–35**, 85, 87, 91, 93

Todhunter, Major H. W., 106–7
Toothbrushes, 182, 187
Tracer bullets, 107
Training manuals, 13, 25–7, 49–51, 53, 55–6, 58, 68–9, 73, 76–8, **95–100**, 120, 130, 134, 165, **179–86**, 191, 196
Training schools, 7, 13, 26, 53, 68, 71, 73, 79, 98–9, 112, 120–29, 133, **159–60**, 170, 179, **186–91**, 196
Travers, Prof. Tim, 11
Trench foot, 21, 28, 79, 86, 110, 182
Trench mortars, 6, 13, 25, 53, 61–2, 73–7, 88, 95, **105–6**, 109, 111, **114–19**, 129, 134, 139, 143, 162, 189, 194, 198–9
Trench raids, 13, 26, 40, **60–62**, 124, 193–4
Trench turrets, 106
Trônes Wood, 174
Tucker microphones, 154
Tudor, Gen. H. H., 146, 180
Tunnels, 86
Turin, 180
Typewriting machines, **180–81**

U-boats, 36, 211
Uniacke, Gen. F. C. C., 83

Vauban, Gen. Sebastian Leprestre de, 48, 205
Vickers guns *see* Machine guns
Vietnam, 8, 49, 157, 192–3, 195, 205
Vincent, Capt., 118
Voice procedure, 172, 199

Wade, Lieut. Aubrey, 157
'War and society', 9, 13, **201–3**, 260
Warlencourt, 144
War Office, 5, 122, 181–2, 202
Watson, Major W. H., 68, 161, 168
Watts, Gen. Sir H. E., 217
'Wave' attacks, **49–50**, **53–7**, 78, 96–7, 123
Wellington, Arthur Duke of, 23, 75, 111, 134
Western Front, *passim*
Wet, Christiaan R. de, 55
Whale Island, 162
Willcocks, Gen. Sir J., 215
Wilson, FM Sir Henry H., 36, 213, **216**
Wilson, Trevor, 15, 59, 150
Wingate, Gen. Orde, 49
Winterbotham, Col. H. St J. L., 152, 154–5
'Wire-less' communication
  by radio, 12, 44, 109, 172–4
  by other instruments, **172–3**
Wolesley, FM Sir Garnett, 131
Woolecombe, Gen. C. L., 216
Woolwich, 105, 109
'Worms', **96–9**
Wormwood Scrubs, 105
Wounded soldiers, 89, 105–6
Wynne, Major G. C., **8–9**, 196

Zero, 93, 117–18, 133, **145**
  see also Secrecy of Zero Hour